Introduction to experimental particle physics

Contributions to the field of experimental particle physics have been accumulating in the literature over the past 40 years and now compose a vast but scattered array of reports and monographs. This book ties together the most important experimental topics into a brief but balanced overview.

The author first gives a review of particle physics and discusses electromagnetic and nuclear interactions. He then goes on to discuss three nearly universal aspects of particle physics experiments: beams, targets, and fast electronics. The second part of the book treats in detail the properties of various types of particle detectors, such as scintillation counters, Cerenkov counters, proportional chambers, drift chambers, sampling calorimeters, and specialized detectors. Wherever possible the author attempts to enumerate the advantages and disadvantages of each detector, and to specify the factors that limit a detector's performance. Finally, the author discusses aspects of specific particle physics experiments, such as properties of triggers, types of measurements, spectrometers, and the integration of detectors into a coherent system.

Throughout the book, the author has attempted to begin each chapter with a discussion of the basic principles involved and follow it by selective examples. Although it is not meant to be a complete survey of experimental particle physics, nevertheless, this book contains much practical information to provide readers with sufficient background in the subject. It will be a useful reference for particle physicists, nuclear physicists, and graduate students studying these topics.

Introduction to experimental particle physics

RICHARD C. FERNOW

Physics Department
Brookhaven National Laboratory

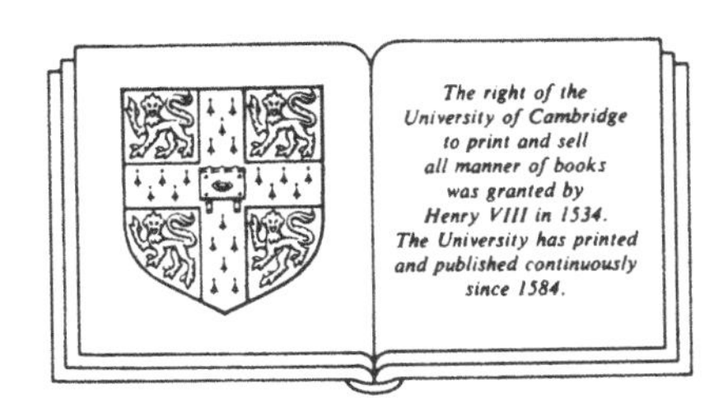

CAMBRIDGE UNIVERSITY PRESS

Cambridge

New York New Rochelle

Melbourne Sydney

Published by the Press Syndicate of the University of Cambridge
The Pitt Building, Trumpington Street, Cambridge CB2 1RP
32 East 57th Street, New York, NY 10022, USA
10 Stamford Road, Oakleigh, Melbourne 3166, Australia

First published 1986
First paperback edition 1989

Library of Congress Cataloging in Publication Data

Fernow, Richard C. (Richard Clinton), 1947–
Introduction to experimental particle physics.
Includes indexes.
1. Particles (Nuclear physics) – Experiments.
2. Particles (Nuclear physics) – Technique.
3. Nuclear counters. 4. Nuclear reactions.
5. Electromagnetic interactions. I. Title.
QC793.2.F47 1986 539.7′21 85–14931

British Library Cataloguing in Publication Data

Fernow, Richard C.
Introduction to experimental particle physics.
1. Particles (Nuclear physics)
I. Title
539.7′21 QC793.2

ISBN 0 521 30170 X hard covers
ISBN 0 521 37940 7 paperback

Transferred to digital printing 2001

Contents

Preface

I have felt for some time that there should be a book that briefly ties together the most important topics in experimental particle physics. The biggest difficulty I have encountered in trying to do this is not that information concerning this subject is lacking, but rather that so much of it exists. Reports on experimental techniques and devices can be found scattered through specialized monographs, conference proceedings, data compilations, review papers, and journal articles. I have had to make enumerable, arbitrary selections in order to produce what I hope is a balanced overview of the subject in a book of reasonable length. I hope that the final product will be useful to graduate students and to others interested in an introduction to the subject and as a reference for practitioners in the field.

The first three chapters give an overview of the subject and discuss the electromagnetic and nuclear interactions of particles. A knowledge of particle interactions is necessary for an understanding of how detectors work, besides being interesting in their own right. The next three chapters are concerned with three nearly universal aspects of particle physics experiments: beams, targets, and fast electronics. Chapters 7 through 12 contain more detailed discussions of various types of detectors. Whenever possible I have attempted to enumerate the advantages and disadvantages of each detector and to specify the factors that limit its performance. The last three chapters are concerned with integrating detectors into a coherent system. A number of examples of specific experiments are given in the last chapter.

Most of the chapters begin with a discussion of basic principles and are followed with selective examples. I have made no attempt to completely survey all the contributions that have been made to each topic. After nearly 40 years the literature is so vast that even if someone contemplated such a project, it would probably require a dozen volumes the size of this one. Although I have included a great deal of practical information, no one should expect to be able to go out and build a detector after reading this book. The successful application of experimental physics almost always requires a period of apprenticeship with an experienced tutor. I do hope, however, that the reader will gain sufficient background to at least start "asking the right questions."

As regards the references, in most cases I have adopted the philosophy of quoting recent articles that I believe contained sufficient details to be useful to an uninitiated reader, rather than making literature searches back to the original papers. Since this book is neither a review paper nor a history, I have preferred this method because the referenced material usually illustrates current applications and techniques and because the reader can always use the references in the cited paper as a starting point for a search if so inclined.

I would like to thank Drs. Suh Urk Chung, Kenneth Foley, Thaddeus Kycia, Thomas Ludlam, David Rahm, Pavel Rehak, Lyle Smith, Mark Sakitt, R.M. Sternheimer, Michael Tannenbaum, and Erich Willen for their helpful suggestions. I would like to thank the many authors who graciously permitted me to reprint figures from their papers in this book. I would also like to thank Ms. Audrey Blake and Jeanne Danko for the nice job they did typing the manuscript. I would like to give special thanks to the people at Cambridge University Press for their cooperation and their faith in this project. Finally, I would like to express my gratitude to my wife Ruth, daughter Jessica, and son Matthew for showing a lot of patience and giving me their support when I needed it.

R. C. Fernow

1
Introduction

Particle physics is the study of the properties of subatomic particles and of the interactions that occur among them. This book is concerned with the experimental aspects of the subject, including the characteristics of various detectors and considerations in the design of experiments. This introductory chapter begins with a description of the particles and interactions studied in particle physics. Next we briefly review some important material from relativistic kinematics and scattering theory that will be used later in the book. Then we give a brief preview of the various aspects of particle physics experiments, before discussing each topic in greater detail in subsequent chapters. Finally, we give a short discussion of some of the tasks involved in analyzing the data from an experiment.

1.1 Particle physics

Particle physics is the branch of science concerned with the ultimate constituents of matter and the fundamental interactions that occur among them. The subject is also known as high energy physics or elementary particle physics. Experiments over the last 40 years have revealed whole families of short-lived particles that can be created from the energy released in the high energy collisions of ordinary particles, such as electrons or protons. The classification of these particles and the detailed understanding of the manner in which their interactions leads to the observable world has been one of the major scientific achievements of the twentieth century.

The notion that matter is built up from a set of elementary constituents dates back at least 2000 years to the time of the Greek philosophers. The ideas received a more quantitative basis in the early nineteenth century with the molecular hypothesis and the development of chemistry. By the end of the century most scientists accepted the idea that matter was constructed from aggregates of atoms. The discovery of radioactivity and the analysis of low energy scattering experiments in the early decades of this century revealed that atoms themselves had a structure. The experiments showed that the positive charge and most of the atomic mass was concentrated in a dense nucleus surrounded by a cloud of electrons.

The discipline of nuclear physics developed in the 1930s, particularly after the discovery of the neutron and the invention of particle accelerators. With sufficient energy the nucleus could be broken apart into its constituent protons and neutrons. At the same time physicists developed new particle detectors, such as Geiger tubes and cloud chambers, to study the properties of cosmic ray particles. The modern discipline of particle physics evolved in the late 1940s from a fusion of high energy nuclear physics and cosmic ray physics.

The chief concerns of this book are a description of the manner in which particles interact in matter, the properties of the detectors used to measure these interactions, and the fundamental considerations involved in designing a particle physics experiment. Two other very important aspects of the subject are data analysis and the interpretation of data using elementary particle theory. A brief survey of data analysis is given in the last section of this chapter. Fortunately, for particle theory an excellent introductory treatment is already available [1].

1.2 Particles and interactions

At the present, as best we can tell, four types of interactions are sufficient to explain all phenomena in physics. The interactions and their approximate relative strengths at distances $\sim 10^{-18}$ cm are [2]

1. strong nuclear, 1;
2. electromagnetic, 10^{-2};
3. weak nuclear, 10^{-5}; and
4. gravitational, 10^{-39}.

The gravitational force controls the interactions between massive bodies separated by large distances. However, the gravitational force between particles, where a typical mass is 10^{-27} kg, is so feeble that it does not appear to have a significant effect on elementary particle interactions. Thus, for particles the electromagnetic force dominates for distances down to 10^{-13} cm, where the nuclear forces begin to become important. The strong nuclear force is responsible for the binding of particles into nuclei, while the weak nuclear force is responsible for processes such as nuclear beta decay.

The electromagnetic interactions of particles can be calculated using the theory of quantum electrodynamics (QED). This is probably the most successful theory in all of physics and is capable of making extremely precise predictions. Recently a model has been developed that successfully treats the weak and electromagnetic interactions as the low energy manifestations of the breakdown of a unified electroweak interaction. A prediction of this model, which has recently been verified, is the existence of massive particles known as the $W^{\pm}$ and Z gauge bosons. Other grand unified models have been developed that assert that the electroweak and strong nuclear interactions have resulted from the breakdown of a single interaction. One consequence of these models is that the proton should have a small but finite probability of decaying.

Hundreds of new particles have been discovered in the study of high energy interactions. Many ways have been devised to group them into families with similar characteristics. One way to classify particles is by the type of interactions in which they participate. The leptons are particles that are not affected by the strong interaction. The electron, muon, and neutrino are examples of leptons. At present leptons appear to be truly elementary particles. They have no measured internal structure and are sometimes referred to as pointlike particles.

Particles that are affected by the strong interaction are known as hadrons. There are two main classes of hadrons. The baryons are hadrons with a half-integral value for the spin quantum number. The mesons, on the other hand, are hadrons with integral values of the spin quantum number. The pions are examples of mesons.

The lowest lying (least massive) baryons are the proton and the neutron. These two common constituents of nuclei are often referred to collectively as nucleons. The hyperons are unstable baryons that decay via the weak interaction and have a nonzero value for the internal quantum number known as strangeness. The lowest lying hyperon is the Λ particle. The decay chain of all unstable baryons ends with a final state containing a proton.

One of the early theories of the strong interaction, known as SU(3), predicted a relation among the baryon masses. Using this relation and the masses of the then-known baryons, it was possible to predict the existence of a hyperon with three units of strangeness, called the Ω^-. Figure 1.1 shows the historic bubble chamber photograph that proved the existence of the Ω^- hyperon. Its discovery marked an important milestone in our understanding of elementary particles.

The largest group of hadrons are referred to as resonances. These particles can decay via the strong interaction and thus have lifetimes on the order of 10^{-23} sec. Even traveling at the speed of light, this lifetime is much too short for the particles to travel a measurable distance in the lab. Thus, the properties of the resonances must be inferred from the properties of their longer-lived decay products.

Unlike the leptons, the hadrons are believed to have an internal structure. In the currently favored model of strong interactions (quantum chromodynamics, or QCD) hadrons are built up from pointlike spin $\frac{1}{2}$ objects known as quarks. The quarks are unlike other particles in several respects. The magnitude of their charge is one-third or two-thirds of the electron's charge, and free quarks have never been observed in scattering experiments. In the QCD model a quark attempting to leave the interior of a hadron would cause new quark–antiquark pairs to be created. The quarks and antiquarks would then recombine in such a way as to form new hadrons. Very energetic quarks would form a narrow spray of hadrons known as a jet.

Another remarkable feature of nature is the existence of antimatter. For every particle there is an antiparticle with the same mass and spin, but with opposite values for the charge and some of the internal quantum

Figure 1.1 The first bubble chamber photograph of the decay of an Ω^- hyperon. The picture was taken by a group headed by N. Samios at the 80-in. chamber at Brookhaven National Laboratory in 1964. (Courtesy of Brookhaven National Laboratory.)

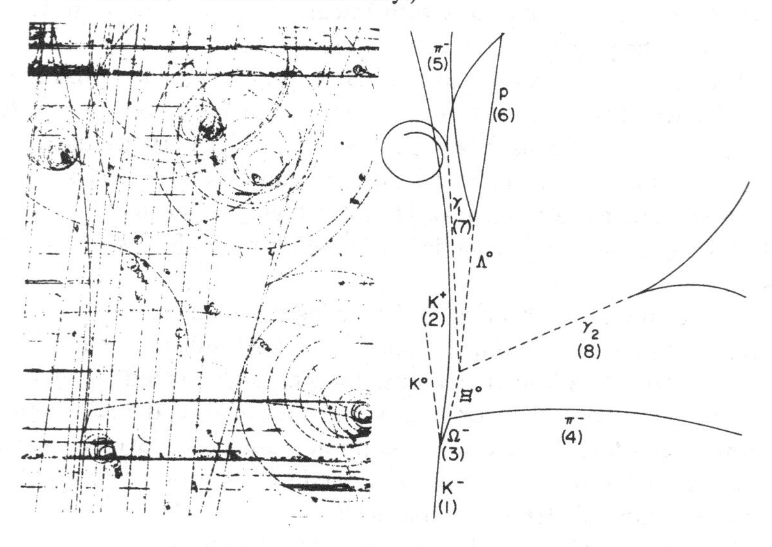

numbers. A familiar example is the positron, which is the antiparticle of the electron.

Besides the leptons and the hadrons there is a third group of particles known as the gauge bosons. These integral spin particles are responsible for transmitting the basic interactions. The most well known example is the photon, which mediates the electromagnetic interaction. The weak interaction is thought to be mediated by the $W^{\pm}$ and Z vector bosons. According to QCD, the carriers of the strong interaction are massless particles known as gluons, while the gravitational interaction is thought to be mediated by spin 2 objects known as gravitons.

1.3 Relativistic kinematics

The mechanics of particle interactions must obey the laws of special relativity [3]. The velocity $\mathbf{v}$ of a particle is frequently specified in terms of the dimensionless quantity

$$\boldsymbol{\beta} = \mathbf{v}/c \tag{1.1}$$

where c is the speed of light in vacuum. The momentum and energy of the particle are given by

$$\mathbf{p} = mc\gamma\boldsymbol{\beta} \tag{1.2}$$

and

$$E = mc^2\gamma \tag{1.3}$$

where m is the mass of the particle measured in the reference frame in which it is at rest, and the auxiliary function γ is defined as

$$\gamma = (1 - \beta^2)^{-1/2} \tag{1.4}$$

The high energy behavior of various phenomena is frequently plotted as a function of γ. In these cases it may be convenient to rewrite the velocity and momentum in the form

$$\beta = [(\gamma^2 - 1)/\gamma^2]^{1/2} \tag{1.5}$$

and

$$p = mc(\gamma^2 - 1)^{1/2} \tag{1.6}$$

It is customary to measure energies in multiples of the electron volt (eV), typically MeV or GeV, at high energies. Then from Eq. 1.2 the unit of momentum is MeV/c, and from Eq. 1.3 the unit for mass is MeV/c^2. The constant c is frequently set to 1 to simplify relativistic calculations.

The energy and momentum of a particle in a second coordinate system moving with constant velocity $-\boldsymbol{\beta}_0$ with respect to the original (primed) system is governed by the Lorentz transformation equations [3]

$$\mathbf{p} = \mathbf{p}' + \boldsymbol{\beta}_0\gamma\left(\frac{\gamma}{\gamma+1}\boldsymbol{\beta}_0 \cdot \mathbf{p}' + \frac{E'}{c}\right)$$
$$\frac{E}{c} = \gamma\left(\frac{E'}{c} + \boldsymbol{\beta}_0 \cdot \mathbf{p}'\right) \tag{1.7}$$

For the special case when the transformation takes place along the z axis, the transformation equations simplify to

$$p_x = p'_x$$
$$p_y = p'_y$$
$$p_z = \gamma\left(p'_z + \beta_0\frac{E'}{c}\right) \tag{1.8}$$
$$\frac{E}{c} = \gamma\left(\frac{E'}{c} + \beta_0 p'_z\right)$$

The quantities $(E/c, \mathbf{p})$ can be interpreted as the components of a vector in a 4-dimensional space and are referred to as the energy–momentum 4-vector. The first quantity in the parentheses is denoted the 0th component.

Another important 4-vector is $(ct, \mathbf{x})$, where $\mathbf{x}$ is the position and t is time. The components of all 4-vectors obey transformation laws analogous to Eq. 1.8. An important consequence of the Lorentz transformation applied to this 4-vector is time dilation. Suppose that an interval of time τ elapses in a coordinate system where some particle is at rest. Time intervals in this frame are referred to as proper times. The corresponding time interval in a coordinate system moving with velocity $-\beta$ with respect to the particle (or equivalently in the frame where the particle has velocity $+\beta$) is

$$t = \gamma\tau \tag{1.9}$$

Thus, time intervals measured in a frame where the particle is moving are increased by the factor γ over the proper time intervals.

We shall identify 4-vectors by using a tilde, for example, $\tilde{a}$. The scalar product of two 4-vectors $\tilde{a}$ and $\tilde{b}$ is defined in the metric we are using as

$$\tilde{a} \cdot \tilde{b} = a_0 b_0 - \mathbf{a} \cdot \mathbf{b} \tag{1.10}$$

It follows immediately that the square of a 4-vector is

$$\tilde{a} \cdot \tilde{a} = a_0^2 - |\mathbf{a}|^2$$

As an example, consider the decay of an unstable particle into two particles with 4-momenta $\tilde{p}_1$ and $\tilde{p}_2$. The effective mass M of the system is defined to be

$$M^2 = (\tilde{p}_1 + \tilde{p}_2)^2$$
$$= \tilde{p}_1^2 + \tilde{p}_2^2 + 2\tilde{p}_1 \cdot \tilde{p}_2$$
$$= m_1^2 + m_2^2 + 2(E_1E_2 - p_1 p_2 \cos\theta) \tag{1.11}$$

where θ is the angle between the 3-vectors $\mathbf{p}_1$ and $\mathbf{p}_2$. The effective mass is a powerful tool for studying the properties of short-lived particles.

1.4 Summary of particle properties

Each of the particles mentioned previously has a unique set of properties that distinguish the particle and describe how it is affected by the fundamental interactions. These properties include

1. charge,
2. mass,
3. spin,
4. magnetic moment,
5. lifetime, and
6. branching ratios.

In addition, a full description of a particle must include the values for a set of internal quantum numbers, such as baryon number and strangeness [4]. The values of the internal quantum numbers determine which particles may be produced together in various reactions and how unstable particles can decay.

If we choose as a time interval the mean lifetime of a particle in its rest frame, then the particle lifetime in the LAB frame is generally longer due to the time dilation effect. The mean distance traveled in the LAB from production to decay is

$$\lambda_D = (p/mc)c\tau \tag{1.12}$$

Note that this grows linearly with the particle's momentum.

Suppose that N_0 unstable particles with mean decay length λ_D have been created at $x = 0$. The number of particle decays occurring in some small interval dx around the distance x is proportional to the number of particles at x and to the fractional size of the interval. Thus,

$$dN(x) = -N(x)\, dx/\lambda_D$$

from which it follows that

$$N(x) = N_0 \exp(-x/\lambda_D) \tag{1.13}$$

Thus, the decay lengths of unstable particles have an exponential distribution with a slope that depends on λ_D and hence on the $c\tau$ value of the particle. This can be useful sometimes in determining the identity of a decay sample.

We summarize in Table 1.1 the properties of the particles most com-

Table 1.1. *Properties of quasistable particles*

		Mass (MeV/c^2)	Spin ($\hbar$)	Magnetic moment	$c\tau$ (cm)	Major decay modes	Branching ratio (%)
Gauge bosons							
γ	photon	0	1	0	stable	—	—
Leptons							
ν_e	e neutrino	~0	$\frac{1}{2}$	0	stable	—	—
ν_μ	μ neutrino	~0	$\frac{1}{2}$	0	stable	—	—
e^-	electron	0.5110	$\frac{1}{2}$	$1.001\mu_B$	stable	—	—
μ^-	muon	105.7	$\frac{1}{2}$	$1.001\ (e\hbar/2m_\mu c)$	6.59×10^4	$e\bar{\nu}\nu$	100
Mesons							
π^0	pion	135.0	0	0	2.5×10^{-6}	2γ	98.8
$\pi^\pm$	pion	139.6	0	0	780.4	$\mu\nu$	100
$K^\pm$	kaon	493.7	0	0	370.9	$\mu\nu$	63.5
						$\pi^\pm\pi^0$	21.2
						$\pi^\pm\pi^+\pi^-$	5.6
K^0_S	K short	497.7	0	0	2.675	$\pi^+\pi^-$	68.6
						$2\pi^0$	31.4
K^0_L	K long	497.7	0	0	1554	$\pi e\nu$	38.7
						$\pi\mu\nu$	27.1
						$3\pi^0$	21.5
						$\pi^+\pi^-\pi^0$	12.4

Baryons							
p	proton	938.3	$\frac{1}{2}$	$2.793\mu_N$	stable	—	—
n	neutron	939.6	$\frac{1}{2}$	$-1.913\mu_N$	2.7×10^{13}	$pe^-\bar{\nu}$	100
Λ	lambda	1115.5	$\frac{1}{2}$	$-0.613\mu_N$	7.89	$p\pi^-$	64.2
						$n\pi^0$	35.8
Σ^+	sigma	1189.4	$\frac{1}{2}$	$2.379\mu_N$	2.40	$p\pi^0$	51.6
						$n\pi^+$	48.4
Σ^0	sigma	1192.5	$\frac{1}{2}$		1.7×10^{-9}	$\Lambda\gamma$	100
Σ^-	sigma	1197.3	$\frac{1}{2}$	$-1.10\mu_N$	4.44	$n\pi^-$	100
Ξ^0	cascade	1314.9	$\frac{1}{2}$	$-1.25\mu_N$	8.69	$\Lambda\pi^0$	100
Ξ^-	cascade	1321.3	$\frac{1}{2}$	$-0.69\mu_N$	4.92	$\Lambda\pi^-$	100
Ω^-	omega	1672.5	$\frac{3}{2}$		2.46	ΛK^-	68.6
						$\Xi^0\pi^-$	23.4
						$\Xi^-\pi^0$	8.0

Source: Particle Data Group, Rev. Mod. Phys. 56: S1, 1984; L. Pondrom, in G. Bunce (ed.), *High Energy Spin Physics—1982,* AIP Conf. Proc. No. 95, 1983, p. 45.

monly encountered in particle physics. Listed are most particles that are not known to decay or that decay via the weak interaction and have a $c\tau$ value greater than 1 cm. We will refer to this group as the quasistable particles. We have also included the neutral members of the pion and sigma families, which decay by electromagnetic processes and thus have a much shorter lifetime than the other listed particles. Table 1.1 does not include the antiparticles, which have identical values for the listed properties, except for the charge and magnetic moment, which are opposite.

The mass of the photon is believed to be identically zero. Although the neutrino masses are very small, there is no compelling theoretical reason why they should be exactly zero. The spins of all particles are found to be multiples of $\hbar/2$, where $\hbar$ is Planck's constant divided by 2π. All neutrinos discovered to date have been "left handed." This means that the neutrino's spin is directed in the opposite direction from its momentum. Antineutrinos are right handed. The photon is the only particle in Table 1.1 to have a spin of 1, while the Ω^- is the only particle with spin $\frac{3}{2}$.

Particles with nonzero spin and nonzero mass have a magnetic moment associated with them. The natural unit for measuring magnetic moments is [5]

$$\mu = e\hbar/2Mc \tag{1.14}$$

where M is the particle's mass. When M equals the electron mass, μ is known as the Bohr magneton. When M is the proton mass, μ is called the nuclear magneton. Table 1.1 shows that the electron magnetic moment is $\sim m_p/m_e$ times larger than the baryon moments.

Apart from the free neutron, the longest lived of the unstable particles is the muon, with a $c\tau = 6.59 \times 10^4$ cm or $\tau = 2.2\ \mu$s. Also listed are the major decay modes of the decaying particles and the corresponding fractions (branching ratios) for each mode.

1.5 Scattering

Most of our knowledge about the interactions between particles has come from the analysis of scattering experiments. Consider the scattering of a beam particle (b) off a target particle (t) in the laboratory (LAB) frame, as shown in Fig. 1.2a. In high energy scattering a particle or group of particles is frequently found to be produced with a momentum comparable to p_b and with a direction close to the beam direction. Such a particle is referred to as the forward or scattered particle (1). In contrast, a second particle or group of particles is frequently found with lower momentum and at a larger angle with respect to the beam direction. This particle is referred to as the backward or recoil particle (2). Such reactions are believed to proceed through the exchange of other (virtual) particles as shown in Fig. 1.2b. Two-body scattering takes its simplest form when viewed in the center of momentum (CM) frame, as shown in Fig. 1.2c.

A useful Lorentz invariant quantity related to the total energy involved in an interaction is

$$\begin{aligned} s &= (\tilde{p}_b + \tilde{p}_t)^2 \\ &= m_b^2 + m_t^2 + 2(E_b E_t - \mathbf{p}_b \cdot \mathbf{p}_t) \end{aligned} \tag{1.15}$$

If we evaluate these quantities in the CM coordinate system ($\mathbf{p}_b = -\mathbf{p}_t$), we find

$$\text{CM:} \qquad s = (E_b^* + E_t^*)^2 = W^2 \tag{1.16}$$

Quantities evaluated in the CM frame will be marked with an asterisk

Figure 1.2 (a) The $2 \rightarrow 2$ body scattering process in the LAB frame. The target particle is at rest. (b) The one-particle exchange diagram. (c) The $2 \rightarrow 2$ body scattering process in the CM system. The initial state particles have opposite momenta, as do the final state particles.

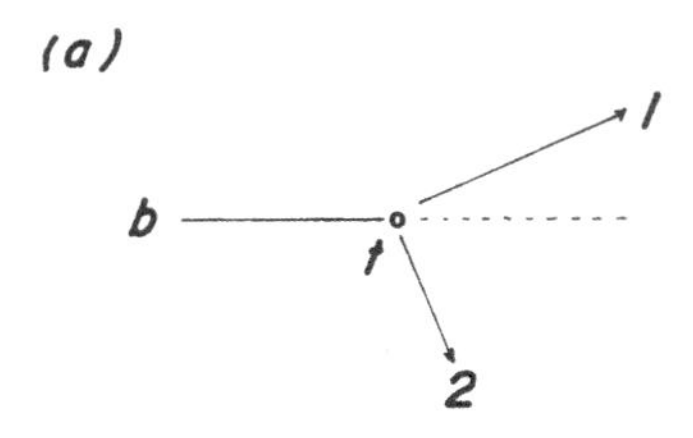

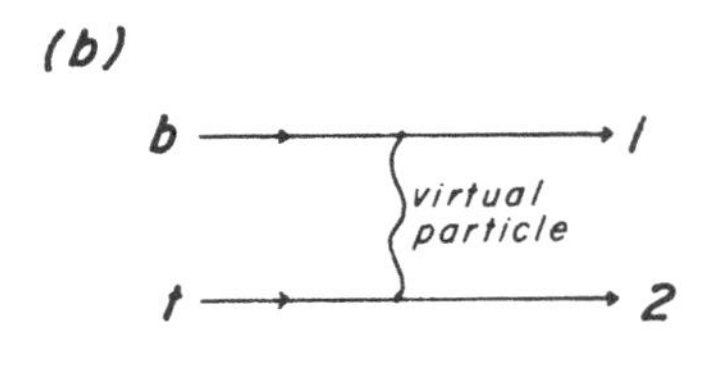

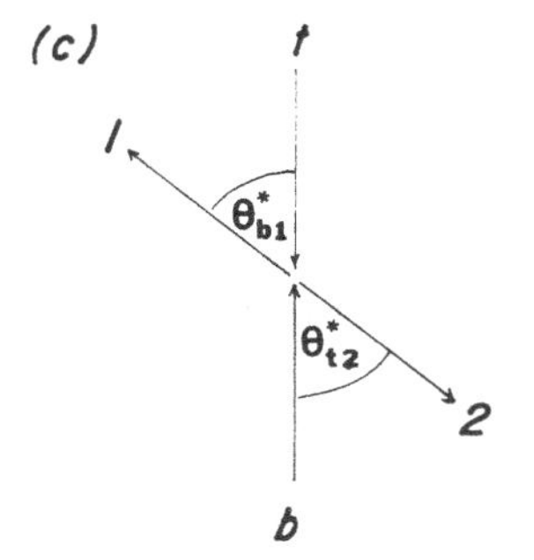

superscript. Thus, s gives the square of the total energy available in the CM system. Expressed in terms of LAB quantities ($p_t = 0$),

$$\text{LAB:} \qquad s = m_b^2 + m_t^2 + 2m_t E_b \tag{1.17}$$

At high energies, where we can neglect the mass terms, the total CM energy should grow like $(E_b)^{1/2}$.

A second Lorentz invariant quantity, which is related to the scattering angle, is the 4-momentum transfer from the beam to the forwardly scattered particle (or system)

$$\begin{aligned} t &= (\tilde{p}_b - \tilde{p}_1)^2 \\ &= m_b^2 + m_1^2 - 2(E_b E_1 - \mathbf{p}_b \cdot \mathbf{p}_1) \end{aligned} \tag{1.18}$$

The differential cross section $d\sigma/d\Omega$ depends on the frame in which the polar angle θ is measured. The cross section $d\sigma/dt$ on the other hand is Lorentz invariant since both σ and t are invariant. We can relate $d\sigma/dt$ to the CM angle by differentiating Eq. 1.18:

$$\frac{d\sigma}{dt} = \frac{d\sigma}{2\, p_b^* p_1^*\, d(\cos\theta^*)}$$

If there are no polarization effects, the differential cross section is independent of the azimuthal angle ϕ, and

$$\frac{d\sigma}{dt} = \frac{\pi}{p_b^* p_1^*} \frac{d\sigma}{d\Omega^*} \tag{1.19}$$

The relations between the kinematic variables of the particles are particularly simple in the CM system. If the incident energy and the masses of the particles are given and the dynamics of the scattering process is independent of azimuth, we have [6]

$$\begin{aligned} \theta_{t2}^* &= \pi - \theta_{b1}^* \\ p_b^* = p_t^* &= \frac{\Lambda^{1/2}(s, m_b^2, m_t^2)}{2(s)^{1/2}} \\ p_1^* = p_2^* &= \frac{\Lambda^{1/2}(s, m_1^2, m_2^2)}{2(s)^{1/2}} \end{aligned} \tag{1.20}$$

where the auxiliary function Λ is defined as

$$\Lambda(a, b, c) = a^2 + b^2 + c^2 - 2ab - 2ac - 2bc \tag{1.21}$$

The relations between the kinematic variables in the LAB frame are discussed in Appendix E.

It is possible to obtain a simple relation between the polar scattering angle θ^* in the CM frame and the polar angle θ in the LAB frame. We have

$$\tan\theta = p_T/p_z$$

where p_T (p_z) is the transverse (longitudinal) component of the particle's momentum. Suppose that the CM frame moves along z in the LAB. The velocity of the CM frame is

$$\beta_0 = \frac{p_0 c}{E_0} = \frac{p_b c}{(p_b^2 c^2 + m_b^2 c^4)^{1/2} + m_t c^2} \tag{1.22}$$

Now, p_T is invariant in the transformation from the CM to the LAB frame. However, the quantity p_z must obey the transformation law in Eq. 1.8. Thus, we have

$$\tan\theta = \frac{p_T^*}{\gamma_0(p_z^* + \beta_0 E^*/c)}$$

Dividing through by the magnitude of the particle's momentum p^*, we obtain

$$\tan\theta = \frac{\sin\theta^*}{\gamma_0(\cos\theta^* + \beta_0/\beta^*)} \tag{1.23}$$

Finally we derive an important result for the maximum energy that can be transferred from an incident particle to a target particle. Consider an incident particle with mass M and momentum p that has a headon collision with a target particle that has mass m and is initially at rest. In the CM frame the recoiling target particle has momentum and energy

$$p_r^* = mc\beta_0\gamma_0$$
$$E_r^* = mc^2\gamma_0$$

where β_0 is the velocity of the CM frame relative to the LAB frame. For cases with $M \gg m$

$$\beta_0 \simeq \frac{pc}{(p^2c^2 + M^2c^4)^{1/2}} = \beta$$

where β is the incident particle's velocity in the LAB. The energy of the recoil particle in the LAB can be determined from Eq. 1.8 by making a Lorentz transformation back to the LAB.

$$E_r \simeq \gamma c\left(\frac{E_r^*}{c} + \beta p_r^*\right)$$

Before the scattering occurred, this particle had only its rest energy mc^2. Thus, after substituting for p_r^* and E_r^*, we find that the maximum energy transfer is

$$\Delta E_{max} \simeq 2mc^2\beta^2\gamma^2 \tag{1.24}$$

Note that since the energy transfer is proportional to γ^2, the recoiling particle can receive a substantial amount of energy from an energetic particle. The exact expression is [4]

$$\Delta E_{\max} = 2mc^2\beta^2\gamma^2 \left(1 + 2\gamma \frac{m}{M} + \frac{m^2}{M^2}\right)^{-1} \tag{1.25}$$

1.6 Particle physics experiments

It is an unfortunate fact of life that nature only begrudgingly reveals the secrets of her elementary particles. Huge experiments involving hundreds of people, years of effort, and the expenditure of millions of dollars may be necessary to measure the properties of new particles or the characteristics of particle interactions. Sometimes the elapsed time from an experiment's conception through its organization, construction, running at the accelerator, and data analysis to the publication of the results is so long that the original goals of the experiment are less important than other topics subsequently developed. In this and the following section we will attempt to give a cursory overview of a particle physics experiment. The subsequent chapters in the book will then treat each of the major topics in more detail.

Most detectors make use of the electromagnetic interactions of particles in matter (Chapter 2). For charged particles heavier than the electron, these interactions tend to be nondestructive because, apart from a small energy loss and a small momentum transfer, the particle is otherwise undisturbed. For these particles ionization of atomic electrons in the detector medium is the dominant source of energy loss. For high energy electrons the energy loss is mainly due to the production of photons through the bremsstrahlung process, while for high energy photons the main source of energy loss is through pair production. Thus, the interactions of high energy electrons and photons are destructive, since the initial particle is destroyed and replaced by a shower of lower energy particles.

The momentum transfer to charged particles is due to the Coulomb interaction of the particle with the nuclei in the medium. This momentum transfer causes small angular changes in the particle's trajectory. The nuclear interaction is important for neutral particles other than the photon and for high energy or large angle processes (Chapter 3).

Most experiments use a beam of particles produced and accelerated to high energy at a particle accelerator (Chapter 4). The main exceptions are experiments searching for evidence of nucleon decay, which look for a signal from a large volume of matter, free quark searches, and cosmic ray experiments. The beam from an accelerator is either directed into a fixed target or collided with a second counterrotating beam of particles. The target for fixed target experiments is usually a small piece of metal or liquid hydrogen (Chapter 5). Spin-polarized targets and gas jets may also be used for special applications.

In a particle physics experiment detectors of various kinds are placed downstream of the fixed target or surrounding the collision point of colliding beams. Particles created in the collisions have electromagnetic or nuclear interactions in the detectors they pass through. The interaction usually creates an analog signal of some kind, which must be measured or converted into standardized pulses using fast pulse electronics (Chapter 6).

Detectors and other electronic apparatus are required for various purposes in every experiment. The tasks required for most experiments include

1. tracking,
2. momentum analysis,
3. neutral particle detection,
4. particle identification,
5. triggering, and
6. data acquisition.

Each detector has particular features for which it excels [7, 8]. The requirements for any given task are generally detrimental to others, so that experimental design requires careful optimization.

The spatial locations of the detector interactions may be combined by computer software to determine the trajectories of the particles. This is referred to as tracking. The most important characteristic of a tracking chamber is the spatial resolution, which measures the accuracy to which the position of the particle trajectory may be localized. Other important characteristics are the response time and the deadtime of the detector. The response time represents the time required to produce a signal after the passage of a particle. It includes the intrinsic time for the interaction between the particle and the detector medium and the time required to collect the photons or charges that were produced by the interaction. A second particle entering the detector during the response time will have its response mixed with the first. In some cases this presents a limitation on the maximum input event rate [7]. The recovery or deadtime is the length of time that must elapse following the passage of a particle before the detector can return to the condition it was in before the arrival of the particle. This time limits the rate at which the experiment can trigger the device. The event rate and average particle multiplicity influence the spatial and temporal resolution required in tracking detectors.

The excited atoms in certain materials can deexcite by emitting light. This is the basis of the scintillation counter (Chapter 7). The light must be efficiently collected and directed onto a photomultiplier tube. This tube first converts the light into an electrical signal and then amplifies the signal to a useful level. Scintillation counters are used for triggering and for particle identification using the time of flight technique.

Charged particles traveling in a dielectric medium with a velocity greater than the speed of light in the medium emit a form of radiation known as Cerenkov light (Chapter 8). This light can also be collected and converted using a photomultiplier tube. Cerenkov counters are used for particle identification.

Particles passing through a chamber containing a gas (or liquid argon) can decompose atoms into electrons and positive ions. If an electric field is present in the chamber, the two charged species drift apart. Near the positive electrode the electrons can acquire sufficient energy to create new ion pairs and a large electrical pulse can develop. This is the basis of the proportional chamber (Chapter 9). Large numbers of wires can be used in parallel to form a multiwire proportional chamber, which is useful for triggering and particle tracking.

In a drift chamber (Chapter 10), instead of detecting the collected charge from a chamber wire, one measures the time from some reference that the electrons take to drift to the wire. Drift chambers are most commonly used for tracking with good resolution. If the pulse height of the signal is also recorded, it is possible to use dE/dx for particle identification.

As mentioned, photons and electrons can create an electromagnetic shower. The characteristics of the shower can be measured with a sampling calorimeter (Chapter 11). A calorimeter usually consists of alternating layers of absorber and detectors. Showers initiated by high energy hadrons may likewise be measured with a hadron calorimeter.

There are a number of other detectors that are normally only used for special applications (Chapter 12). These include emulsions (excellent spatial resolution), bubble chambers and streamer chambers (large solid angle acceptance), transition radiation detectors (high energy particle identification), and silicon detectors (vertex information). Table 1.2 lists the most common uses for a number of detectors.

Every experiment needs a signal (trigger) to indicate when the spatial and temporal correlation of detector signals has determined that a potentially interesting event may have occurred (Chapter 13). The trigger may look for the characteristics of a certain type of particle or for a large deposition of energy in a calorimeter. The system of detectors must efficiently signal the occurrence of the interesting events, even when they are accompanied by a large background of more common reactions. Microprocessors may be employed to make more complicated decisions based on a property of the particle or correlations between particles.

All the detectors in an experiment must be carefully integrated into a detector system (Chapter 14). If a set of tracking chambers is used in conjunction with a magnet, the resulting spectrometer may be used to measure the momentum of charged particles. System design involves a series of compromises on the size and location of the various detectors, the type and strength of magnetic field, the acceptance, segmentation, and rate handling capability. In large experiments careful attention must be given to calibration of the detector signals and to online monitoring of their performance.

Enormous amounts of analog and digital data are generated by the detectors in a large experiment. For example, Fig. 1.3 shows a display of drift chamber information from a high energy $\bar{p}p$ interaction. This data must be channeled via data acquisition systems into an online computer and then some storage medium such as magnetic tape for later data analysis. The data recording rate must be carefully matched with the trigger rate.

Sometimes it is necessary to know the identity (i.e., the mass) of at least some of the particles resulting from an interaction. Two separate kinematic measurements are necessary for particle identification. Usually one is provided by the momentum measurement. The second measurement

Table 1.2. *Detector uses*

Detector	Common uses	Chapter
Scintillation counter	tracking, fast timing, triggering	7
Cerenkov counter	particle identification, triggering	8
Proportional chamber	tracking, triggering	9
Drift chamber	tracking, particle identification	10
Sampling calorimeters	neutral particle detection, triggering	11
Bubble chamber	vertex detector, tracking	12.1
Emulsion	high resolution vertex detection	12.2
Spark chamber	tracking	12.3
Streamer chamber	vertex detector, tracking	12.4
Transition radiation detector	high energy particle identification	12.5
Semiconductor detector	vertex detector	12.6
Flashtube hodoscope	tracking	12.6
Spark counter	high resolution timing	12.6

then requires a specialized detector whose response is proportional to the velocity or energy of the particle.

Finally, after one has selected an appropriate beam of particles and a target, an arrangement of magnetic field and detectors, a trigger, and a data acquisition system, one can do an experiment. Certain types of experiments are fundamentally important, such as those that measure the properties of particles or the total or elastic scattering cross sections (Chapter 15).

1.7 Data analysis

Computers play an essential role in particle physics experiments. We have already mentioned the use of online computers, which monitor the experiment and control the acquisition of data from the detectors. A second major use of computers is to process the data tapes through a series of programs that eventually yield the physics results the experiment was designed to obtain. This function is performed offline in the sense that the processing occurs independently of the experiment, although the same computers and much of the same software may be involved in both tasks. Much of this software is experiment, detector, or computer dependent. Therefore, we will only give a brief survey of the analysis tasks likely to be found in most experiments.

Table 1.3 outlines some common tasks for offline analysis. Of course, some experiments will not require all of these tasks, while others will need additional levels of processing. In general, as the processing level increases, software from different experiments tends to become more alike.

Figure 1.3 Particle trajectories in a $\bar{p}p$ collision with $W = 540$ GeV. The trajectories were determined from drift chamber hits. The electron track, indicated by the arrow, was identified using an electromagnetic calorimeter. This event was one of the first examples of a W vector boson decay. (After G. Arnison et al., Phys. Lett. 122B: 103, 1983.)

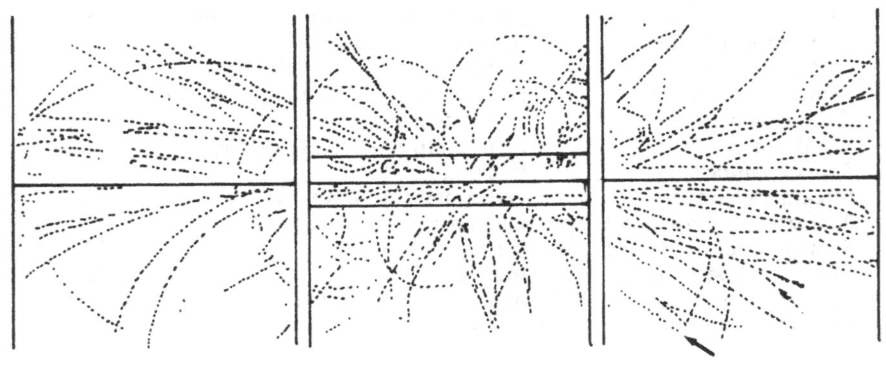

1.7.1 Preprocessing

The information from an experiment can appear in many different forms. It may, for example, include chamber wire numbers, drift times, photomultiplier tube signals, or scaler counts. All of this information must be written in some format on the permanent storage medium, which we take to be magnetic tape. The experimental signals may be channeled to the tape through a standard interface, such as CAMAC, or it may proceed through homemade electronics.

The information on the raw data tapes is usually organized into groups depending on the source of the information. Thus, for instance, beam chamber information may be in one group and drift chamber times in another. The data is usually packed as densely as possible in order to minimize the amount of tape required to record each event.

The first job for the preprocessor is to convert these stored records into a more manageable form. As a result, the preprocessor is the most experiment and computer dependent set of software. Data formatting routines unpack the experimental information and fill appropriate arrays that serve as input for the subsequent processing.

One of the most important software jobs is tracking. Thus, a second task of the preprocessing program is to calculate the spatial coordinates of all hits in the tracking chambers. The program applies predetermined calibration constants in order to convert the output of the devices into spatial coordinates. For example, the space–time relation for a drift chamber can be measured by scanning the position of the beam across a drift chamber cell and measuring the drift times. It is also necessary to determine the absolute positions of the chambers in some coordinate system. In order to do this, special alignment runs are performed with tracks whose trajectories have been determined independently. The alignment constants are adjusted until the deviations of the positions given by the chambers from the actual positions (residuals) are mini-

Table 1.3. *Offline analysis chain*

Level	Task	Purpose
1	Preprocessing	decodes raw data tapes, finds spatial coordinates
2	Pattern recognition	finds tracks, rough momentum with approximate field
3	Geometrical fitting	best track parameters using true field
4	Vertexing	associates tracks, particle decays
5	Kinematic fitting	assigns masses, finds missing neutrals
6	Physics analysis	finds effective masses, Dalitz plots, etc.

mized. In some chambers it may also be necessary to perform $\mathscr{E} \times B$ corrections on the apparent coordinates because of the motion of the electrical discharge in the fields.

It may also be possible to filter out certain classes of events at the preprocessing stage if it is known with certainty that they will fail a subsequent level of processing. For example, unless there are a minimum number of hits, pattern recognition will be unable to determine if a track was present. Eliminating such events as soon as possible minimizes the total processing time.

1.7.2 Pattern recognition

In order to do tracking, the programs must first recognize from the arrays of chamber hit positions when it is likely that the pattern of hits was caused by the passage of charged tracks and to determine the best values for the parameters describing the tracks. The first task is referred to as pattern recognition, while the second is known as geometrical fitting or more simply as geometry.

The pattern recognition program must take the arrays of spatial positions and determine when a set of hits represents a track. For 3-dimensional track reconstruction, information must be available for more than one plane. Typically a third set of planes may be used to resolve ambiguities. Pattern recognition is one of the most difficult software tasks. These programs must be carefully optimized for the specific experiment, the quality of the beam, and the performance of the tracking chambers. A large number of problems can arise, and specific algorithms must be available for every eventuality.

Several general methods of pattern recognition have been used for finding tracks [9]. A brute force examination of all possible combinations of hits is too time consuming for anything but the most simple experiments. Track following is a method commonly used with sets of closely spaced chambers. Here one starts with sets of three or four hits as far from the target or interaction region as possible so that the confusion of nearby tracks is minimized. The program then predicts the next few hits by extrapolating the assumed trajectory. If a chamber hit is present within a window determined by the errors on the extrapolation, the process is continued. On the other hand, if after taking into account the chamber efficiencies no more hits are found, the track may be abandoned. A vectorlike variation of this technique can be used when the chambers consist of closely spaced planes that measure more than one dimension of the trajectory. Then each chamber module gives a vector on the track, and one can search neighboring modules for corresponding vectors.

Another technique forms a track "road" by picking initial points near the beginning and end regions of chambers. If the track is curved, a third point near the center is also required. The program then uses a simple model of the trajectory and the measured position errors to define the road through the chambers and checks to see if additional hits lie on the road.

In some cases it is possible to use a global method of pattern recognition. If all the points on a given track have approximately the same value for some function of the coordinates, the tracks can be recognized by making a histogram of the function. The points belonging to a given track will cluster together. For example, the quantity y/x is the same for all points on a track in a field free region.

Some of the problems encountered in pattern recognition are illustrated in Fig. 1.4, which shows the pattern of hits in a set of chambers. Figure 1.4a shows the pattern measured perpendicular to the direction of

Figure 1.4 Pattern of chamber hits in two views.

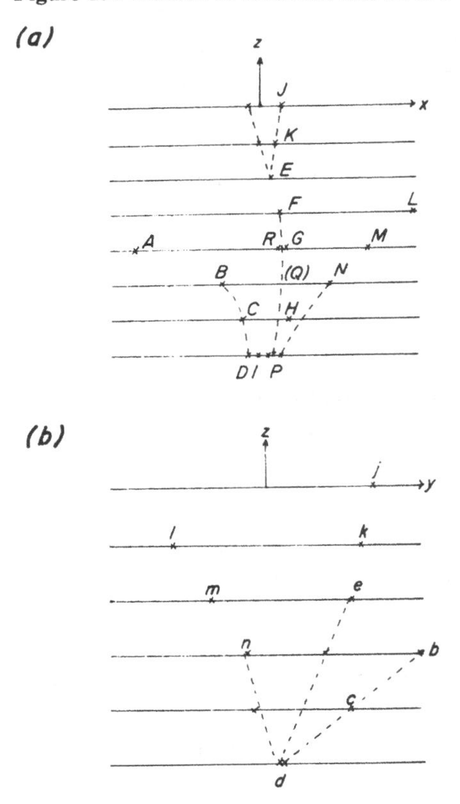

the magnetic field (y), while Fig. 1.4b shows measurements in the plane containing the field. Assuming the field is approximately uniform, trajectories in the view perpendicular to the field form circular arcs. Now there must be some minimum requirements for what constitutes a track. Chambers may have spurious noise hits (A), while the chambers closest to the target may have many closely spaced hits (D, I, P). The position of each hit is only known to the accuracy of the chamber resolution. This makes it difficult to determine whether possible short track combinations such as BCD are really tracks. Examination of the hit patterns in the y-measuring planes may give additional confidence. In this view tracks lie along approximately straight lines. However, unless one is using specially constructed, 3-dimensional tracking chambers, the measurements in x and y occur at different values of z, and there is not an exact one-to-one correspondence between measurements in the two views. Since points b and c in Fig. 1.4b point back to the target, it is likely that BCD is a track, and that it appears so short because it is produced at a large angle with respect to the measuring planes.

The opposite problem from chamber noise is chamber inefficiency. Some tracks may have a missing hit (Q). Sometimes a track has a large angle multiple scattering or interaction in the chamber (E). This causes the apparent trajectory to appear as two broken segments. Another problem that can occur near the edge of a large magnet is nonuniformity in the field. This can cause a trajectory (LMN) to deviate smoothly from a circle in the xz view and from a line in the yz view.

1.7.3 Geometrical fitting

After the pattern recognition programs have determined which sets of chamber hits belong to tracks, it is necessary to obtain accurate measurements of the track parameters. Particle trajectories in a magnetic field $\mathbf{B}$ must satisfy the equation of motion

$$\frac{d^2\mathbf{x}}{ds^2} = \frac{q}{pc}\frac{d\mathbf{x}}{ds} \times \mathbf{B} \tag{1.26}$$

where q is the particle's charge, p is its momentum, and s is the distance along the trajectory. Neglecting energy loss, the solution of Eq. 1.26 is a straight line for a field free region and a helix for a uniform magnetic field. For these cases five parameters must be specified to define the trajectory: the coordinates x_0 and y_0 at some reference plane $z = z_0$, the magnitude of the momentum, and two angles to specify the direction.

Figure 1.5 shows the parameters for a helical trajectory in a uniform magnetic field $\mathbf{B} = B\hat{e}_y$. The axis of the helix is parallel to the direction of the field. The dip angle λ measures the orientation of the momentum above or below the xz plane. The angle ϕ_0 is the azimuthal angle in the xz plane of the projection of the starting point with respect to the x axis. The projection of the trajectory onto the xz plane is a circle with radius of curvature given by

$$\rho = \frac{p \cos \lambda}{qB} \tag{1.27}$$

The projected curvature

$$k = 1/\rho \tag{1.28}$$

is usually used as a track parameter instead of p because the error in k is constant for constant position measurement errors [9, 10]. Any point on the helix can be expressed as a function of the arc length traversed from the reference point as

$$\begin{aligned} x(s) &= (1/k)[\cos(\phi_0 + ks \cos \lambda) - \cos \phi_0] + x_0 \\ y(s) &= s \sin \lambda + y_0 \\ z(s) &= (1/k)[\sin(\phi_0 + ks \cos \lambda) - \sin \phi_0] + z_0 \end{aligned} \tag{1.29}$$

If the energy loss is appreciable, as it is for bubble chambers, the curvature is a function of s.

Figure 1.5 Definition of helical track parameters.

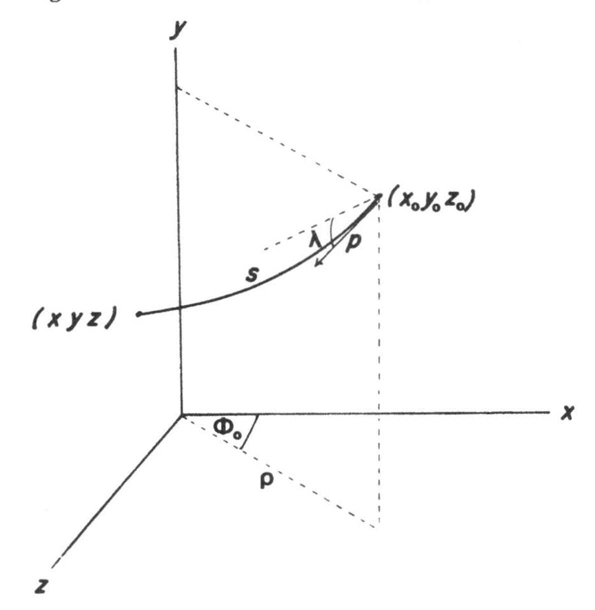

The curvature also varies if the magnetic field is nonuniform, in which case the trajectory may have to be broken up into segments. The trajectory through each segment is then either represented by a low order polynomial (spline fit) or determined by numerically integrating the equations of motion [9]. This is usually done by rewriting the x and y components of Eq. 1.26 in the form

$$\begin{aligned} x'' &= \frac{q}{pc}\frac{ds}{dz}[x'y'B_x - (1 + x'^2)B_y + y'B_z] \\ y'' &= \frac{q}{pc}\frac{ds}{dz}[(1 + y'^2)B_x - x'y'B_y - x'B_z] \end{aligned} \tag{1.30}$$

where the primes refer to derivatives with respect to z and

$$ds/dz = (1 + x'^2 + y'^2)^{1/2}$$

If the positions (x_n, y_n) and directions (x'_n, y'_n) of the trajectory are known at z_n, we can estimate their values at a nearby position z_{n+1} using Eq. 1.30 to give the second derivatives [11]. If we define $h_n = z_{n+1} - z_n$ and expand in a Taylor's series around z_n, we find

$$\begin{aligned} x_{n+1} &= x_n + x'_n h_n + 1/2 x''_n h_n^2 \\ x'_{n+1} &= x'_n + h_n x''_n \end{aligned} \tag{1.31}$$

with similar equations for y_{n+1} and y'_{n+1}. Then by reevaluating the second derivatives at z_{n+1}, one can continue stepping through the inhomogeneous magnetic field.

Once some model of the track trajectory has been adopted, the track parameters are determined by making a least squares fit to the measured spatial coordinates. Suppose we have a set of N chambers at fixed values of z. Then a χ^2 function can be defined by

$$\chi^2 = \sum_{i=1}^{N} \left[\frac{x_i - f(\omega; z_i)}{\sigma_i}\right]^2 \tag{1.32}$$

where x_i is the measured coordinate for the plane at z_i, σ_i is the error on the measurement, and $f(\omega; z_i)$ is the projection of the trajectory determined by the parameters ω onto the measurement plane. For a helical trajectory ω is the set $\{x_0, y_0, \phi_0, \lambda, k\}$. The parameter values are adjusted until the χ^2 function is minimized and

$$\frac{\partial \chi^2}{\partial \omega_j} = 0 \tag{1.33}$$

for each parameter ω_j. The least squares fitting procedure also produces a complete covariance matrix for the parameter errors.

1.7.4 Vertexing

The vertexing programs attempt to ascertain if isolated tracks originated from a common point. Bubble chamber analysis programs generally combine this task with kinematic fitting. All the tracks in a true event should originate from a production vertex or result from the decay of another particle that did.

In experiments using chambers for tracking, the geometrical tracks usually have to be extrapolated from the first chambers back toward the interaction region. This may require taking a series of small steps or making an initial pass with large tolerances if the magnetic field is inhomogeneous. The program then computes the distance of closest approach of the tracks, and tests if it is smaller than some minimum distance. Once two or more tracks are found that appear to be associated, the position of the actual vertex can be estimated by minimizing a χ^2 function defined in terms of the distances of the tracks to the assumed vertex point and the errors on the extrapolated tracks due to the errors on the track parameters. Alternatively, one could do an overall fit of the associated tracks to their chamber hits with the constraint that all the tracks must originate from a point.

In multivertex events the downstream decay vertices are found first. For vees the sum of the measured track momenta gives the momentum of the decaying particle. It can then be checked if the decaying particle is associated with the beam and other tracks at a production vertex. The tolerance that must be allowed for associating tracks is sensitive to the quality of the track measurements. Within the errors tracks may appear to come from more than one vertex. Thus, the physics questions under study may influence how the tracks are assigned to vertices.

1.7.5 Kinematic fitting

The complete kinematic description of an event requires that we specify the mass of each particle in addition to its momentum. For some tracks additional information may be available from particle identification detectors, such as Cerenkov counters or dE/dx chambers. The masses of the other tracks in an event are generally ambiguous. In this case one can assign various mass hypotheses to each of the tracks and perform a kinematic fit to the overall event. Given the mass assignments, the 4-vectors of the initial and final state systems are determined, and a true event must satisfy the laws of conservation of energy and momentum.

The fit proceeds by minimizing a χ^2 function defined in terms of the difference between the track parameters and their geometry values as well as the errors on the track parameters [12]. Alternatively, one could consider the residuals of the tracks from the measured chamber hits. The energy-momentum constraints are added to the function using the method of Lagrangian multipliers. Since these constraints are nonlinear in the track parameters, the function must be minimized iteratively.

If all the tracks in an event have been well measured, there are four constraints ($4C$) on the overall fit. If there is a missing track, three constraints are lost in order to determine its momentum, and only one constraint ($1C$) remains on the fit. Multivertex events can be combined either by first fitting the downstream vertices and then working back toward the production vertex or by first fitting each vertex independently and then refitting them all simultaneously.

1.7.6 Physics analysis

It is obvious that this stage of the data analysis is totally experiment dependent. However, much of the software that is used at this stage is applicable to many different problems. Software should be available for making histograms and scatter plots of the data. When the 4-vectors of the particles are either fitted or assumed, effective masses, t distributions, Dalitz plots, and missing masses are commonly calculated. A number of cuts are usually applied to the data to obtain a clean sample of the particular types of events of interest.

Another important task at this stage is to understand the normalization for the data collected in the experiment. This is usually done by generating events using Monte Carlo (statistical) techniques and then propagating the created tracks through a simulation of the experimental apparatus. This allows one to find how the acceptance for various quantities in the experiment depend on known properties of the created tracks.

References

[1] D. Perkins, *Introduction to High Energy Physics,* 2nd ed., Reading: Addison-Wesley, 1982.
[2] G. t'Hooft, Gauge theories of the forces between elementary particles, Sci. Amer., June: 104-38, 1980. A fifth type of interaction has been proposed to explain the phenomenon of CP violation in K decays. At this point it is not clear if this is necessary.
[3] R. Hagedorn, *Relativistic Kinematics,* New York: Benjamin, 1964.
[4] Particle Data Group, Review of particle properties, Rev. Mod. Phys. 56: S1-S304, 1984.
[5] L. Pondrom, Magnetic moments of baryons and quarks, in G. Bunce (ed.), *High Energy Spin Physics—1982,* AIP Conference Proceeding No. 95, New York: AIP, 1983, pp. 45-57.

[6] W.S.C. Williams, *An Introduction to Elementary Particles,* 2nd ed., New York: Academic, 1971, pp. 491–4.
[7] C. Fabjan and H. Fischer, Particle detectors, Rep. Prog. Phys. 43: 1003–63, 1980.
[8] K. Kleinknecht, Particle detectors, Phys. Rep. 84: 85–161, 1982.
[9] H. Grote, Data analysis for electronic experiments, in C. Verkerk (ed.), Proc. of the 1980 CERN School of Computing, CERN Report 81-03, 1981, pp. 136–81.
[10] T. Fields, Magnets for bubble and spark chambers, in R. Shutt (ed.), *Bubble and Spark Chambers,* New York: Academic, 1967, pp. 1–50.
[11] J. Hart and D. Saxon, Track and vertex fitting in an inhomogeneous magnetic field, Nucl Instr. Meth. 220: 309–26, 1984.
[12] M. Alston, J. Franck, and L. Kerth, Conventional and semiautomatic data processing and interpretation, in R. Shutt (ed.), *Bubble and Spark Chambers,* New York: Academic, 1967, pp. 51–139.

Exercises

1. Suppose we express β for a particle in terms of the sine of some parametric angle θ, that is, $\beta = \sin\theta$. How are γ and p/m given in terms of θ?

2. What is the mean lifetime of a 100-GeV muon in the LAB frame?

3. For a 3-body final state with particles of masses m_1, m_2, and m_3, show that the lower limit for the effective mass of two of the particles is $m_1 + m_2$, and the upper limit is $W - m_3$, where W is the total energy in the CM frame.

4. Suppose in a hyperon production experiment we want a mean decay region of 1 m following the target. Find the required momentum for the produced hyperon for Λ, Σ^-, Ξ^-, Ω^-, and Σ^0.

5. What is the CM momentum of the π^- in the reaction $\pi^- p \rightarrow K^0 \Lambda$ at a CM energy of 3 GeV? What is the CM momentum of the Λ? Can the K^0 be emitted in the backward hemisphere in the LAB?

6. What is the maximum energy transfer to an electron from a 100-GeV pion?

7. Suppose that the velocity of a certain particle in the CM frame β^* is less than the velocity β_0 of the CM in the LAB frame. Show that there is a maximum angle θ_{max} at which the particle may be emitted in the LAB given by

$$\tan\theta_{max} = \frac{\beta^*}{\gamma_0(\beta_0^2 - \beta^{*2})^{1/2}}$$

8. Make a rough flow chart for a pattern recognition program that takes into account the problems illustrated in Fig. 1.4.
9. Derive Eq. 1.26 starting from the Lorentz force equation. Derive Eq. 1.30.
10. Show that Eq. 1.29 is a solution to Eq. 1.26 for the case when $\mathbf{B} = B\hat{e}_y$.

2

Electromagnetic interactions

Before a particle can be detected, it must first undergo some sort of interaction in the material of a detector. Processes that result from the electromagnetic interaction are the most important for particle detection. In this chapter we will consider four major topics. The first is the loss of energy by charged particles heavier than the electron due to the excitation or ionization of atomic electrons. We will calculate the most probable value of the ionization energy loss and the distribution of the fluctuations in that quantity. Second, we will consider the interactions of electrons. These include ionization losses and the loss of energy due to photon emission (bremsstrahlung). The third topic is the interaction of photons with matter. The most important of these are the photoelectric effect, the Compton effect, and pair production. Lastly, we will examine Coulomb scattering of charged particles with the atomic nucleus, which is responsible for multiple scattering.

There are additional processes that, although they are electromagnetic in nature, are more appropriately discussed in other sections of the book. These include scintillation light (Chapter 7), Cerenkov light (Chapter 8), ionization in gases (Chapter 9), electromagnetic showers (Chapter 11), and transition radiation (Chapter 12). Strong and weak nuclear interactions of particles in matter are discussed in Chapter 3.

A rigorous treatment of electromagnetic effects requires calculations using QED. This theory describes the interactions in terms of the exchange and emission of photons. Since these calculations require special techniques and tend to be rather lengthy, we shall be content to present simple arguments for the processes under consideration and only quote QED results [1].

2.1 Energy loss in matter

Let us begin by considering the loss of kinetic energy of an incident charged particle due to its Coulomb interaction with charged particles in matter. We first give a semiclassical argument that demonstrates the physical causes of the energy loss. Let the incident particle have mass M, charge $z_1 e$, and velocity v_1. We assume it is interacting with a particle in the material with mass m and charge $z_2 e$ and that the material particle is essentially at rest. We restrict ourselves to cases where only small momentum transfers are involved, so that the trajectory of the incident particle is not appreciably altered and the material particle only has a small recoil. The trajectory of the incident particle defines the axis of a cylinder as shown in Fig. 2.1. We consider the interaction with a particle in the cylindrical shell a distance b from the axis. The distance b is referred to as the impact parameter for the interaction.

The moving charge creates an electric and magnetic field at the location of the material particle. Since the material particle is assumed to have only a small velocity, the magnetic interaction is not important. By symmetry the net force acting on the material particle is perpendicular to the cylinder. The transverse electric field is

$$\mathscr{E}_\perp = z_1 eb/r^3 \tag{2.1}$$

in the rest frame of the incident particle. The electric field observed in the LAB changes with time. Suppose that the incident particle reaches its point of closest approach at $t = 0$. At time t the transverse electric field in the LAB frame is given by [2]

$$\mathscr{E}_\perp = \frac{\gamma z_1 eb}{(b^2 + \gamma^2 v_1^2 t^2)^{3/2}} \tag{2.2}$$

Figure 2.1 A cylindrical sheet of matter surrounding a particle trajectory.

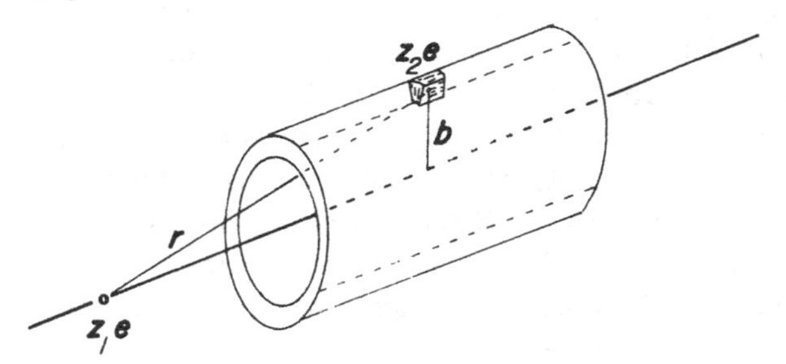

The momentum acquired by the bound particle is

$$\Delta p = \int F\,dt$$

$$= \int_{-\infty}^{\infty} \frac{(z_2 e)\gamma z_1 e b\,dt}{(b^2 + \gamma^2 v_1^2 t^2)^{3/2}} \tag{2.3}$$

$$\Delta p = \frac{2 z_1 z_2 e^2}{v_1 b}$$

The incident particle will have collisions with both the nuclei and the electrons of the atoms. Since the bound particle is assumed to have only a small velocity, the energy transfer can be written

$$\Delta E = \frac{(\Delta p)^2}{2m} = \frac{2 z_1^2 z_2^2 e^4}{b^2 v_1^2 m} \tag{2.4}$$

We see that the energy transfer is inversely proportional to the square of the incident particle velocity and to the square of the impact parameter. Thus, most of the energy transfer is due to close collisions. We have $m = m_e$ and $z_2 = 1$ for electrons and $m = Am_p$ and $z_2 = Z$ for nuclei. With Z electrons in an atom and $A \approx 2Z$,

$$\frac{\Delta E\text{ (electrons)}}{\Delta E\text{ (nucleus)}} = \frac{Z}{m_e}\left(\frac{Z^2}{2Zm_p}\right)^{-1} \approx 4000$$

so we see that the atomic electrons are responsible for most of the energy loss. We will let $m = m_e$ for the rest of this chapter.

Now let us calculate the total energy lost by the incident particle per unit length in the material. We have just seen that most of the energy loss is due to interactions with the atomic electrons. There are $n_e \times 2\pi b\,db\,dx$ electrons in the cylindrical shell of Fig. 2.1, where

$$n_e = Z_2 n_a = Z_2 \frac{N_A \rho}{A} \tag{2.5}$$

is the number of electrons per unit volume (cf. Appendix D). Summing over the total energy transfer in each b interval, the total energy loss per unit length is

$$dE/dx = 2\pi n_e \left(\frac{2Z_1^2 e^4}{m v_1^2}\right) \int_{b_{min}}^{b_{max}} \frac{db}{b}$$

$$= \frac{4\pi n_e Z_1^2 e^4}{m v_1^2} \ln \frac{b_{max}}{b_{min}} \tag{2.6}$$

The limiting values of the impact parameter are determined by the range of validity of the various assumptions that were made in deriving Eq. 2.6. We have assumed that the interaction takes place between the electric field of the incident particle and a free electron. However, the electron is actually bound to an atom. The interaction may be considered to be with a free electron only if the collision time is short compared to the characteristic orbital period of electrons in the atom. Examination of Eq. 2.2 shows that the transverse electric field in the LAB is very small except near $t = 0$. The full width at half maximum of the $\mathscr{E}(t)$ distribution is $b/v\gamma$ times a constant of order 1, so we take [2]

$$t_{\text{coll}} \approx \frac{b}{v_1 \gamma} \tag{2.7}$$

An upper limit for the impact parameter then is

$$b_{\text{max}} \approx \frac{\gamma v_1}{\omega} \tag{2.8}$$

where ω is a characteristic orbital frequency. A lower limit of validity for b is obtained from the requirement that ΔE cannot exceed the maximum allowed energy transfer for a head-on collision. Thus, equating Eq. 2.4 evaluated at b_{min} with Eq. 1.24, we find that

$$b_{\text{min}} \approx \frac{Z_1 e^2}{\gamma m v_1^2} \tag{2.9}$$

If we substitute Eqs. 2.8 and 2.9 into Eq. 2.6, we obtain

$$\frac{dE}{dx} = \frac{4\pi n_e Z_1^2 e^4}{m v_1^2} \ln \frac{m v_1^3 \gamma^2}{Z_1 e^2 \omega} \tag{2.10}$$

A more rigorous classical calculation, originally due to Bohr, treats the atom as a harmonically bound charge for distant collisions. However, the results of this calculation differ numerically from Eq. 2.10 by a negligible amount [2].

Now let us determine the classical electromagnetic cross section for an incident particle to lose an amount of energy W. Consider a ring of width db at an impact parameter b from an atom. Every incident particle passing through the annular region undergoes a certain deflection. By definition, the differential cross section is the area of the ring and

$$\frac{d\sigma(b)}{db} db = 2\pi b \, db$$

We use Eq. 2.4 to relate b to the energy transferred to an atomic electron, which we assume is equal to the energy W lost by the incident particle. We find that

$$\frac{d\sigma}{db}\,db = \frac{2\pi Z_1^2 Z_2^2 e^4}{m v_1^2}\,\frac{dW}{W^2}$$
$$= \frac{d\sigma(W)}{dW}\,dW$$

Thus, the classical cross section for obtaining an energy loss W is

$$\frac{d\sigma}{dW} = \frac{2\pi Z_1^2 Z_2^2 e^4}{m v_1^2 W^2} \tag{2.11}$$

When electromagnetic scattering represents the dominant source of energy loss, a pure beam of monoenergetic, charged, stable particles heavier than the electron travels approximately the same range R in matter. For example, a beam of 1 GeV/c protons has a range of about 200 g/cm^2 in lead (17.6 cm). Because of their light mass, the paths of electrons in matter have much larger deviations. A plot of the number of heavy charged particles in a beam is shown in Fig. 2.2 as a function of the depth into the material. Also shown is the local value of dE/dx. The small decrease in intensity that occurs at all depths is caused by nuclear or large angle scattering processes. An interaction that removes particles of a given type from a beam leads to an exponential decrease in the intensity of those particles. Most of the ionization loss occurs near the end of the path, where the velocity is smallest. This increase in the energy loss is referred to as the Bragg peak. The depth at which half the initial particles remain is called the mean range. This is related to the energy loss by

$$R(E) = \int_E^0 \frac{1}{-dE/dx}\,dE \tag{2.12}$$

The range represents the distance traversed along the trajectory of the

Figure 2.2 Number of heavy charged particles in a beam and dE/dx as a function of depth in the absorber (R is the mean range).

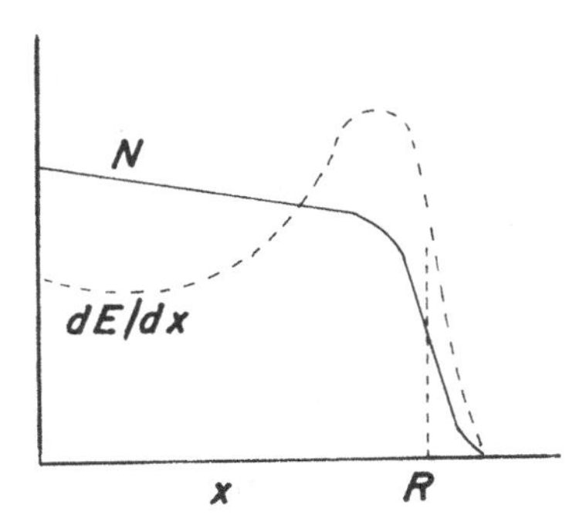

particle itself and differs from the thickness of the absorber because of multiple scattering.

We can derive an important scaling law for the range. The energy loss formula can be written

$$\frac{dE}{dx} = Z_1^2 f(v) = Z_1^2 g\left(\frac{E}{M}\right)$$

where f and g are functions. Then, according to Eq. 2.12, the range is given by an equation of the form

$$R = \int \frac{1}{Z_1^2 g(E/M)} M \frac{dE}{M}$$
$$R\left(\frac{E}{M}\right) = \frac{M}{Z_1^2} h\left(\frac{E}{M}\right) \tag{2.13}$$

where h is a universal function of E/M. To illustrate the usefulness of Eq. 2.13, suppose that the range of some particle, a proton, for example, has been measured as a function of v or of E/M. Then the range of another particle, an alpha particle, for example, with energy E_α can be related to the proton range by

$$\frac{Z_p^2}{M_p} R_p\left(\frac{E_\alpha}{M_\alpha}\right) = \frac{Z_\alpha^2}{M_\alpha} R_\alpha\left(\frac{E_\alpha}{M_\alpha}\right)$$
$$R_\alpha\left(\frac{E_\alpha}{M_\alpha}\right) = \frac{M_\alpha}{M_p} \frac{Z_p^2}{Z_\alpha^2} R_p\left(\frac{E_\alpha}{M_\alpha}\right)$$

The range–energy relation can often be expressed empirically in the form

$$R(E) = (E/E_0)^n \tag{2.14}$$

For example, the range in meters for low energy protons in air can be approximated with $n = 1.8$ and $E_0 = 9.3$ MeV [3].

The energy losses and ranges of a number of incident particles in a variety of absorber materials are shown in Fig. 2.3. The range and dE/dx are expressed in terms of the amount of mass traversed (g/cm^2) instead of the linear distance. All the dE/dx curves show the $1/v^2$ drop for small momentum and a region of minimum ionization for higher momentum. Similar curves are given for particles in liquid hydrogen in Fig. 2.4.

2.2 Quantum treatment of the energy loss

The semiclassical treatment of the energy loss given in the preceding section treats the quantum nature of the particles in an ad hoc fashion. A proper treatment must take into account (1) the fact that

Figure 2.3 Mean range and energy loss of charged particles in solids. Calculations use the Bethe–Bloch equation with density effect corrections. Refer to the cited reference for a discussion of assumptions and qualifications. (Particle Data Group, Rev. Mod. Phys. 56: S1, 1984.)

PARTICLE DETECTORS, ABSORBERS, AND RANGES

Mean Range and Energy Loss in Lead, Copper, Aluminum, and Carbon

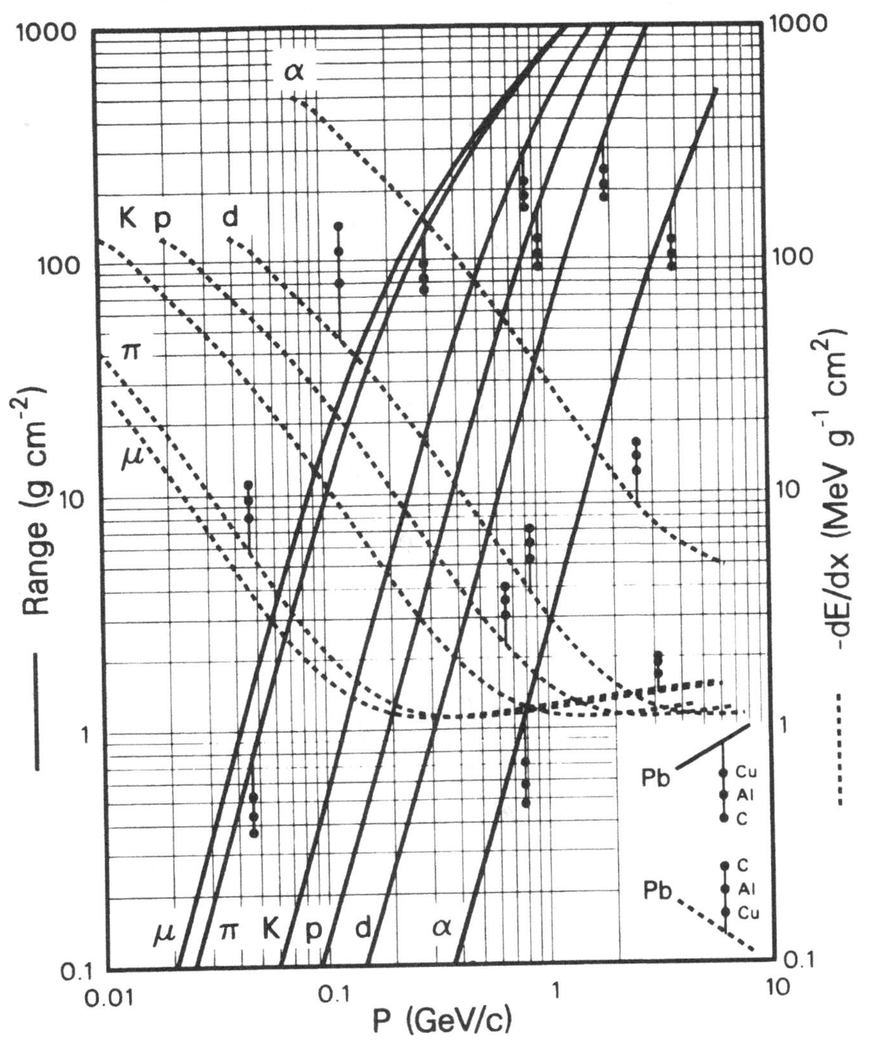

Figure 2.4 Mean range and energy loss of charged particle in liquid hydrogen. Calculations use the Bethe–Bloch equation with density effect corrections. Refer to the cited reference for a discussion of assumptions and qualifications. (Particle Data Group, Rev. Mod. Phys. 56: S1, 1984.)

PARTICLE DETECTORS, ABSORBERS, AND RANGES

Mean Range and Energy Loss in Liquid Hydrogen

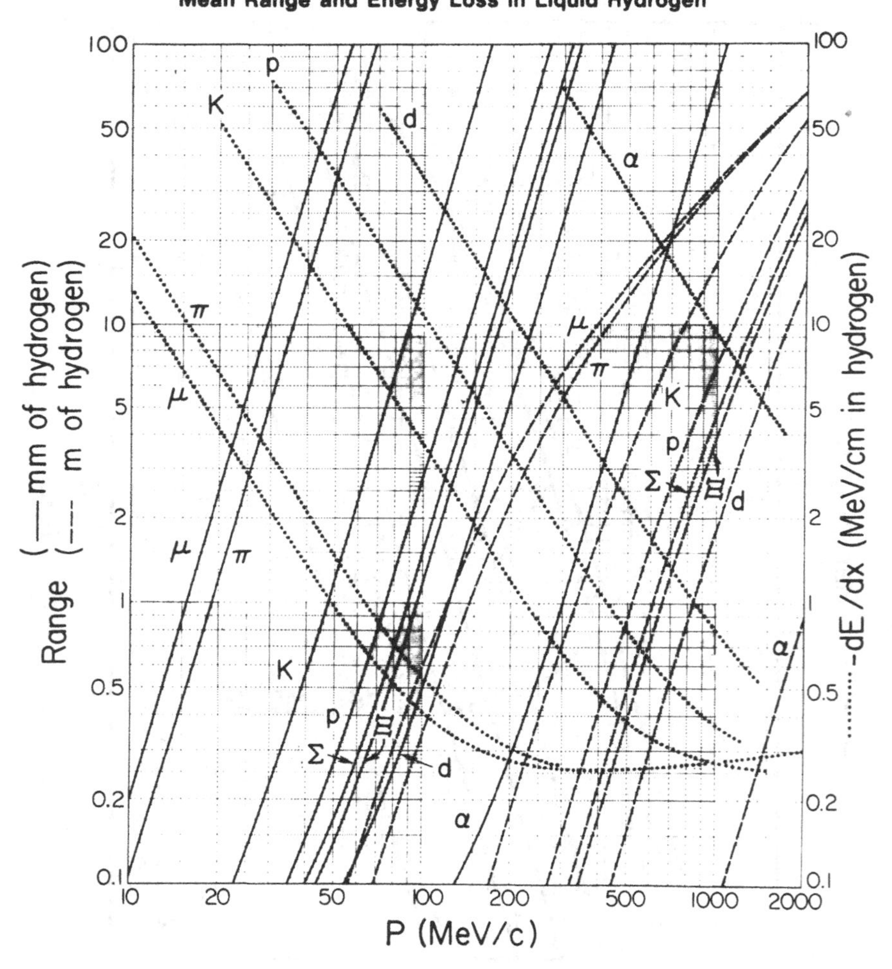

energy transfers to the atomic electrons only occur in discrete amounts and (2) the wave nature of the particles. For very close collisions the classical specification of a particle as an object with a well-defined position and momentum conflicts with the uncertainty principle. In the early 1930s Bethe and Bloch treated the problem of energy loss in the framework of quantum mechanics. We will summarize some important aspects of their treatment in this section [4, 5].

Bethe's theory classifies atomic collisions according to the amount of momentum or energy transfer to the bound electron. This is an observable quantity in contrast to the impact parameter used in the semiclassical derivation. However, one can associate the small momentum transfer processes with a large impact parameter (distant collisions) and the large momentum transfers with small impact parameters (close collisions).

In distant collisions the incident particle interacts with the atom as a whole. There will be a certain probability that the energy lost by the incident particle will cause excitation of an electron to a higher energy level or will cause ionization. Bethe calculated the probabilities for these transitions using first-order perturbation theory. The incident particle is treated as a plane wave. The spin and magnetic moment of the atomic electrons are properly treated if Dirac wavefunctions are used. The perturbation is the Coulomb potential plus a coupling to the photon field. The total contribution to the energy loss comes from summing all excitation energies, each weighted by the cross section for that excitation. Thus,

$$\left.\frac{dE}{dx}\right|_{W<\eta} = n_a \sum_n \int E_n \, d\sigma_n \tag{2.15}$$

where η is a limiting energy transfer (~ 50 KeV) for which the assumptions used in deriving Eq. 2.15 are valid [1]. The expression obtained after evaluating Eq. 2.15 depends on the atomic properties through the mean ionization potential I.

For close collisions the interaction can be considered to be with free electrons, and atomic properties are not involved. The energy loss due to close collisions can be written

$$\left.\frac{dE}{dx}\right|_{W>\eta} = n_e \int_\eta^{W_{max}} W \frac{d\sigma}{dW}(E, W) \, dW \tag{2.16}$$

Note that $(d\sigma/dW)(E, W)$ is the cross section for an incident particle with energy E to lose an amount of energy W in the collision with a free electron. The cross section depends on the type of incident particle. For spin 0 particles heavier than the electron, the differential cross section is given by [1]

$$\frac{d\sigma}{dW}(E, W) = 2\pi \frac{e^4}{mc^2} \frac{1}{\beta^2 W^2} \left(1 - \beta^2 \frac{W}{W_{\max}}\right) \tag{2.17}$$

while for heavy spin $\frac{1}{2}$ incident particles

$$\frac{d\sigma}{dW}(E, W) = 2\pi \frac{e^4}{mc^2} \frac{1}{\beta^2 W^2} \left[1 - \beta^2 \frac{W}{W_{\max}} + \frac{1}{2}\left(\frac{W}{E + mc^2}\right)^2\right] \tag{2.18}$$

When $W \ll W_{\max}$, both of these cross sections reduce to that for Coulomb scattering, Eq. 2.11. Thus, for small energy transfers the cross sections only depend on the energy of the recoiling electron and the velocity of the incident particle. Spin only plays an important role when $W \sim E$.

The total energy loss is the sum of the contributions from close and distant collisions. The result for the energy loss of a heavy, spin 0 incident particle is [5]

$$\frac{dE}{dx} = \frac{4\pi n_e Z_1^2 e^4}{m v_1^2} \left(\ln \frac{2mv^2\gamma^2}{I} - \beta^2\right) \tag{2.19}$$

It is important to note that the final expression for the energy loss does not depend on the intermediate energy transfer η used to separate the classes of collisions.

It is useful to break up the constant in front of Eq. 2.19 into separate factors relating to the incident particle, the material medium, and the intrinsic properties of the electron. First recall that the electron's charge is related to the so-called classical radius of the electron by

$$r_e = e^2/(mc^2) \tag{2.20}$$

Using Eqs. 2.5 and 2.20, the fixed constants and electron properties can be combined into the constant

$$\begin{aligned} D_e &= 4\pi r_e^2 mc^2 \\ &= 5.0989 \times 10^{-25} \text{ MeV-cm}^2 \end{aligned} \tag{2.21}$$

Equation 2.19 can then be written in the convenient form

$$\frac{dE}{dx} = D_e \left(\frac{Z_1}{\beta_1}\right)^2 n_e \left[\ln \frac{2mc^2\beta^2\gamma^2}{I} - \beta^2\right] \tag{2.22}$$

Now let us consider the important features of Eq. 2.22. The energy loss depends quadratically on the charge and velocity of the incident particle, but not on its mass. Thus, for a beam of particles with a given charge, the energy loss is a function of the velocity only. The energy loss depends on the material linearly through the electron density factor n_e and logarithmically through the mean ionization potential I.

As the velocity of the particle increases from near zero, dE/dx falls due to the $1/v^2$ factor. All incident particles have a region of minimum ionization with $dE/dx \sim 2$ MeV/g cm^2 for $\beta\gamma \approx 3$. As β continues to increase, the ln γ^2 factor in Eq. 2.22 begins to dominate and dE/dx starts to increase. This is referred to as the region of relativistic rise. In the semiclassical picture the relativistic deformation of the Coulomb field of the incident particle increases the upper limit for impact parameters involved in the collision (see Eq. 2.8).

The mean ionization potential per electron depends on the atomic number of the atom. Bloch used the Thomas–Fermi model of the atom to show that I should vary linearly with Z. An approximate expression is

$$I/Z \simeq 10 \text{ eV} \tag{2.23}$$

which is generally valid for $Z \gtrsim 20$. Numerical values for the mean ionization potential of some materials are given in Table 2.1.

Table 2.1. *Electromagnetic properties of elements*[a]

Material	Z	n_a ($\times 10^{23}$/cm^3)	n_e ($\times 10^{23}$/cm^3)	I (eV)	L_R (cm)	X_R (g/cm^2)	Density (g/cm^3)
H_2	1	0.423	0.423	21.8	891	63.05	0.0708
He	2	0.188	0.376	41.8	755	94.32	0.125
Li	3	0.463	1.39	40.0	155	82.76	0.534
Be	4	1.23	4.94	63.7	35.3	65.19	1.85
B	5	1.32	6.60	76	22.2	52.69	2.37
C	6	1.146	6.82	78	18.8	42.70	2.27
N_2	7	0.347	2.43	85.1	47.0	37.99	0.808
O_2	8	0.429	3.43	98.3	30.0	34.24	1.14
Ne	10	0.358	3.58	137[b]	24.0	28.94	1.20
Al	13	0.603	7.84	166	8.89	24.01	2.70
Si	14	0.500	6.99	173	9.36	21.82	2.33
Ar	18	0.211	3.80	188[b]	14.0	19.55	1.40
Fe	26	0.849	22.1	286	1.76	13.84	7.87
Cu	29	0.845	24.6	322	1.43	12.86	8.92
Zn	30	0.658	19.6	330	1.75	12.43	7.14
Kr	36	0.155	5.59	352[b]	5.26	11.37	2.16
Ag	47	0.586	27.6	470	0.85	8.97	10.5
Sn	50	0.371	18.5	488	1.21	8.82	7.31
W	74	0.632	46.8	727	0.35	6.76	19.3
Pt	78	0.662	51.5	790	0.31	6.54	21.45
Au	79	0.577	45.6	790	0.34	6.46	18.88
Pb	82	0.330	27.0	823	0.56	6.37	11.34
U	92	0.479	44.1	890	0.32	6.00	18.95

[a] Values are for solid and liquid states unless noted.
[b] Gaseous state.
Source: Particle Data Group, Rev. Mod. Phys. 56: S1, 1984, S53; S. Ahlen, Rev. Mod. Phys. 52: 121, 1980, Table 6; Y. Tsai, Rev. Mod. Phys. 46: 815, 1974, Table 3.6; *Handbook of Chemistry and Physics,* 64th ed., Boca Raton: CRC Press, 1983, p. B65; R.M. Sternheimer, M.J. Berger, and S.M. Seltzer, Atomic Data and Nuclear Data Tables 30: 261, 1984, Table 1.

The relativistic rise does not continue indefinitely. The theory we have described so far treats the interaction of the incident particle with an isolated atom. However, for dense materials where the interatomic spacing is small, the upper limit on allowed impact parameters may encompass many atoms. In this case interactions among the atomic electrons can cause a screening of the projectile's electric field. Fermi developed a theory of dielectric screening that explains the reduction of energy loss for distant collisions [4]. This phenomenon is known as the density effect since it is affected by the density of the medium. It causes the energy loss in the region of relativistic rise to only increase like $\ln \gamma$ instead of $\ln \gamma^2$ and causes the loss to become constant at very large γ. The constant ionization loss at large γ is referred to as the Fermi plateau.

Taking the density effect into account, the energy loss formula can be written

$$\frac{dE}{dx} = D_e \left(\frac{Z_1}{\beta_1}\right)^2 n_e \left[\ln \frac{2mc^2\beta^2\gamma^2}{I} - \beta^2 - \frac{\delta(\gamma)}{2}\right] \tag{2.24}$$

Figure 2.5 Density effect correction parameter δ for several materials. (The parameter was calculated using the formulas and coefficients given in R.M. Sternheimer, M.J. Berger, and S.M. Seltzer, Atomic Data and Nuclear Data Tables 30: 261, 1984.)

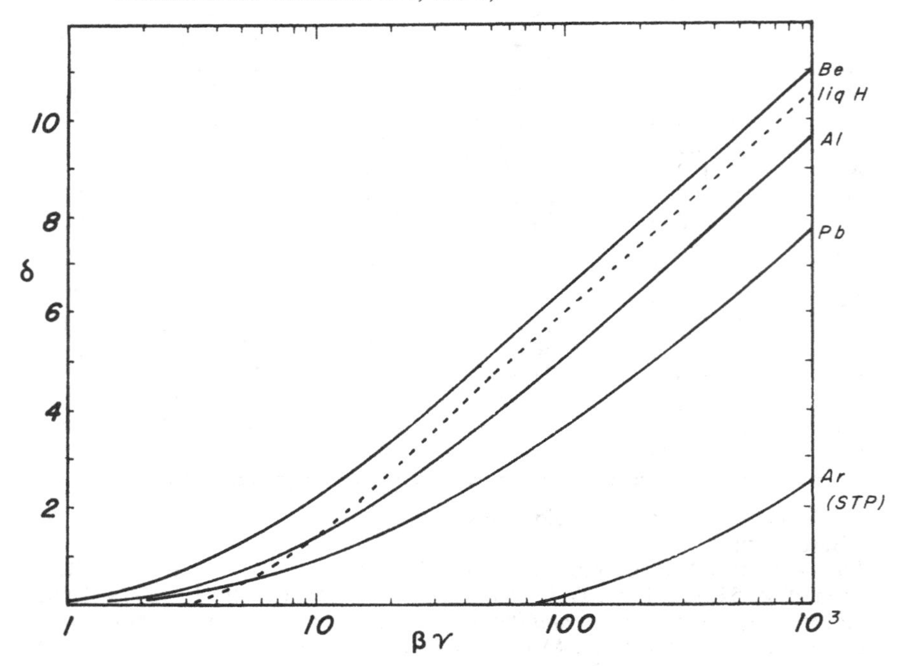

where $\delta(\gamma)$ is a correction due to the density effect. A successful model for calculating δ in terms of atomic properties has been developed by Sternheimer. Formulas giving the correction in terms of the ionization potential and the plasma frequency of the material can be found in Sternheimer's papers [6]. Values of δ for several materials are shown in Fig. 2.5. The density effect correction changes the calculated ionization loss by ~15% for particles with $\beta\gamma = 100$ in metals. Note that nonconductors have a sharp threshold in $\beta\gamma$.

Measurements of dE/dx in propane are shown in Fig. 2.6 as a function of $\beta\gamma$ and the gas pressure [7]. The measurements were made by collecting the charge liberated by ionization in a proportional chamber. Note that the plateau value of dE/dx at large $\beta\gamma$ decreases with increasing pressure due to the density effect. Measurements [8] of the ionization losses of high energy protons and pions in high pressure hydrogen gas have also shown that the energy loss in the Fermi plateau remains constant for γ values as high as 1800.

One should keep in mind that the derivation of the energy loss expression given in Eq. 2.19 makes use of a number of simplifying assumptions.

Figure 2.6 Measured mean energy losses in propane as a function of pressure and $\beta\gamma$. The energy losses are normalized to those for 3-GeV/c protons. (After A. Walenta, J. Fischer, H. Okuno, and C. Wang, Nuc. Instr. Meth. 161: 45, 1979.)

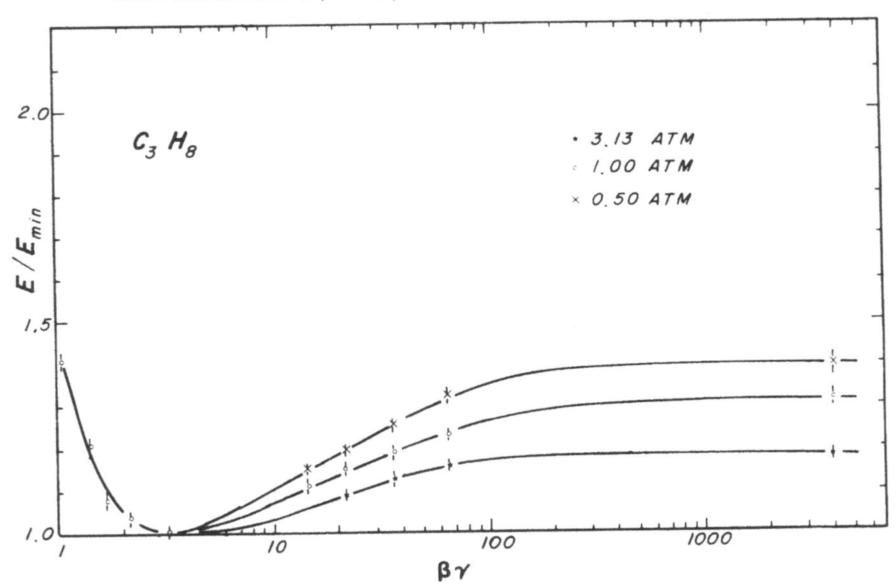

At the low β extreme atomic shell corrections are necessary when the velocity of the incident particle becomes comparable to the velocity of the bound electrons. On the other hand, at large γ radiation, kinematic, and incident particle structure corrections may be necessary [5]. One should also remember that excitation and ionization are not the only causes of energy loss. At large γ other contributing processes include Cerenkov radiation, transition radiation, bremsstrahlung, and pair production.

2.3 Fluctuations in energy loss

The amount of energy lost by a charged particle that has traversed a fixed thickness of absorber will vary due to the statistical nature of its interactions with individual atoms in the material. The value of dE/dx calculated in the preceding section is an averaged value. We have seen in Eq. 2.11 that collisions with small energy transfers are more likely than those with large transfers. Thus, the most probable energy loss will be shifted to the lower half of the range of possible energy transfers. The large energy transfer events are associated with the production of high energy recoil electrons (and from nuclear interactions). The result is that the energy loss distribution will be asymmetric with a tail on the high energy side.

For historical reasons the high energy recoil electrons are called delta rays. The number of delta rays produced with energy greater than E_1 in a thickness x is

$$N(E \geqslant E_1) = \int_{E_1}^{E_{max}} \xi \frac{dE}{E^2} \tag{2.25}$$

where

$$\xi = \frac{2\pi n_e Z_1^2 e^4}{mv^2} x \tag{2.26}$$

and E_{max} is the maximum possible energy transfer. Thus,

$$N(E \geqslant E_1) = \xi \left(\frac{1}{E_1} - \frac{1}{E_{max}} \right) \tag{2.27}$$

So long as $E_1 \ll E_{max}$, we see that the number of energetic delta rays falls off inversely with the energy and that the parameter ξ is the energy above which there will be, on the average, one delta ray produced. As such, it represents a "typical" value of the energy loss in the material.

The probability that an incident particle with energy E will lose energy between W and $W + dW$ while traversing an infinitesimal thickness dx of absorber is (see Appendix D)

$$\phi(W)\,dW\,dx = n_a \frac{d\sigma(W)}{dW}\,dW\,dx \tag{2.28}$$

where $d\sigma/dW$ is the differential cross section for the incident particle to lose energy W in a single collision with an absorber atom. The total probability of a collision in the thickness dx, regardless of the energy transfer, is $q\,dx$ where

$$q = n_a \int_0^\infty d\sigma/dW\,dW \tag{2.29}$$

The quantity q is called the primary ionization rate.

Although the probability for an energy loss W in an infinitesimal absorber layer is given trivially by Eq. 2.28, the calculation of the corresponding probability for a finite thickness can be very complicated. Consider a beam of N particles all having energy E. Let $\chi(W, x)\,dW$ be the probability that a particle loses an energy between W and $W + dW$ after crossing a thickness x of absorber. The form of χ may be determined by considering how it changes if the particles traverse an additional infinitesimal thickness dx in the absorber. The number of particles with energy losses between W and $W + dW$ increases because some particles with energy loss less than W at x will undergo a collision in dx that increases its loss to between W and $W + dW$. On the other hand, the number of particles with energy losses between W and $W + dW$ decreases because some particles in the correct interval will undergo a collision in dx and increase the total energy loss above $W + dW$. We assume that successive collisions are statistically independent, that the absorber medium is homogeneous, and that the total energy loss is small compared to the particles' incident energy. Then we can express the change in the number of particles as

$$N\chi(W, x + dx)\,dW - N\chi(W, x)\,dW = N \int_0^\infty \chi(W - \epsilon, x)\phi(\epsilon)\,dW\,dx\,d\epsilon - N\chi(W, x)\,dW\,q\,dx \tag{2.30}$$

Thus χ satisfies the equation

$$\frac{\partial\chi(W, x)}{\partial x} = \int_0^\infty \phi(\epsilon)\chi(W - \epsilon, x)\,d\epsilon - q\chi(W, x) \tag{2.31}$$

A number of investigators have determined solutions to Eq. 2.31. The differences in the treatments arise chiefly from different assumptions made about the single collision energy transfer probability $\phi(W)$. Landau used the classical free electron cross section given in Eq. 2.11. He assumed that a typical energy loss was (1) large compared to the binding energy of the electrons in the material, yet (2) small compared to the maximum possible energy loss. With these assumptions the function χ can be factorized into the form [9, 10]

$$\chi(W, x) = \frac{1}{\xi} f_{\mathrm{L}}(\lambda)$$

where

$$\lambda = \frac{1}{\xi}\left[W - \xi\left(\ln\frac{\xi}{\epsilon'} + 1 - C_{\mathrm{E}}\right)\right]$$

$$\ln \epsilon' = \ln \frac{(1-\beta^2)I^2}{2mv^2} + \beta^2 \tag{2.32}$$

$$C_{\mathrm{E}} = 0.577 \quad \text{(Euler's constant)}$$

and ξ is given by Eq. 2.26. The quantity ϵ' is the low energy cutoff of possible energy losses. It was chosen by Landau so that the mean energy loss agreed with the Bethe–Bloch theory.

The universal function $f_{\mathrm{L}}(\lambda)$ can be expressed in terms of the integral

$$f_{\mathrm{L}}(\lambda) = \frac{1}{\pi}\int_0^\infty \exp[-u(\ln u + \lambda)] \sin \pi u \, du \tag{2.33}$$

The most probable value of the energy loss is given by [5]

$$W_{\mathrm{mp}} = \xi\left(\ln\frac{\xi}{\epsilon'} + 0.198 - \delta\right) \tag{2.34}$$

where δ is the density effect correction used in Eq. 2.24. The full width at half maximum of the distribution is

$$\mathrm{FWHM} = 4.02\xi \tag{2.35}$$

It is convenient to use the quantity ξ/E_{max} to classify various theories of energy losses. Landau's second assumption mentioned above requires $\xi/E_{max} \lesssim 0.01$. In this case the number of delta rays with energies near E_{max} is very small, and single large energy loss events give an asymmetric high energy tail to the energy loss distribution. Landau's first assumption breaks down in very thin absorbers such as gases, where the typical loss may be comparable to the electron binding energy in the gas atoms. Experimental energy loss distributions for gases are broader than predicted by the Landau theory. Accurate treatment of the energy loss requires that the theory take into account the presence of discrete atomic energy levels [9].

The distributions approach a Gaussian for $\xi/E_{max} \gtrsim 1$. In this case the number of delta rays observed with energy near E_{max} is large. The most probable energy loss is also large, and the high energy loss events tend to average out. The width of the Gaussian is given simply by [5]

$$\sigma_W^2 = 4\pi Z_1^2 Z_2 n_a e^4 x \tag{2.36}$$

Vavilov has developed a theory for the intermediate case $0.01 \lesssim \xi/E_{max} \lesssim 1$ using the physical upper limit for the maximum energy transfer [11].

The pulse height spectra of high energy protons and electrons in a gaseous proportional chamber are shown in Fig. 2.7. It is clear that the most probable value of the energy loss is skewed to the low energy side of the asymmetric energy loss distribution.

Because of fluctuations in energy loss, a beam of particles of fixed energy will have a distribution of ranges in a thick absorber. This phenomenon is known as straggling. The two fluctuations are related by

$$\langle (E - \overline{E})^2 \rangle = \left(\frac{dE}{dx}\right)^2 \langle (R - \overline{R})^2 \rangle \tag{2.37}$$

Figure 2.7 Measured pulse height distributions for 3-GeV/c protons and 2-GeV/c electrons in a 90% Ar + 10% CH_4 gas mixture. (After A. Walenta, J. Fischer, H. Okuno, and C. Wang, Nuc. Instr. Meth. 161: 45, 1979.)

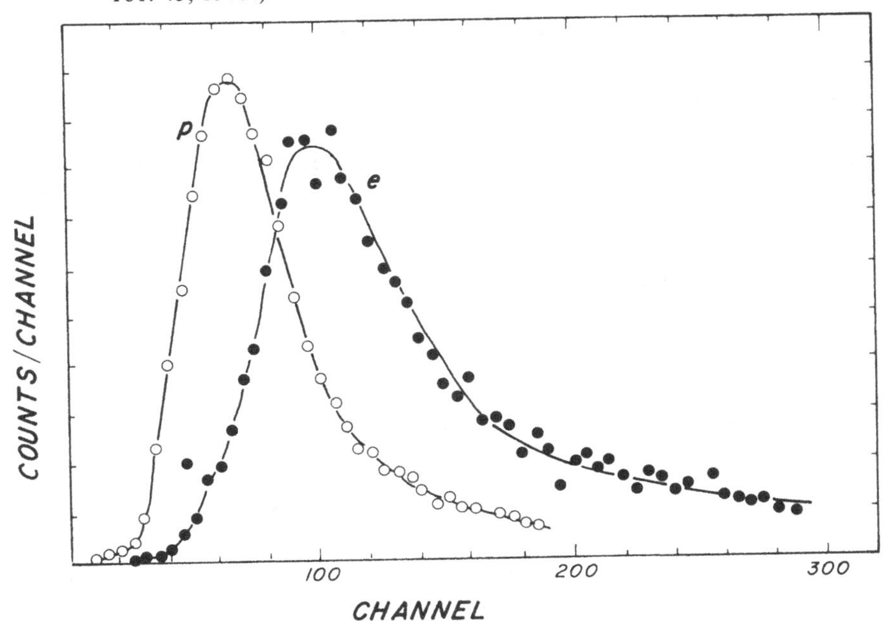

Calculations of straggling have shown that the range distributions for protons in various metals are nearly Gaussian [5]. For a pure, monoenergetic beam of particles the fractional straggling σ_R/R increases with increasing Z of the absorber. The fractional straggling in a given absorber decreases with increasing kinetic energy and approaches a value $\sigma_R/R \sim \frac{1}{2}(m/M)^{1/2}$ at high energy, where M is the mass of the incident particle.

2.4 Energy loss of electrons and positrons

Electrons and positrons lose energy by ionization just as the heavier charged particles do. However, because of their small mass, they also have significant losses due to the production of radiation. For lead the fractional energy loss due to bremsstrahlung exceeds that due to ionization for electron energies above 10 MeV. Other significant sources of energy loss for low energy electrons are elastic scattering and positron annihilation. For high energy electrons the bremsstrahlung and pair production processes lead to the production of electromagnetic showers.

Electrons and positrons have similar electromagnetic interactions in matter. Most of the statements we will make about electrons apply to positrons as well. When this is not the case, we will explicitly say so.

2.4.1 Ionization energy loss

The portion of the ionization loss resulting from distant collisions is the same for all incident particles. On the other hand, the portion resulting from close collisions depends on the form of the free electron cross section of the incident particle (Eq. 2.16). The differential cross section for the relativistic scattering of an electron from a free electron was first calculated by Moller. The cross section for finding a scattered electron with kinetic energy between W and $W + dW$ is [12]

$$\frac{d\sigma}{dW} = \frac{2\pi e^4}{mv^2}\left[\frac{1}{W^2} - \frac{1}{W(E-W)}\frac{mc^2(2E+mc^2)}{(E+mc^2)^2} + \frac{1}{(E-W)^2} + \frac{1}{(E+mc^2)^2}\right] \tag{2.38}$$

where E is the kinetic energy of the incident electron. The energy loss for relativistic electrons using this cross section is

$$\frac{dE}{dx} = \frac{2\pi n_e e^4}{mc^2}\left(2\ln\frac{2mc^2}{I} + 3\ln\gamma - 1.95\right) \tag{2.39}$$

The energy loss for positrons differs slightly since the Bhabha differential cross section must be used in place of the Moller cross section. The corresponding expression for the energy loss of singly charged, heavy particles traveling at relativistic velocities can be found by evaluating Eq. 2.19 for $\beta \to 1$

$$\frac{dE}{dx} = \frac{2\pi n_e e^4}{mc^2}\left(2\ln\frac{2mc^2}{I} + 4\ln\gamma - 2\right) \tag{2.40}$$

Comparison of Eqs. 2.39 and 2.40 shows that to first order all singly charged particles with $\beta \sim 1$ lose energy by collisions at approximately the same rate. The second terms indicate that the rate of relativistic rise for electrons will be slightly smaller than for heavier particles.

Because of their small mass, electrons follow a very irregular path through matter. For this reason, the range for electrons has a wider distribution than the range for heavier particles. The mean range for 1-MeV electrons varies from about 0.22 g/cm^2 in hydrogen to 0.78 g/cm^2 in lead [13].

2.4.2 Bremsstrahlung

The dominant energy loss mechanism for high energy electrons is the production of electromagnetic radiation. This is usually referred to as synchrotron radiation for circular acceleration and bremsstrahlung for motion through matter. Conservation of energy requires that $E_i = E_f + k$, where E_i (E_f) is the initial (final) energy of the electron and k is the energy of the produced photon. The time rate of energy loss depends quadratically on the acceleration experienced by the particle through the well-known relation

$$dE/dt = (2e^2/3c^3)a^2 \tag{2.41}$$

where a is the acceleration.

A semiclassical calculation of the bremsstrahlung cross section for a relativistic particle gives [2]

$$\frac{d\sigma}{dk} \simeq 5\frac{e^2}{\hbar c} Z_1^4 Z_2^2 \left(\frac{mc}{Mv_1}\right)^2 \frac{r_e^2}{k} \ln\frac{Mv_1^2\gamma^2}{k} \tag{2.42}$$

We note that the cross section depends inversely on the square of the incident particle mass M. It is for this reason that up to the present date radiation energy loss has only been important for electrons and very high energy muons. The cross section depends on the medium through the factor Z_2^2, implying that heavy elements are most efficient at causing energy loss by radiation. Recall that ionization energy loss was proportional to Z_2. Finally note that the cross section falls off with increasing photon energy roughly as $1/k$.

Interactions of the incident particle with the Coulomb field of the nucleus go like Z_2^2. There is also a contribution from the atomic electrons that goes like Z_2. Thus, for all but the lightest elements the bremsstrahlung cross section is dominated by interactions with the nucleus. However, the atomic electrons cause another important effect by screening the nuclear charge. Classically, when the impact parameter is larger than the atomic radius, we expect the cross section to fall sharply since the effective charge seen by the incident particle is greatly reduced. This case is referred to as complete screening. The effect is also true quantum mechanically since one may define an "effective distance" of the electron from the nucleus $\hbar/q$, where q is the momentum transfer from the electron to the nucleus.

The results of QED calculations for the bremsstrahlung process can be found in a review by Tsai [14]. At least one virtual photon must be exchanged to the target system in order to conserve 4-momentum. The lowest-order QED diagrams shown in Fig. 2.8 do not depend on the sign of the charge of the lepton. Thus, we expect the same cross section from incident electrons and positrons. The interaction of the virtual photon with the target system depends on properties of the target, such as its mass, internal structure, spin, and screening of the nuclear charge by atomic electrons.

Bethe and Heitler made a quantum mechanical calculation of the bremsstrahlung cross section for an electron in the field of an infinitely heavy, pointlike, spinless nucleus [15]. The calculation makes use of the Born approximation, which is valid if

$$\frac{2\pi Z_2 e^2}{\hbar v} \ll 1 \tag{2.43}$$

where v can be the velocity of the electron before or after the photon emission. This relation is generally satisfied for energetic particles, except possibly for the heavy elements.

Figure 2.8 Lowest-order Feynman diagrams for bremsstrahlung.

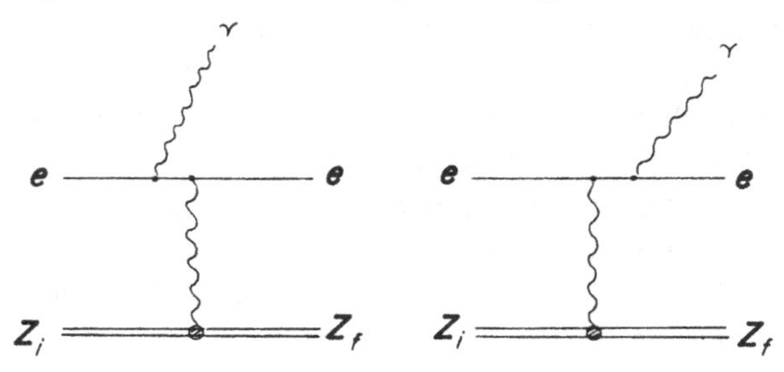

The effects of screening are determined by the parameter [12]

$$\Gamma = \frac{100mc^2}{Z_2^{1/3}} \frac{k}{E_i E_f}$$
$$= \frac{100mc^2}{Z_2^{1/3}} \frac{k/E_i}{E_i(1 - k/E_i)} \tag{2.44}$$

This parameter results from dividing the Thomas–Fermi radius of the atom [15]

$$r_a = \frac{a_0}{Z_2^{1/3}} = \frac{\hbar}{\alpha mc Z_2^{1/3}} \tag{2.45}$$

by the maximum allowed value of $\hbar/q$ [12]. The quantity a_0 is the ground state radius of the hydrogen atom in the Bohr theory and

$$\alpha = e^2/\hbar c \simeq 1/137 \tag{2.46}$$

is the fine structure constant. The Bohr radius a_0 is related to the classical radius of the electron by $r_e = \alpha^2 a_0$. The case of complete screening corresponds to $\Gamma \approx 0$. Figure 2.9a shows the region of allowed photon energies as a function of the incident electron energy. The contour with $\Gamma = 0.1$ for lead is shown together with the region of complete screening.

Measurements of the angular distribution of bremsstrahlung photons agree with the predictions of the Bethe–Heitler calculations [16]. If the angles are measured with respect to the incident electron direction, the differential cross section is given approximately by

$$\frac{d\sigma}{(dp_f/p_f)\, d\Omega_f d\Omega_\gamma} = \frac{8Z_2^2\alpha^3}{\pi^2} \frac{p_f(p_i^2 + p_f^2)}{kp_i^4\theta_f^2\theta_\gamma^4(\theta_f + \theta_\gamma)^2} \frac{r^2}{(1 + r^2)^2} \tag{2.47}$$

where i (f) refers to the incoming (outgoing) electron and $r = Q_\perp/Q_\parallel$ is the ratio of the 3-momentum transfer to the nucleus perpendicular and parallel to the incident electron direction.

At high energy the mean angle of photon emission is

$$\bar{\theta}_\gamma \simeq \frac{mc^2}{E} \tag{2.48}$$

independent of the photon energy. Thus, most of the radiation lies inside a narrow cone around the electron's momentum vector. The cone becomes more and more narrow as the energy is increased. Bremsstrahlung photons are in general polarized with the polarization vector normal to the plane formed by the photon and incident electron [3]. The photon polarization is influenced by the polarization of the incident electrons.

The cross section integrated over angles in the case of complete screen-

Figure 2.9 (a) The kinematic region of complete screening for bremsstrahlung and the contour for lead with $\Gamma = 0.1$. (b) The complete screening bremsstrahlung cross sections for aluminum and lead versus the variable $f = k/(E_i - mc^2)$.

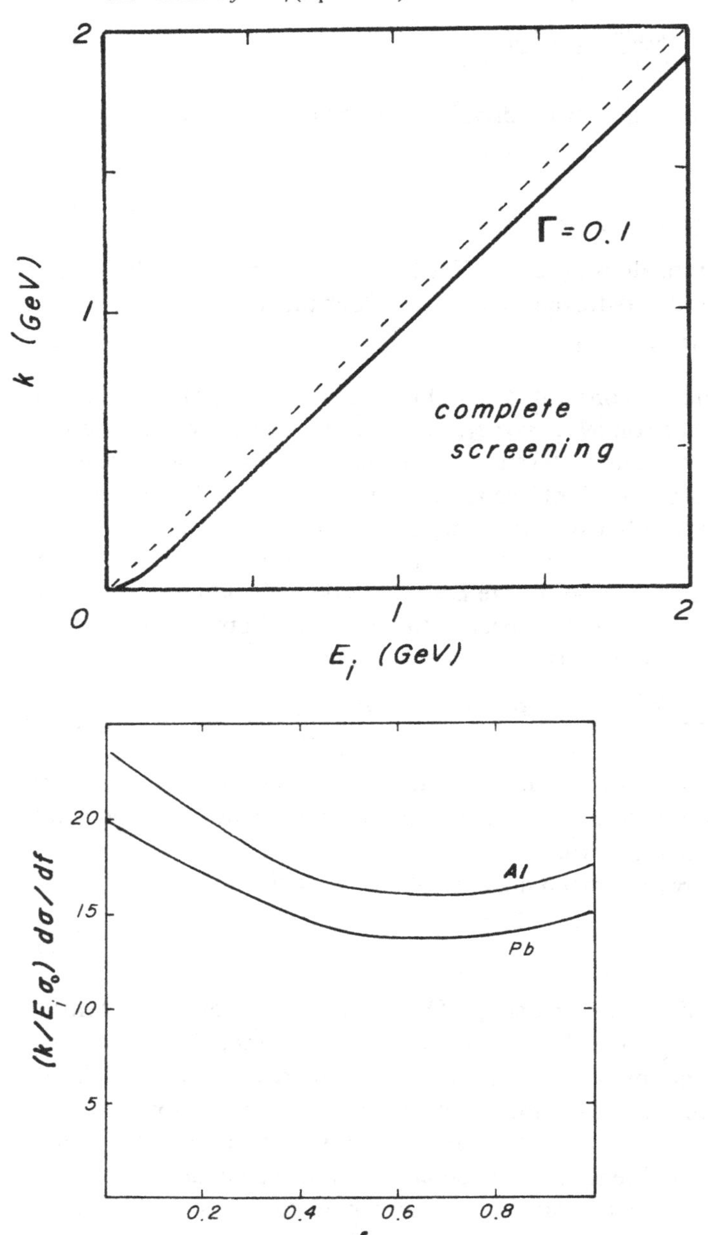

ing ($\Gamma \ll 1$) is [12]

$$\frac{d\sigma}{dk} = \frac{4\sigma_0}{k}\left[\left(1 + w^2 - \frac{2}{3}w\right)\ln\frac{183}{Z_2^{1/3}} + \frac{w}{9}\right] \tag{2.49}$$

where $w = E_f/E_i$ and $E_i \gg mc^2$, and

$$\sigma_0 = \alpha Z_2^2 r_e^2 \tag{2.50}$$

The contribution of the atomic electrons can be included by substituting $Z_2^2 \rightarrow Z_2(Z_2 + 1)$. Note that Eq. 2.49 only depends on the fractional energy w and not on E_i itself. It also depends on the material principally through the factor Z_2^2 in σ_0.

Expressions for the cross section for the case $\Gamma > 0$ have been defined using auxiliary functions and can be found in the literature [1, 12]. For $\Gamma = 0.1$ the complete screening approximation in lead differs from the more accurate partial screening calculations by ~3%. It can be seen that to this level of approximation and so long as $E_i \gg mc^2$, one may assume complete screening, except for the production of very energetic photons that carry off almost all of the electron's energy.

The factor k in the denominator of Eq. 2.49 implies that the cross section for low energy photon production increases without limit. Of course, this infrared divergence does not actually occur. These equations are not valid in the limit $k \rightarrow 0$.

For the purpose of illustrating the energy dependence of the bremsstrahlung cross sections, it is convenient to use as the independent variable

$$f = \frac{k}{E_i - mc^2}$$

which is the fraction of the incident kinetic energy of the electron given to the photon. Multiplying the cross section by k gives the photon intensity distribution. The complete screening cross sections for aluminum and lead are shown in Fig. 2.9b. For very energetic photon production (large f) screening corrections are negligible. The most important feature of Fig. 2.9b is that the energy distribution does not fall off much faster than $1/k$. Hence there is a significant probability of obtaining a photon of any energy up to the maximum allowed. The deposited energy spectrum $d\sigma/df$ is fairly uniform. The function plotted in Fig. 2.9b only changes a small amount for different elements. Hence the major material dependent factor is the Z^2, which comes from σ_0.

The energy loss due to radiation of an electron traversing some material is as follows:

$$\left.\frac{dE}{dx}\right|_{\text{rad}} = \int_0^{k_{\max}} k n_{\text{a}} \frac{d\sigma}{dk}\,dk$$

where $k_{\max} = E_{\text{i}} - mc^2$ is the maximum allowed photon energy. It is convenient to separate out the incident energy and write the energy loss as

$$\left.\frac{dE}{dx}\right|_{\text{rad}} = n_{\text{a}} E_{\text{i}} \sigma_{\text{rad}} \tag{2.51}$$

where

$$\sigma_{\text{rad}} = \frac{1}{E_{\text{i}}} \int_0^{k_{\max}} k \frac{d\sigma}{dk}\,dk$$

For the case of complete screening and $E_{\text{i}} \gg mc^2/\alpha Z^{1/3}$

$$\sigma_{\text{rad}} = 4\sigma_0 [\ln(183 Z_2^{-1/3}) + \tfrac{1}{18}] \tag{2.52}$$

Note that σ_{rad} in this case is independent of the electron's energy. Thus, according to Eq. 2.51 the energy loss due to radiation is proportional to E_{i}. The collision energy loss, on the other hand, increased like $\ln E_{\text{i}}$. In addition, the radiative energy loss is proportional to Z_2^2, whereas the collision loss was proportional to Z_2. If we rewrite Eq. 2.51 in the form $dE/E = dx\, n_{\text{a}}\sigma_{\text{rad}}$, we see that $n_{\text{a}}\sigma_{\text{rad}}$ must be the inverse of some constant length L_{R}, where

$$L_{\text{R}}^{-1} = 4\sigma_0 n_{\text{a}} \ln(183 Z_2^{-1/3}) \tag{2.53}$$

and we have neglected the small term $\frac{1}{18}$ in Eq. 2.52. If we also neglect the collisional energy loss, we find that radiation losses cause the mean energy of the electron to decrease exponentially so long as the restrictions on Eq. 2.52 are satisfied.

Values of L_{R} for various materials are listed in Table 2.1. Radiation lengths for molecules can be determined from the atomic values by weighting the terms by the appropriate atomic weight [14]. It is often convenient to express lengths as a multiple of the radiation length in the material.

The ratio of the radiation energy loss to the collision energy loss is given approximately by [12]

$$\left.\frac{dE}{dx}\right|_{\text{rad}} \Big/ \left.\frac{dE}{dx}\right|_{\text{coll}} = \frac{Z_2 E_{\text{i}}}{1600 mc^2}$$

We see that at high incident electron energies, the energy loss is almost totally due to the production of radiation. The energy at which the loss due to radiation just equals the loss due to ionization is sometimes called the critical energy

$$E_{crit} = \frac{1600}{Z_2} mc^2 \tag{2.54}$$

Another process by which fast moving charged particles can lose energy is through direct pair production [15]. The electromagnetic field of the charged particle may be considered as a flux of virtual photons. In the Coulomb field of the nucleus or an electron the virtual photons can decay into an electron and positron, just as real photons do in the process of pair production.

We have seen that the dominant source of energy loss for high energy electrons is through bremsstrahlung. Figure 2.9b showed the energy distribution of the photons as a function of the fraction of the initial kinetic energy given to the photon. We saw that there was a substantial probability that the emitted photon will carry off a large fraction of the electron's energy. Thus, we expect that the distribution of electron energy losses will be quite wide compared to that of a heavy charged particle. The Landau theory discussed in Section 2.3 is only applicable to low energy electrons, whose energy losses are dominated by collision processes. At high energies the probability due to bremsstrahlung that the energy of an electron will decrease by the factor $e^{-\zeta}$ while traversing a thickness x of material is [15]

$$w(x, \zeta) = \frac{e^{-\zeta} \zeta^{bx-1}}{\Gamma(bx)} \tag{2.55}$$

where

$$b \sim 21 n_a \sigma_0$$

and Γ is the gamma function. This relation is only valid for small thicknesses of low Z material. Otherwise, the electron has a large probability of initiating an electromagnetic shower (see Section 11.1).

2.4.3 Positron annihilation

The ultimate fate of most positrons in matter is annihilation with an electron into photons. The most likely process is

$$e^+ + e^- \rightarrow \gamma + \gamma$$

Annihilation into a single photon is possible if the electron is bound to a nucleus, but the cross section for this process is at most 20% of that for two photons [15]. The two-photon annihilation cross section for a positron with LAB energy E_+ is given by

$$\sigma_{ann} = \pi r_e^2 \frac{1}{\gamma + 1} \left[\frac{\gamma^2 + 4\gamma + 1}{\gamma^2 - 1} \ln(\gamma + \sqrt{\gamma^2 - 1}) - \frac{\gamma + 3}{\sqrt{\gamma^2 - 1}} \right] \tag{2.56}$$

where

$$\gamma = E_+/mc^2$$

This cross section peaks near $\gamma = 1$. Thus, it is most likely that a high energy positron will lose energy by collision and radiation until the velocity becomes small, at which point it will annihilate into photons.

The electron and the positron can also form a temporary bound state called positronium. This system is analogous to the hydrogen atom, although the energy level spacing is reduced by a factor of 2 due to the reduced mass of the electron – positron system. Annihilation occurs when there is an overlap between the electron and positron wavefunctions and is most likely in the *S* state. The two particles can form both singlet (spins antiparallel) and triplet (spins parallel) states. The most common decay is the singlet state decay at rest into two collinear 0.511-MeV photons, which occurs with a mean lifetime of $\sim 10^{-10}$ sec. The triplet state decays into three photons with a mean lifetime of $\sim 10^{-7}$ sec.

2.5 Interactions of photons

We have seen that the collisional interactions of heavy charged particles tend to be small perturbations that, apart from removing a small amount of energy and causing a small change in the particle's trajectory, leave the particle basically undisturbed. Thus, the number of particles in a beam remains roughly constant until the velocity has been reduced to a small value. Photon interactions are different since in general there is a large probability that an interacting photon will be removed from the beam.

Consider a collimated, monoenergetic beam of N photons. The number removed from the beam while crossing a thickness dx of material is

$$dN = -\mu N\, dx \tag{2.57}$$

The constant of proportionality μ is known as the linear attenuation coefficient and is related to the probability that a photon will be scattered or absorbed in the material [17]. As a consequence, the intensity of the original photons will decrease exponentially with depth in matter.

There are three major electromagnetic processes by which photons interact with matter:

1. photoelectric effect,
2. Compton effect, and
3. pair production.

Figure 2.10 shows the contributions of these three processes to the total photon interaction cross section for lead. For photon energies below 500 keV the interactions are almost totally due to the photoelectric effect, while for photon energies above 50 MeV they are primarily due to pair production off the nucleus. The Compton effect plays an important role in the intermediate energy range. Two other effects indicated in Fig. 2.10 are coherent (Rayleigh) scattering and photonuclear absorption. Rayleigh scattering is a process in which photons scatter from the atomic electrons without exciting or ionizing the atom. Photonuclear absorption is actually a nuclear interaction where the photon is absorbed by the nucleus. It is most important in the region of the "giant resonance" (10–25 MeV) and is frequently accompanied by the emission of a neutron.

2.5.1 *Photoelectric effect*

The photoelectric effect can be considered to be an interaction between the photon and the atom as a whole. Incident photons whose energy k exceeds the binding energy E_b of an electron in the atom may be absorbed, and an atomic electron ejected with kinetic energy $T = k - E_b$. The photoelectron is emitted near 90° from the incident photon direction

Figure 2.10 Contributions to the photon interaction cross section in lead. τ, photoelectric effect; σ_{COH}, Rayleigh scattering; σ_{INCOH}, Compton scattering; $\sigma_{PH.N}$, photonuclear absorption; K_n, pair production off the nucleus; K_e, pair production off atomic electrons. (J. Hubbell, H. Gimm, and I. Overbo, J. Phys. Chem. Ref. Data 9: 1023, 1980.)

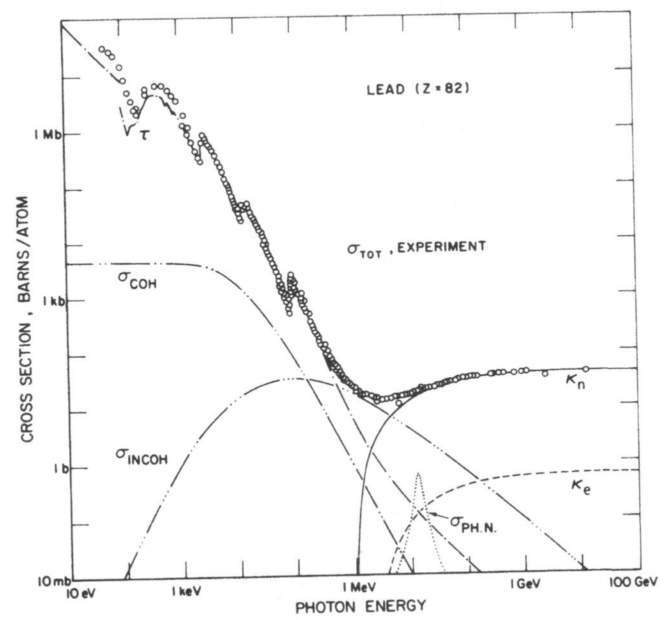

for very low energy, unpolarized photons [15]. The emission angle becomes more and more forward as the photon energy increases. However, the directionality can be quickly randomized due to multiple scattering.

There are sharp discontinuities in the photoelectric spectrum corresponding to the binding energies of the atomic shells, the most prominent of which is due to the innermost, or *K*, shell. The cross section increases by a large factor when the photon energy exceeds the binding energy of electrons in the shell. The ejected photoelectron may be accompanied by fluorescence or additional (Auger) electrons. The photoelectric cross section at high energies falls roughly as Z_2^5/k and only plays a negligible role in high energy interactions.

2.5.2 Compton effect

The Compton effect involves the scattering of an incident photon with an atomic electron. Consider a photon with energy k_0 scattering from an electron considered to be at rest and producing a scattered photon with energy k together with the recoiling electron, as shown in Fig. 2.11. The laws of conservation of energy and momentum require that

$$k_0 + mc^2 = k + T + mc^2$$
$$\mathbf{k}_0 = \mathbf{k} + \mathbf{p}c$$

These equations can be solved in terms of the scattering angle θ to obtain the frequency of the scattered photon [15],

$$\omega = \frac{\omega_0}{1 + \epsilon(1 - \cos\theta)} \qquad (2.58)$$

the kinetic energy of the recoil electron,

$$T = mc^2 \frac{\epsilon^2(1 - \cos\theta)}{1 + \epsilon(1 - \cos\theta)} \qquad (2.59)$$

Figure 2.11 The Compton scattering process.

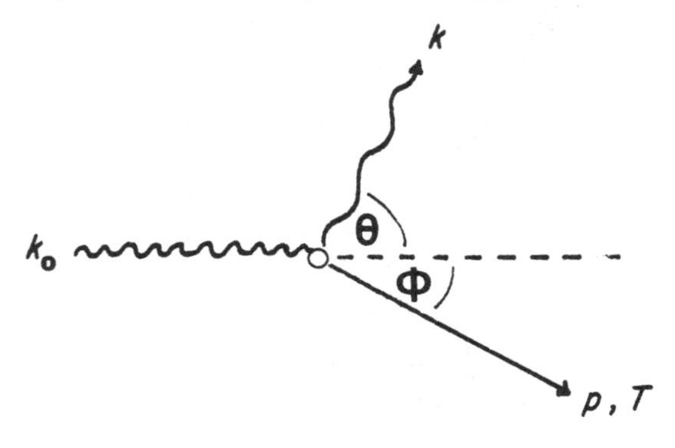

and the electron recoil angle

$$\cos\phi = (1+\epsilon)\left[\frac{1-\cos\theta}{2+\epsilon(\epsilon+2)(1-\cos\theta)}\right]^{1/2} \tag{2.60}$$

where

$$\epsilon = \hbar\omega_0/mc^2$$

We can express the difference in the wavelength of the incident and scattered photons as

$$\lambda - \lambda_0 = h/mc(1-\cos\theta) \tag{2.61}$$

The scattered photon has a longer wavelength than the incident one. The quantity h/mc is referred to as the Compton wavelength of the electron.

The quantum mechanical derivation of the cross section for Compton scattering was performed by Klein and Nishima. The differential cross section averaged over the polarization of the incident photon is [12]

$$\frac{d\sigma}{d\Omega} = \frac{1}{2}r_e^2\left(\frac{h\omega}{h\omega_0}\right)^2\left(\frac{h\omega_0}{h\omega}+\frac{h\omega}{h\omega_0}-\sin^2\theta\right) \tag{2.62}$$

If we use Eq. 2.58 to eliminate ω from this equation, we obtain the angular distribution

$$\frac{d\sigma}{d\Omega} = \frac{1}{2}r_e^2\frac{(1+\cos^2\theta)}{[1+\epsilon(1-\cos\theta)]^2}\left\{1+\frac{\epsilon^2(1-\cos\theta)^2}{(1+\cos^2\theta)[1+\epsilon(1-\cos\theta)]}\right\} \tag{2.63}$$

This distribution is shown as a function of energy and scattering angle in Fig. 2.12. The angular distribution of scattered radiation from classical electrodynamics is [2]

$$I = I_0(1+\cos^2\theta)$$

independent of frequency. The quantum results should approach the classical one as $h \rightarrow 0$ and thus as $\epsilon \rightarrow 0$. At high energies the scattered photons are produced mainly in the forward direction. The scattered photons may also be polarized.

The energy distribution may be obtained by substituting for θ in Eq. 2.62. The result is [12]

$$\frac{d\sigma}{d(h\omega)} = \frac{\pi r_e^2}{\epsilon h\omega}\left[1+\left(\frac{h\omega}{h\omega_0}\right)^2 - \frac{2(\epsilon+1)}{\epsilon^2}+\frac{1+2\epsilon}{\epsilon^2}\frac{h\omega}{h\omega_0}+\frac{1}{\epsilon^2}\frac{h\omega_0}{h\omega}\right] \tag{2.64}$$

where the scattered photon energy must satisfy the inequality

$$\frac{1}{1+2\epsilon} \leqslant \frac{h\omega}{h\omega_0} \leqslant 1$$

For small ϵ there is only a small spread of scattered photon energies, close to the incident photon energy. For higher energies the number of low energy scattered photons increases and the distribution becomes very broad.

The total cross section for photon scattering from classical electrodynamics is the Thomson cross section [3]

$$\sigma_{Th} = \tfrac{8}{3}\pi r_e^2 \simeq 0.67 \times 10^{-24}\ \text{cm}^2$$

The quantum mechanical result is obtained by integrating Eq. 2.63 over all angles giving [12]

$$\sigma_{Comp} = \sigma_{Th}\frac{3}{8\epsilon}\left\{\left[1 - \frac{2(\epsilon+1)}{\epsilon^2}\right]\ln(2\epsilon+1) + \frac{1}{2} + \frac{4}{\epsilon} - \frac{1}{2(2\epsilon+1)^2}\right\} \tag{2.65}$$

At high energy the cross section is given approximately by

$$\sigma_{Comp} \simeq \sigma_{Th}\frac{3}{8\epsilon}\left(\ln 2\epsilon + \frac{1}{2}\right)$$

Thus, at high energy the Compton scattering per atom will fall off roughly like $Z_2/\hbar\omega_0$.

Figure 2.12 The angular distribution for Compton scattering. The parameter $\epsilon = \hbar\omega_0/mc^2$.

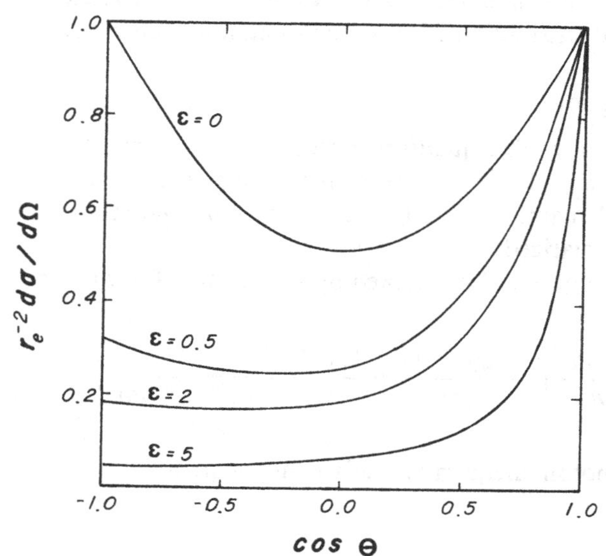

2.5.3 Pair production

The third interaction of photons in matter and the most important at high energies is pair production. The intense electric field near the nucleus can cause the photon to decay into an electron and a positron. The threshold energy for this process is $2mc^2$. The nucleus must be there to satisfy conservation of momentum, but it acquires very little recoil energy. Pair production may also take place near an atomic electron. The threshold in this case is $4mc^2$, and the recoil electron acquires significant kinetic energy. In a track sensitive detector this would appear as a triplet of tracks.

The bremsstrahlung and pair creation processes are intimately related in QED. Examination of the Feynman diagrams in Figs. 2.8 and 2.13 shows that the effects differ only in the directions of the incident and outgoing particles. Both processes proceed to lowest order through the exchange of a single virtual photon. As a consequence, both effects are most important when the momentum transfer is small.

Bethe and Heitler made a quantum mechanical calculation of the cross section for pair creation in the Born approximation. The matrix elements can be determined from those used in the bremsstrahlung calculation with some simple substitutions. This comes about in the Dirac theory since one can regard the inverse of pair production as a bremsstrahlung process in which a positive energy electron emits a photon and falls into a negative energy state [12].

Screening of the nucleus by the atomic electrons is again an important effect. Let the incident photon have energy k and the created electron (positron) have energy E_- (E_+) so that $k = E_+ + E_-$. The screening is measured using the parameter

$$\begin{aligned}\Gamma' &= \frac{100mc^2}{Z_2^{1/3}} \frac{k}{E_+ E_-} \\ &= \frac{100mc^2}{Z_2^{1/3}} \frac{1}{kw_+(1-w_+)}\end{aligned} \tag{2.66}$$

Figure 2.13 Lowest-order Feynman diagrams for pair production.

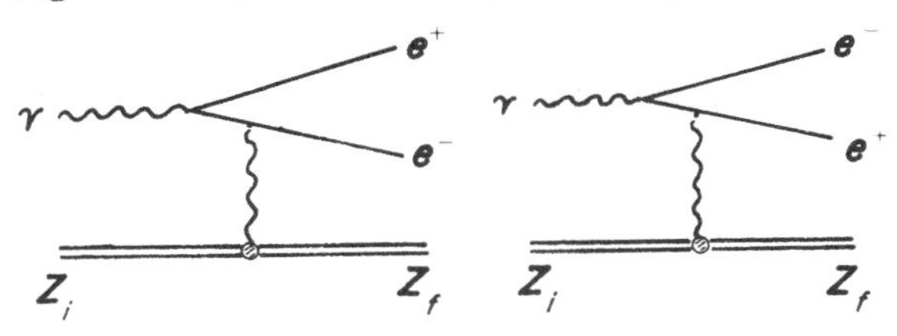

where we have used the energy ratios $w_\pm = E_\pm/k$. Note that the screening parameter is symmetric in w_+ and $w_- = 1 - w_+$ and decreases in general with increasing k. For a given value of k the screening parameter is smallest for symmetric pairs, the limit where $\Gamma' \sim 0$ corresponding to complete screening. Figure 2.14a shows the contour with $\Gamma' = 0.1$ for lead and the region of complete screening on a graph of E_+ versus k.

The differential cross section for the production of a positron with energy between E_+ and $E_+ + dE_+$ and an electron of energy $E_- = k - E_+$ from a photon with energy k in the limit of complete screening is [12]

$$\frac{d\sigma}{dE_+} = \frac{4\sigma_0}{k}\left[\left(w_+^2 + w_-^2 + \frac{2}{3}w_+w_-\right)\ln\left(\frac{183}{Z_2^{1/3}}\right) - \frac{1}{9}w_+w_-\right] \tag{2.67}$$

where σ_0 was given by Eq. 2.50. Note that Eq. 2.67 is symmetric between electron and positron energies. Actually at very low energies where the Born approximation is no longer valid, nuclear repulsion tends to make the positrons more energetic than the electrons. The cross section for the case $\Gamma' > 0$ can be expressed in terms of the same auxiliary functions used for bremsstrahlung [12]. The corrections to the complete screening approximation for lead when $\Gamma' = 0.1$ is $\sim 3\%$.

The pair production cross section is conveniently expressed as a function of the variable

$$g = \frac{E_+ - mc^2}{k - 2mc^2}$$

which is the fraction of the total available kinetic energy taken by the positron. Figure 2.14b shows the complete screening cross sections for lead and aluminum normalized to σ_0 and plotted as a function of the variable g. The cross sections are fairly uniform, indicating that positrons are likely to be produced with any allowed energy. Thus, in general, the electron and the positron in the pair do not have the same energy. Since the plotted functions vary little for different elements, the major material dependence is again the Z_2^2 factor that comes from σ_0. When $g = 0$ or $g = 1$, the complete screening approximation is no longer valid. The correct theory causes the cross section to fall off rapidly for these cases.

At high energy the electron and positron tend to be produced at small angles with respect to the incident photon direction. The mean production angle of an electron or positron with energy E is approximately [12]

$$\theta = \frac{mc^2}{E} \tag{2.68}$$

The total pair production cross section is obtained by integrating Eq.

Figure 2.14 (a) The kinematic region of complete screening for pair production and the contour for lead with $\Gamma' = 0.1$. (b) Complete screening pair production cross sections for aluminum and lead versus the variable $g = (E_+ - mc^2)/(k - 2mc^2)$.

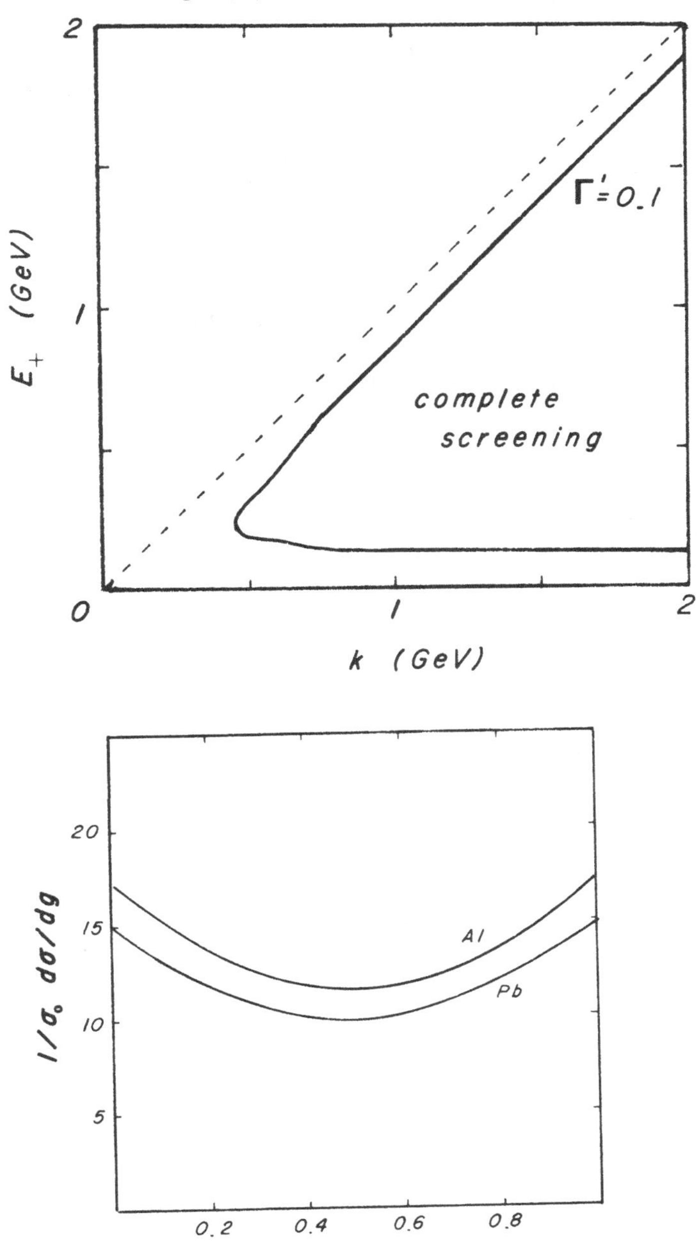

2.67 over all possible positron energies. The total cross section increases rapidly as the photon energy increases, approaching the asymptotic value

$$\sigma_{\text{pair}} = 4\sigma_0\left[\tfrac{7}{9}\ln(183Z_2^{-1/3}) - \tfrac{1}{54}\right] \tag{2.69}$$

for $k \gg 137mc^2Z^{-1/3}$. Note that the cross section is approximately proportional to Z^2 and is independent of the incident photon energy. If we compare Eq. 2.69 with Eq. 2.52, we see that the cross section for pair production is roughly seven-ninths that for bremsstrahlung.

Total cross sections for pair production off of the nucleus and atomic electrons in lead were shown in Fig. 2.10. Pair production off the nucleus contributes over 95% of the photon total cross section at 60 MeV. The majority of the remaining 5% is due to Compton scattering. The assumptions used in the Born approximation break down for high Z elements. The actual cross section for lead is about 10% smaller than the calculated one [12].

Figure 2.15 Measurements of the pair production cross section as a function of the average photon energy $\bar{k}$ and the atomic number Z of the absorber. Solid lines are QED calculations. (J. Eickmeyer et al., Phys. Rev. D 21: 3001, 1980.)

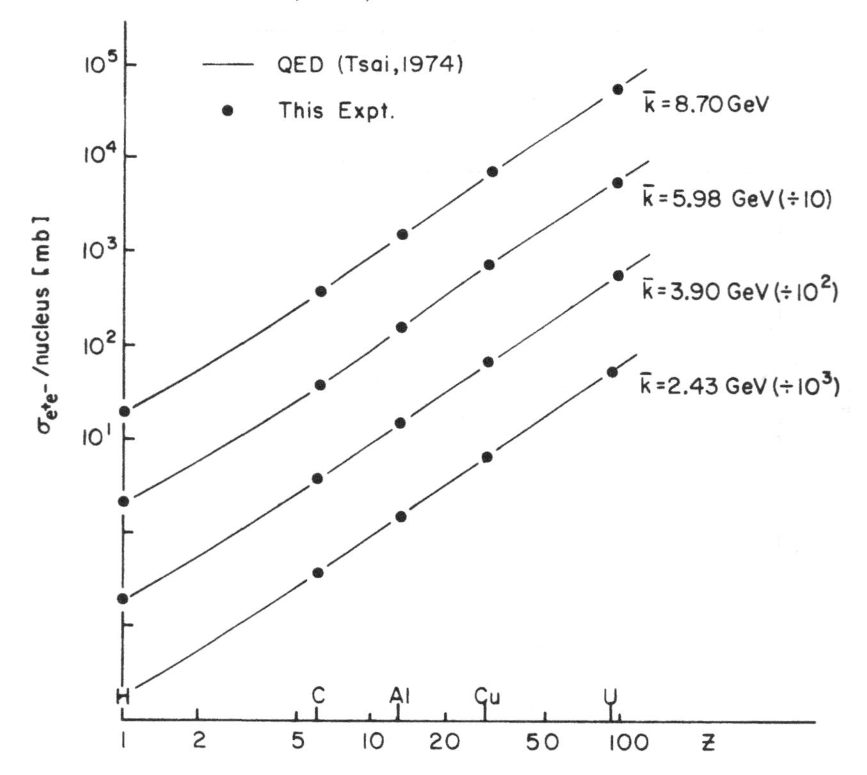

A summary of the results of QED calculations of pair production is contained in the review of Tsai [14]. Figure 2.15 presents a comparison of these calculations for the total cross section with experimental measurements of high energy pair production by Eickmeyer et al. [18]. The calculations are in excellent agreement with the measurements for all elements and photon energies over the range 2.4–8.7 GeV.

In a homogeneous medium the intensity of a beam of collimated, monoenergetic photons decreases exponentially

$$I(x) = I_0 \exp(-\mu x)$$

where μ is the linear attenuation coefficient. The coefficient μ is frequently divided by the density to obtain the mass attenuation coefficient μ/ρ, which has the dimensions cm^2/g. The mass attenuation coefficient is related to the total photon interaction cross section σ_{tot} by

$$\begin{aligned} \frac{\mu}{\rho} &= \frac{N_A}{A}\sigma_{tot} \\ &\simeq \frac{N_A}{A}(\sigma_{pE} + Z\sigma_{Comp} + \sigma_{pair}) \end{aligned} \tag{2.70}$$

where A is the atomic weight of the material. Figure 2.16 shows the mass

Figure 2.16 Photon mass attenuation coefficients for H, C, Fe, and Pb as a function of photon energy. (Data from J. Hubbell, H. Gimm, and I. Overbo, J. Phys. Chem. Ref. Data 9: 1023, 1980; J. Hubbell, National Bureau of Standards report NSRDS-NBS 29, 1969.)

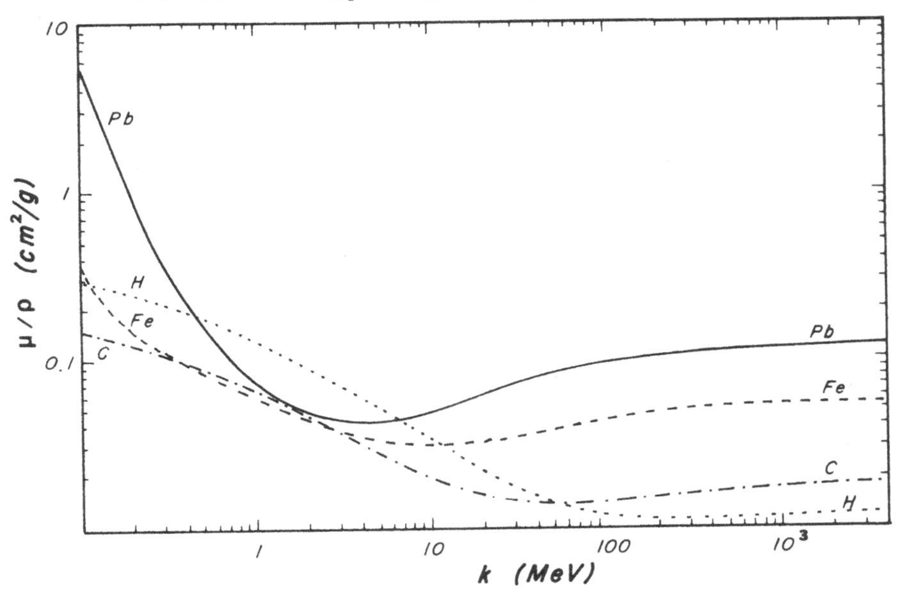

attenuation coefficients for a number of materials as a function of the incident photon energy.

2.6 Elastic scattering

We have seen that the dominant source of energy loss for heavy charged particles is the inelastic excitation and ionization of atomic electrons. The energy loss resulted from the Coulomb interaction between the incident particles and the atomic electrons. So far we have considered the particle's trajectory to be a straight line through the absorber. However, the force on the incident particle due to the charged particles in the absorber can also introduce small deflections in the trajectory.

Consider a charged particle traversing a thickness of material. We saw in Section 2.1 that the Coulomb field of the incident particle gives the charged particles in the medium a momentum perpendicular to the original direction of

$$p_\perp = \frac{2Z_1Z_2e^2}{bv}$$

The incident particle receives an oppositely directed momentum of the same magnitude. It will therefore be deflected through a small angle

$$\theta \simeq \frac{p_\perp}{p} = \frac{2Z_1Z_2e^2}{bvp} \tag{2.71}$$

The differential cross section for scattering with an impact parameter between b and $b + db$ is $2\pi b\, db$. Using Eq. 2.71, we can relate this to the cross section for scattering between θ and $\theta + d\theta$,

$$d\sigma = 2\pi\left(\frac{2Z_1Z_2e^2}{pv}\right)^2 \frac{d\theta}{\theta^3}$$

In the small angle approximation the solid angle $d\Omega \simeq 2\pi\theta\, d\theta$. It follows that the small angle form of the elastic differential cross section can be written in the form

$$\frac{d\sigma}{d\Omega} = 4Z_1^2Z_2^2r_e^2\left(\frac{mc}{\beta p}\right)^2 \frac{1}{\theta^4} \tag{2.72}$$

The θ^{-4} dependence of the cross section shows that a particle is much more likely to undergo a small angle scatter than a large one.

The exact form of the scattering differential cross section for a Coulomb potential in both classical and quantum mechanics is the Rutherford scattering formula

$$\frac{d\sigma}{d\Omega_R} = \frac{1}{4}\, Z_1^2Z_2^2r_e^2\left(\frac{mc}{\beta p}\right)^2 \frac{1}{\sin^4(\theta/2)} \tag{2.73}$$

This cross section is valid for a spin 0 incident particle. The cross section for the interaction of a spin $\frac{1}{2}$ incident particle in the Coulomb field of a nucleus is the Mott cross section, given by [1]

$$\frac{d\sigma}{d\Omega_{\mathrm{M}}} = \frac{d\sigma}{d\Omega_{\mathrm{R}}}\left(1 - \beta^2 \sin^2 \frac{\theta}{2}\right) \tag{2.74}$$

We see that the additional term, which arises from the use of spinor wavefunctions, is only important for large β and large θ. For small angle scattering the Mott cross section reduces to the Rutherford case.

The elastic differential cross section for the scattering of two relativistic electrons is known as the Moller cross section and is given in the CM frame by [19]

$$\frac{d\sigma}{d\Omega} = \frac{\alpha^2}{4s}\left[\frac{10 + 4x + 2x^2}{(1-x)^2} + \frac{10 - 4x + 2x^2}{(1+x)^2} + \frac{16}{(1-x)(1+x)}\right] \tag{2.75}$$

where $x = \cos\theta^*$ and s is the total CM energy squared. The corresponding cross section for positrons and electrons was derived by Bhabha. A QED calculation of the relativistic cross section gives

$$\frac{d\sigma}{d\Omega} = \frac{\alpha^2}{4s}\left[\frac{10 + 4x + 2x^2}{(1-x)^2} - \frac{2(1+x)^2}{1-x} + (1 + x^2)\right] \tag{2.76}$$

The excellent agreement of the calculation with measurements [20] of the cross section between 14 and 34 GeV is shown in Fig. 2.17.

The Rutherford scattering cross section is not valid at very small or very large angles. For very small angles, which correspond to very large impact parameters, the Coulomb potential of the nucleus is screened by the presence of the atomic electrons. The effective potential drops sharply for separation distances that exceed the Thomas–Fermi radius of the atom given in Eq. 2.45. This has the effect of modifying the small angle Rutherford cross section to [2]

$$\frac{d\sigma}{d\Omega} = 4Z_1^2 Z_2^2 r_e^2 \left(\frac{mc}{\beta p}\right)^2 \frac{1}{(\theta^2 + \theta_{\min}^2)^2} \tag{2.77}$$

The cross section for scattering at angles less than some angle $\theta_{\min}$ will be very small. The form of Eq. 2.77 shows that instead of diverging at $\theta = 0$ the cross section levels off to a constant value.

We can estimate $\theta_{\min}$ classically using Eq. 2.71 with b evaluated at the atomic radius r_a

$$\theta'_{\min} \simeq 2Z_1 Z_2^{4/3} \frac{mc^2}{pv}\left(\frac{r_e}{a_0}\right) \tag{2.78}$$

A quantum mechanical limit on θ arises since the incident trajectory must be localized to within $\Delta x \sim r_a$ to obtain a reasonable probability of scattering. Then, by the uncertainty principle, the incident momentum is uncertain by an amount $\Delta p \sim \hbar/r_a$ and the scattering angle by $\Delta\theta \sim \hbar/pr_a$. Scattering for angles smaller than $\Delta\theta$ is smeared, causing the cross section to flatten out. Using Eq. 2.45 for the atomic radius, we find

$$\theta_{\min} = \alpha Z_2^{1/3} \frac{mc}{p} \tag{2.79}$$

In general, the more restrictive of the two limits should be used.

The Rutherford scattering law also breaks down when the particle's wavelength λ becomes comparable with the size r_n of the nucleus. In

Figure 2.17 Angular distributions for Bhabha scattering. Solid lines are results of QED calculations. (R. Brandelik et al., Phys. Lett. 117B: 365, 1982.)

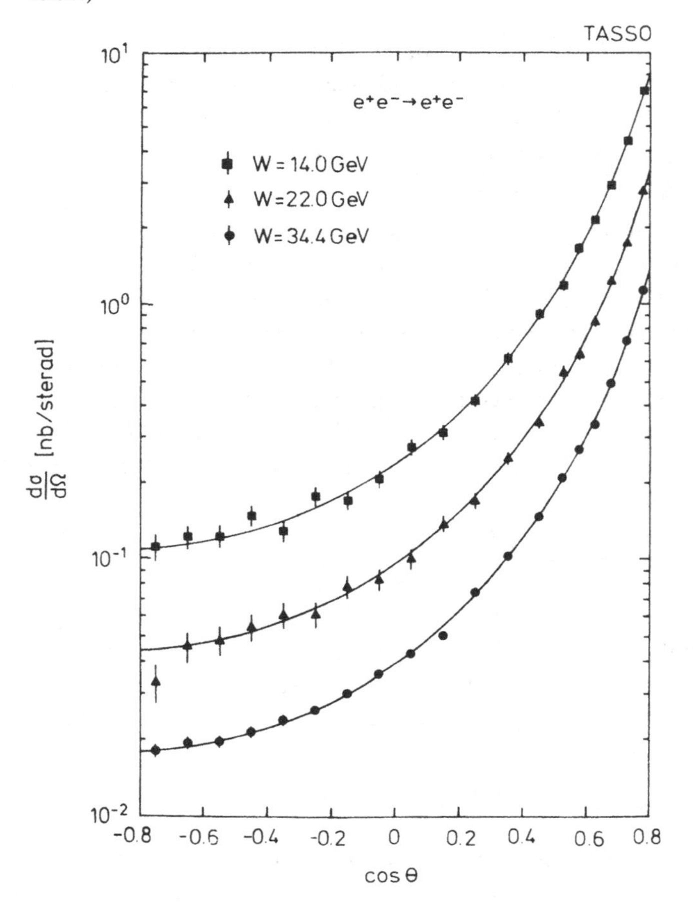

analogy with the first minimum of a diffraction pattern from an object of size r_n, the scattering is predominantly confined to angles smaller than

$$\theta_{max} \simeq \frac{\lambda}{r_n} \simeq \frac{\hbar}{pr_n}$$

Using the simple relation [1, 2]

$$r_n \simeq \tfrac{1}{2} r_e A^{1/3} \tag{2.80}$$

we find that

$$\theta_{max} \simeq \frac{2A^{-1/3}}{\alpha} \frac{mc}{p} \tag{2.81}$$

Other corrections that are sometimes important include the mass of the target and particle identity.

The total elastic scattering cross section can be estimated by integrating Eq. 2.72 from θ_{min} to θ_{max},

$$\sigma = \int \frac{d\sigma}{d\Omega} \sin\theta \, d\theta \, d\phi$$
$$\simeq \pi r_e^2 \, 4 \left(\frac{Z_1 Z_2^{2/3}}{\alpha\beta} \right)^2 \tag{2.82}$$

Note that the total elastic cross section falls off like β^{-2}.

2.7 Multiple scattering

We have seen in the previous section that there is a significant probability that a charged particle will undergo a Coulomb scattering collision while traversing a block of matter. Suppose that downstream of the block we observe how many particles are traveling at an angle θ with respect to the incident beam direction. Figure 2.18 shows the trajectories of two particles. The first particle only makes a single scatter at the angle θ. On the other hand, since the cross section for Rutherford scattering grows rapidly for decreasing scattering angles, it is also possible for a particle to leave the block at the angle θ after making a large number of small angle collisions. This latter case is referred to as multiple scattering. In a single event one cannot tell whether a particle observed at some angle has had a single scatter or has undergone multiple scattering. What one can do instead is to determine distribution functions for various processes, so that the probability of a given process resulting in a particle at the angle θ can be calculated.

We have seen in Eq. 2.72 that the Coulomb scattering cross section is proportional to the squares of the incident particle and target charges. To first order the probability for scattering due to the nucleus goes like Z_2^2, while the contribution of the Z_2 atomic electrons goes like Z_2. Thus, except for the lightest elements, multiple scattering is dominated by Coulomb scattering off the nuclei.

Since each individual small angle scatter is a random process, we expect the mean scattering angle of a beam of particles with respect to the incident direction to be zero. On the other hand, the rms scattering angle will in general be nonzero. The expectation value of θ^2 due to multiple scattering of a particle while crossing a length L of the material is

$$\theta_s^2 = Ln_a \int \theta^2 \frac{d\sigma}{d\Omega}\, d\Omega$$

where we assume the particle's velocity is not appreciably reduced while crossing the material. Using the small angle Rutherford cross section (Eq. 2.72), we obtain the variance of the cumulative angle distribution

$$\theta_s^2 = 8\pi L n_a r_e^2\, Z_1^2 Z_2^2 \left(\frac{mc}{\beta p}\right)^2 \ln \frac{2}{\alpha^2 A^{1/3} Z_2^{1/3}} \qquad (2.83)$$

where we have used Eqs. 2.79 and 2.81 for θ_{min} and θ_{max}. If we take $A \simeq 2Z$, the logarithm factor can be written in the form $2 \ln(173 Z_2^{-1/3})$. This logarithmic dependence is similar to that encountered in the definition of the radiation length L_R in Eq. 2.53. Thus, rewriting Eq. 2.83 in terms of L_R, we get

$$\theta_s^2 = \frac{4\pi}{\alpha} Z_1^2 \frac{m^2c^2}{\beta^2 p^2} \frac{L}{L_R}$$

Note that the expression of the multiple scattering angle in terms of a radiation length is just a convenience based on the fact that both quanti-

Figure 2.18 Scattering in a thick material.

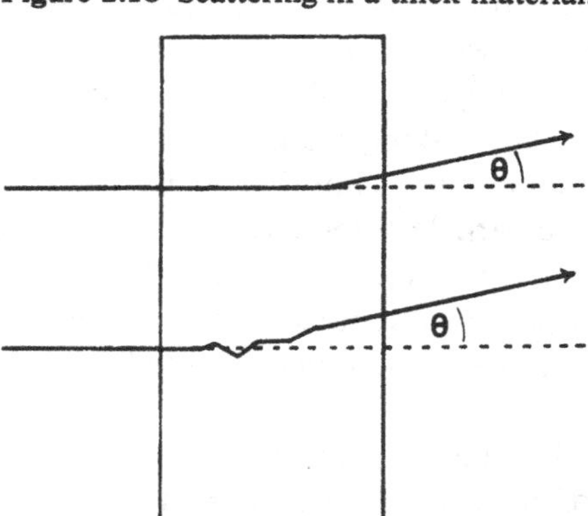

ties have a similar dependence on the properties of the material. Multiple scattering is not a radiation process. Now combining some of the factors into the energy

$$E_s = \sqrt{4\pi/\alpha}\, mc^2 = 21.2 \text{ MeV} \tag{2.84}$$

we obtain the Rossi–Greisen equation for the rms scattering angle

$$\theta_s = \frac{E_s}{pv} Z_1 \sqrt{L/L_R} \tag{2.85}$$

This equation is only accurate if the particle traverses many radiation lengths of the material. Otherwise it tends to overestimate the amount of scattering.

Highland [21] has given a more accurate form of Eq. 2.85. He used the Moliere theory described below to investigate the dependence of E_s on Z and L. This leads to an empirical formula for θ_e, the space angle for which the distribution drops to $1/e$ of its peak value, of

$$\theta_e = \frac{17.5 \text{ MeV}}{pv} Z_1 \sqrt{\frac{L}{L_R}} \left[1 + 0.125 \log_{10}\left(\frac{10L}{L_R}\right)\right] \tag{2.86}$$

It is sometimes convenient to consider the projection of the scattering angle on a plane. Define the z axis of a right-handed coordinate system to be along the direction of motion of the incident particle as shown in Fig. 2.19. If θ is the space angle of the scattered particle and θ_x and θ_y are the projections of the space angle onto the xz and yz planes, respectively, then

$$\cos\theta = (1 + \tan^2\theta_x + \tan^2\theta_y)^{-1/2}$$

Expanding for small θ, we find $\theta^2 \simeq \theta_x^2 + \theta_y^2$, so that on the average the angle projected onto a fixed plane is

$$\langle \theta_{pr}^2 \rangle = \tfrac{1}{2}\langle \theta^2 \rangle \tag{2.87}$$

Up to this point we have only been concerned with the rms or $1/e$ values of the scattering angle. Another important question is the distribution of scattering angles. Let $\psi_{ms}(\theta, x)\theta\, d\theta$ be the probability that a particle will emerge at an angle between θ and $\theta + d\theta$ with respect to the incident direction after traversing a finite thickness x of material. The distribution function is governed by a transport equation analogous to Eq. 2.30

$$\frac{\partial \psi_{ms}}{\partial x}(\theta, x) = \int \psi_{ms}(\theta - \zeta, x) n_a \frac{d\sigma}{d\Omega}(\zeta)\zeta\, d\zeta - \psi_{ms}(\theta, x) \int n_a \frac{d\sigma}{d\Omega}(\zeta)\zeta\, d\zeta \tag{2.88}$$

where we assume that the distribution is independent of the azimuthal scattering angle ϕ.

Moliere has obtained a solution of this equation that, if we take $d\sigma/d\Omega$

to be the small angle Rutherford cross section, becomes [22, 23]

$$\psi_{ms}(\theta, x) = \frac{1}{\theta_c^2} \int_0^\infty y\, dy\, J_0\left(\frac{\theta y}{\theta_c}\right) \exp\left[\frac{y^2}{4}\left(-b + \ln\frac{y^2}{4}\right)\right] \quad (2.89)$$

where J_0 is a Bessel function and θ_c is a characteristic angle given by

$$\theta_c^2 = \frac{4\pi n_a e^4 Z_1^2 Z_2 (Z_2 + 1) x}{(pv)^2} \quad (2.90)$$

The angle θ_c contains the dependence of ψ_{ms} on the macroscopic properties of the scattering medium. Note that θ_c^2 grows linearly with x. The quantity b is defined as

$$b = \ln(\theta_c/\theta_a)^2 + 1 - 2C_E \quad (2.91)$$

where C_E is Euler's constant. The dependence of the scattering on the screening angle θ_a is given by

$$\theta_a^2 \simeq \theta_0^2\left[1.13 + 3.76\left(\frac{Z_1 Z_2 e^2}{\hbar v}\right)^2\right] \quad (2.92)$$

where θ_0 is approximately the minimum scattering angle given by Eq. 2.79,

$$\theta_0 \simeq 1.13\theta_{min} \quad (2.93)$$

Figure 2.19 The projected scattering angles.

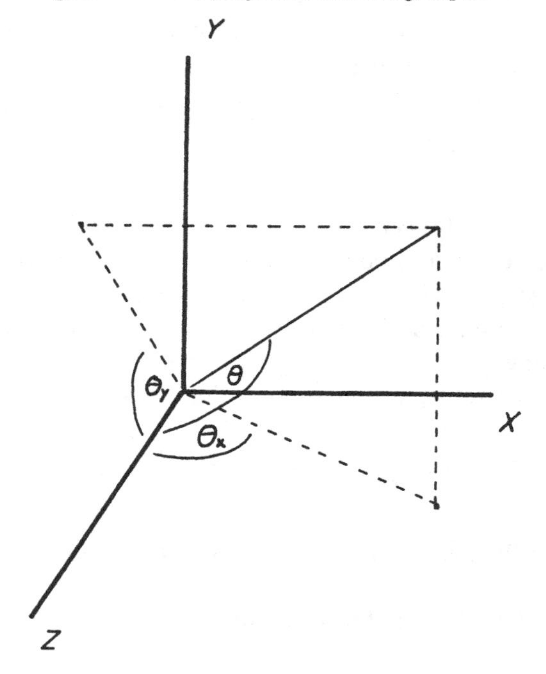

The Moliere theory should be valid when $\theta_c \gg \theta_0$. The scattering angle distribution approaches the Rutherford single scattering distribution for large angles.

The mean number of collisions between an incident particle and the target atoms in the thickness x can be found by dividing the target length by the interaction mean free path

$$N_{coll} = x/\lambda_I = x n_a \sigma \tag{2.94}$$

If we use Eq. 2.82 for the total Coulomb scattering cross section, we find that

$$N_{coll} = 4 x n_a \pi r_e^2 \left(\frac{Z_1 Z_2^{2/3}}{\alpha\beta} \right)^2 \tag{2.95}$$

When $N_{coll} \gtrsim 1000$, the distribution is approximately Gaussian for small scattering angles.

Scattering angle distributions of 15.7 MeV electrons from gold foils are in excellent agreement with the predictions of the Moliere theory [24]. The small angle region is approximately Gaussian, and the tails fit nicely to the large angle, single scattering distributions. At high energies Shen et al. [25] have measured the small angle scattering distributions for $\pi^\pm$, $K^\pm$, and $p^\pm$ incident on H, Be, C, Al, Cu, Sn, and Pb targets. In each case the incident particle traversed about 0.1 radiation length of material. Figure 2.20 shows typical θ^2 distributions for hydrogen and lead. The experimental results for θ_e, the angle at which the distribution drops to $1/e$ of the peak value, agreed with the predictions of the Moliere theory. There was no statistically significant dependence on the type of projectile. The result for θ_e from Eq. 2.86 was found to give results ~ 3% larger than experiment for $Z \geqslant 6$. For $Z < 6$ the error increased, reaching ~ 10% for hydrogen.

The angular spread arising from multiple scattering also introduces a lateral spread in a beam of particles. The mean square lateral displacement, irrespective of angle, is given approximately by [1]

$$\langle y^2 \rangle = \tfrac{1}{6} \theta_s^2 L^2 \tag{2.96}$$

where L is the distance traversed into the scattering medium.

2.8 Other electromagnetic effects

A large number of additional electromagnetic effects have been investigated. In this section we will discuss two interesting phenomena that may have useful applications in certain circumstances: channeling and acoustic radiation.

Until this point we have implicitly assumed that particles were passing through an amorphous material. In such a case each interaction is a random event, so that the results of multiple interactions are not correlated. The situation can be different, however, for crystals of materials such as silicon or germanium [26]. If the incident angle ψ_{inc} of the particle with respect to a low index, crystal axis is on the order of or smaller than the critical angle

$$\psi_{\text{crit}} = \sqrt{\frac{4Z_1Z_2e^2}{p_1v_1d}} \tag{2.97}$$

channeling effects may occur. The quantity d is the interatomic spacing along the channel, while p_1 (v_1) is the momentum (velocity) of the incident particle. For $\psi_{\text{inc}} \lesssim \psi_{\text{crit}}$ the atoms of the crystal appear as an axial string. The correlated collisions with the string cause the particles to be gently attracted or repelled, depending on their charge. The atomic planes can have a similar steering effect.

Channeled positive particles tend to avoid the axial string. As a result, processes that are most important at small impact parameters tend to be suppressed. For example, channeled protons or π^+ show smaller than random energy loss, reduced nuclear absorption, and reduced wide angle

Figure 2.20 The distribution of θ^2 for 50-GeV/c negative hadrons in hydrogen and 70-GeV/c negative hadrons in lead. The lower curves show the distributions taken with the target removed. The solid curves are fits using the Moliere theory. (After G. Shen et al., Phys. Rev. D 20: 1584, 1979.)

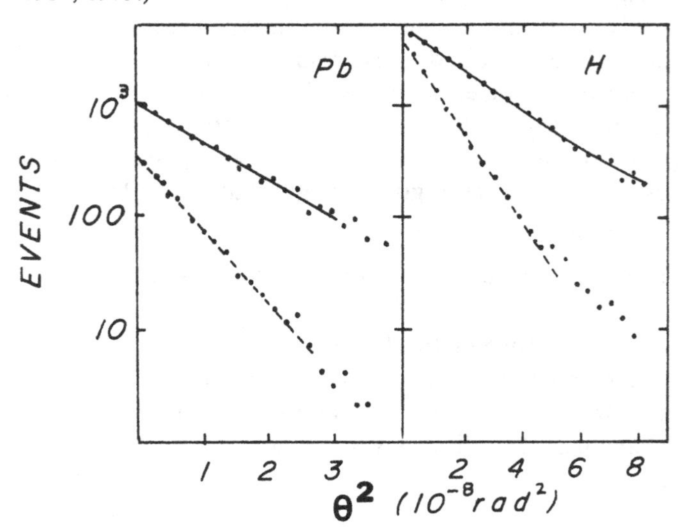

scattering and have a large transmission near $\psi_{inc} = 0°$. Channeled π^-, on the other hand, are attracted to the string and show increased nuclear absorption, increased wide angle scattering, and a reduced transmission for $\psi_{inc} \sim 0°$.

Channeling can also influence multiple scattering [27]. Figure 2.21 shows multiple scattering distributions for 15-GeV/c protons and π^-. The random scattering data agree well with the predictions of the Moliere theory. When $\psi_{inc} < \psi_{crit}$ (Δ) the multiple scattering for protons is reduced, while that for π^- is increased. Interestingly, for $\psi_{inc} \sim 3\psi_{crit}$ the multiple scattering for both charges is larger than for random scattering.

The oscillatory motion of channeled electrons and positrons leads to the emission of channeling radiation [26, 28]. This type of radiation has the same origin as ordinary bremsstrahlung, discussed in Section 2.4, and as coherent bremsstrahlung, which results when the incoming particle has periodic contacts with atoms in the target material. Ordinary bremsstrahlung differs from the other types of radiation because it has a continuous photon emission spectrum. In coherent bremsstrahlung the transverse

Figure 2.21 Scattering angle distributions of 15-GeV/c p and π^- transmitted through a 4.2-mm-thick germanium crystal. ψ_1 is the critical angle. (S. Andersen et al., Nuc. Phys. B 167: 1, 1980.)

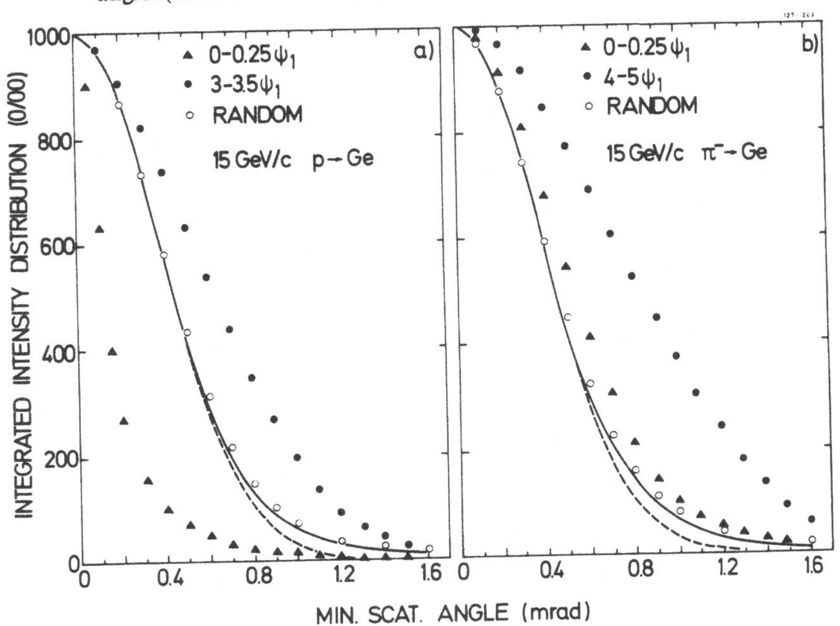

motion of the incident particle is free, while in channeling radiation the transverse motion is bound by the potential of the atomic string.

In the channeling regime the intensity of the emitted radiation is enhanced over the ordinary bremsstrahlung by a factor proportional to $\gamma^{1/2}Z_2^{-2/3}$. Channeled radiation is emitted in the forward direction within a cone with characteristic half-angle $\theta_c \sim \gamma^{-1}$. The emitted photon frequency ω and emission angle θ in the LAB are related by

$$\omega \simeq \frac{2\gamma^2\omega_0}{1 + \gamma^2\theta^2} \tag{2.98}$$

where $\gamma\omega_0$ is the classical oscillation frequency for the electron or positron in its rest frame. Since the channeling radiation is linearly polarized, channeled beams of electrons or positrons may provide a useful source for a polarized photon beam. The emitted photon spectrum from incident electrons tends to be broader and show less structure than that from incident positron beams.

When a beam of particles is passed through a liquid, the deposited energy can cause the affected volume to undergo an adiabatic expansion. This in turn produces a detectable bipolar acoustic pressure wave [29]. The thermoacoustic model predicts that the time dependent pressure wave $P(t)$ satisfies the equation

$$\int_{-\infty}^{\infty} tP(t)\,dt \simeq -\frac{K}{4\pi C_P}\frac{E}{R} \tag{2.99}$$

where K is the volume coefficient of expansion and C_P is the heat capacity of the medium, E is the total deposited energy, and R is the distance to the observation point. The dependence of the acoustic signal on the quantities on the right-hand side of Eq. 2.99 was confirmed by experiments. The signal amplitude was linearly dependent on the deposited energy to within a factor of 2 for deposited energies between 2×10^{15} and 4×10^{20} eV. The signal amplitude in CCl_4 was 24 times as large as the signal in H_2O, in rough agreement with the ratios of the values of K/C_P.

Some of the applications envisioned for acoustic radiation include a beam monitor, heavy ion detector, hadron calorimeter, or a cosmic ray detector. The primary advantage of the technique is that the hydrophones, which detect the acoustic pulses, are much less expensive than photomultiplier tubes or other common detectors. In addition, one can sample large volumes since the attenuation length of sound in liquids is long. The major disadvantage is the very large threshold required to produce a signal. It is estimated that the ultimate detector threshold would be around 10^{13} eV [29].

References

Some other general references not specifically cited below include Enrico Fermi, *Nuclear Physics,* Chicago: University of Chicago, 1950, Chap. 2; R. M. Sternheimer, Interaction of Radiation with Matter, in L. C. Yuan and C.-S. Wu, (eds.), *Methods of Experimental Physics,* New York: Academic, 1961, Vol. 5A, pp. 1–88.

[1] An excellent introduction to the subject matter of this chapter can be found in B. Rossi, *High Energy Particles,* Englewood Cliffs: Prentice-Hall, 1952, Chap. 2. References 2 and 3 also contain much useful information.

[2] J. Jackson, *Classical Electrodynamics,* New York: Wiley, 1962.

[3] E. Segre, *Nuclei and Particles,* 2nd ed., Reading: Benjamin, 1977, Chap. 2.

[4] U. Fano, Penetration of protons, alpha particles, and mesons, Ann. Rev. Nuc. Sci. 13: 1–66, 1963.

[5] S. Ahlen, Theoretical and experimental aspects of the energy loss of relativistic heavily ionizing particles, Rev. Mod. Phys. 52: 121–73, 1980. This reference contains a nice summary of various investigators' contributions to the energy loss problem.

[6] The theory of the density effect correction is given by R.M. Sternheimer, The density effect for the ionization loss in various materials, Phys. Rev. 88: 851–9, 1952; the corrections have been updated using the latest data for the ionization potential in R.M. Sternheimer, S.M. Seltzer, and M.J. Berger, Density effect for the ionization loss of charged particles in various substances, Phys. Rev. B 26: 6067–76, 1982; erratum, B 27: 6971, 1983; and R.M. Sternheimer, M.J. Berger, and S.M. Seltzer, Density effect for the ionization loss of charged particles in various substances, Atomic Data and Nuclear Data Tables 30: 261–71, 1984.

[7] A. Walenta, J. Fischer, H. Okuno, and C. Wang, Measurement of ionization loss in the region of relativistic rise for noble and molecular gases, Nuc. Instr. Meth. 161: 45–58, 1979.

[8] J. Burq, M. Chemarin, M. Chevallier, A. Denisov, T. Ekelof, P. Grafstrom, E. Hagberg, B. Ille, A. Kashchuk, A. Kulikov, M. Lambert, J. Martin, S. Maury, M. Querrou, V. Schegelsky, I. Tkach, and A. Vorobyov, Observation of the ionization energy loss of high energy protons and pions in hydrogen gas, Nuc. Instr. Meth. 187: 407–11, 1981.

[9] R. Talman, On the statistics of particle identification using ionization, Nuc. Instr. Meth. 159: 189–211, 1979.

[10] H. Maccabee and D. Papworth, Correction to Landau's energy loss formula, Phys. Lett. 30A: 241–2, 1969.

[11] The parameters used in Vavilov's and others theories are presented in some detail in H. Bichsel and R. Saxon, Comparison of calculational methods for straggling in thin absorbers, Phys. Rev. A 11: 1286–96, 1975.

[12] H. Bethe and J. Ashkin, Passage of radiation through matter, in E. Segre (ed.), *Experimental Nuclear Physics,* Vol. 1, Part 2, New York: Wiley, 1959.

[13] Particle Data Group, Review of particle properties, Phys. Lett. 111B: 1, 1982.

[14] Y. Tsai, Pair production and bremsstrahlung of charged leptons, Rev. Mod. Phys. 46: 815–51, 1974.

[15] W. Heitler, *The Quantum Theory of Radiation,* 3rd ed., Oxford: Clarendon Press, 1953.

[16] R. Siemann, W. Ash, K. Berkelman, D. Hartill, C. Lichtenstein, and R. Littauer, Wide angle bremsstrahlung, Phys. Rev. Lett. 22: 421–4, 1969.
[17] J. Hubbell, Photon cross sections, attenuation coefficients, and energy absorption coefficients from 10 keV to 100 GeV, National Bureau of Standards report, NSRDS-NBS 29, 1969.
[18] J. Eickmeyer, T. Gentile, S. Michalowski, N. Mistry, R. Talman, and K. Ueno, High energy electron-pair photoproduction from nuclei: Comparison with theory, Phys. Rev. D 21: 3001–4, 1980.
[19] J. Bjorken and S. Drell, *Relativistic Quantum Mechanics*, New York: McGraw-Hill, 1964, Chap. 7.
[20] R. Brandelik, W. Braunschweig, K. Gather, F.J. Kirschfink, K. Lubelsmeyer, H.-U. Martyn, G. Peise, J. Rimkus, H.G. Sander, D. Schmitz, D. Trines, W. Wallraff, H. Boerner, H.M. Fischer, H. Hartmann, E. Hilger, W. Hillen, G. Knop, L. Kopke, H. Kolanoski, R. Wedemeyer, N. Wermes, M. Wollstadt, H. Burkhardt, S. Cooper, J. Franzke, D. Heyland, H. Hultschig, P. Joos, W. Koch, U. Kotz, H. Kowalski, A. Ladage, B. Lohr, D. Luke, H.L. Lynch, P. Mattig, K.H. Mess, D. Notz, J. Pyrlik, D.R. Quarrie, R. Riethmuller, W. Schutte, P. Soding, G. Wolf, R. Fohrmann, H.L. Krasemann, P. Leu, E. Lohrmann, D. Pandoulas, G. Poelz, O. Romer, P. Schmuser, B.H. Wiik, I. Al-Agil, R. Beuselinck, D.M. Binnie, A.J. Campbell, P.J. Dornan, D.A. Garbutt, T.D. Jones, W.G. Jones, S.L. Lloyd, J.K. Sedgbeer, K.W. Bell, M.G. Bowler, I.C. Brock, R.J. Cashmore, R. Carnegie, P.E.L. Clarke, R. Devenish, P. Grossmann, J. Illingworth, M. Ogg, G.L. Salmon, J. Thomas, T.R. Wyatt, C. Youngman, B. Foster, J.C. Hart, J. Harvey, J. Proudfoot, D.H. Saxon, P.L. Woodworth, M. Holder, E. Duchovni, Y. Eisenberg, U. Karshon, G. Mikenberg, D. Revel, E. Ronat, A. Shapira, T. Barklow, T. Meyer, G. Rudolph, E. Wicklund, Sau Lan Wu, and G. Zobernig, Electroweak coupling constants in the leptonic reactions $e^+e^- \rightarrow e^+e^-$ and $e^+e^- \rightarrow \mu^+\mu^-$ and search for scalar leptons, Phys. Lett. 117B: 365–71, 1982.
[21] V. Highland, Some practical remarks on multiple scattering, Nuc. Instr. Meth. 129: 497–9, 1975; erratum, Nucl. Instr. Meth. 161: 171, 1979.
[22] H. Bethe, Moliere's theory of multiple scattering, Phys. Rev. 89: 1256–66, 1953.
[23] W. Scott, The theory of small angle multiple scattering of fast charged particles, Rev. Mod. Phys. 35: 231–313, 1963.
[24] A. Hanson, L. Lanzl, E. Lyman, and M. Scott, Measurement of multiple scattering of 15.7 MeV electrons, Phys. Rev. 84: 634–7, 1951.
[25] G. Shen, C. Ankenbrandt, M. Atac, R. Brown, S. Ecklund, P. Gollon, J. Lach, J. MacLachlan, A. Roberts, L. Fajardo, R. Majka, J. Marx, P. Nemethy, L. Rosselet, J. Sandweiss, A. Schiz, and A. Slaughter, Measurement of multiple scattering at 50 to 200 GeV/c, Phys. Rev. D 20: 1584–8, 1979.
[26] V. Beloshitsky and F. Komarov, Electromagnetic radiation of relativistic channeling particles (The Kumakhov effect), Phys. Rep. 93: 117–197, 1982.
[27] S. Andersen, O. Fich, H. Nielsen, H. Schiott, E. Uggerhoj, C. Vraast Thomsen, G. Charpak, G. Petersen, F. Sauli, J. Ponpon, and P. Siffert, Influence of channeling on scattering of 2–15 GeV/c protons, π^+, and π^- incident on Si and Ge crystals, Nuc. Phys. B 167: 1–40, 1980.
[28] J. Andersen, E. Bonderup, and R. Pantell, Channeling radiation, Ann. Rev. Nuc. Part Sci. 33: 453–504, 1983.
[29] L. Sulak, T. Armstrong, H. Baranger, M. Bregman, M. Levi, D. Mael, J. Strait, T. Bowen, A. Pifer, P. Polakos, H. Bradner, A. Parvulescu, W. Jones, and J. Learned, Experimental studies of the acoustic signature of proton beams traversing fluid media, Nuc. Instr. Meth. 161: 203–17, 1979.

Exercises

1. Consider a 10-GeV/c proton incident on an aluminum target. Estimate the range of valid impact parameters for calculating dE/dx. Assume that $\hbar\omega$ is approximately equal to the ionization potential for the atom.
2. What is the expected mean energy loss of 50-GeV/c protons in beryllium? How much is this result affected by the density effect correction?
3. Consider a 10-GeV/c K^- beam in liquid hydrogen. What is the maximum kinetic energy of delta rays produced by the beam? How many delta rays with kinetic energy greater than 100 MeV are produced in 2 cm?
4. Calculate the Landau distribution function numerically and plot $f_L(\lambda)$ versus λ.
5. Find the most probable energy loss of 100 GeV/c π^- in copper. What is the probability of observing an energy loss of half of this amount and twice this amount?
6. Find the ionization energy loss of a 20-GeV positron in lead.
7. Find the cross section for a 30-GeV electron to undergo bremsstrahlung in a lead target and emerge with an energy of 25 GeV. Is the complete screening hypothesis justified?
8. Consider the Compton scattering of a 1-MeV photon. What is the energy of a photon scattered at 30°? What is the kinetic energy and angle of the recoil electron?
9. Find the total pair production cross section for a 50-GeV photon in gold. What is the differential cross section for producing a 20-GeV electron? Is the complete screening hypothesis justified?
10. Plot the difference between the Moller and Bhabha differential cross sections as a function of $\cos\theta^*$.
11. Make a table showing the minimum and maximum angles of validity for Rutherford scattering and the total Coulomb cross section for 1-GeV protons in carbon, iron, and lead.
12. Find the Rossi–Greisen mean scattering angle θ_s for 6-GeV K^-

after traversing 2 cm of copper. How large is the Highland correction to θ_s?

13. Calculate numerically the Moliere scattering distribution for 10-GeV electrons after passing through 1 radiation length of lead. Plot ψ_{ms} as a fuction of θ.

14. Estimate the critical angle for channeling of a 20-GeV/c π^- in tin.

15. Use the data of Table 2.1 to check the validity of Bloch's expression for the ionization potential.

3

Nuclear interactions

In the preceding chapter we discussed the electromagnetic interaction, which was responsible for the energy loss and small angle scattering of charged particles, and for the production and interactions of photons. However, there are other types of processes where the nuclear interaction may represent the dominant mechanism. These include particle creation reactions, interactions at high energies or large momentum transfers, and interactions of neutral particles other than the photon. In this chapter we will examine some of the basic properties of nuclear interactions. We will not be directly concerned with the physics underlying subnuclear phenomena. Instead, our main concern will be to survey its overall features, principally total cross sections, particle production multiplicities, and angular distributions.

We will first discuss the strong interaction. The group of particles known as hadrons is influenced by this interaction. Next we will briefly discuss the weak interaction, which is responsible for the interactions of neutrinos in matter and for the decay of most quasistable particles.

3.1 Strong interactions

The group of particles known as hadrons are subject to the strong interaction in matter. The neutron is an ideal probe of this interaction since it has no appreciable electromagnetic interactions. We have seen that the cross sections for most electromagnetic interactions are strongly peaked in the forward direction and fall off with increasing energy. Thus, away from the forward direction, the high energy behavior of all hadrons is determined by the strong interaction. In this section we will examine a few important features of strong interactions [1].

The total cross sections for interactions of some common hadrons on a proton target are shown in Fig. 3.1. Most of the total cross sections have a complicated structure at low energies, due in part to resonance production. However, above 5 GeV/c incident lab momentum, the exhibited cross sections show a broad minimum where the value of the total cross section is 20–40 mb. At very high energies all the cross sections are observed to rise slowly [2]. The pp data in Fig. 3.1 show that the increase is consistent with a ln(s) dependence, where s is the square of the total CM energy. Elastic scattering represents 15–20% of the pp total cross section at high energy. Note that the negative variety of each particle has a slightly larger cross section than its positive partner. Figure 3.2 shows that similar behavior is observed in scattering from neutrons. This data is either obtained using a neutron beam or extracted from deuterium target experiments.

The rate of particle interactions I is related to the total cross section σ_T and to the incident beam rate I_0 by

$$I = I_0\left(\frac{\rho N_A t}{A}\right)\sigma_T \tag{3.1}$$

Figure 3.1 Total cross section for πp, Kp, and pp interactions as a function of the total CM energy squared, s. (After G. Giacomelli, Phys. Rep. 23: 123, 1976.)

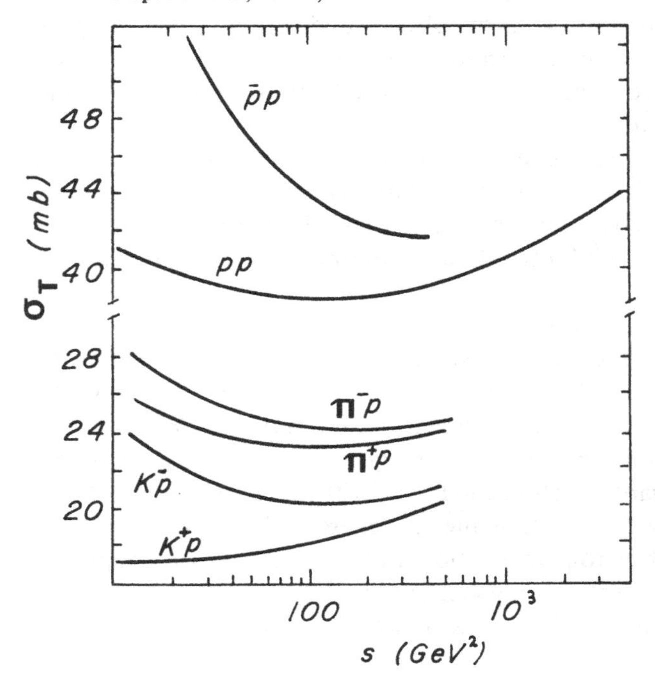

where ρ, t, and A are the density, thickness, and atomic weight of the target material. Another measure of the likelihood of an interaction is given by the mean free path between collisions (collision length)

$$L_{\text{coll}} = \frac{A}{N_A \rho \sigma_T} \tag{3.2}$$

or

$$X_{\text{coll}} = \frac{A}{N_A \sigma_T} \tag{3.3}$$

For calculations of attenuation in matter, a more relevant quantity is the absorption cross section, defined as

$$\sigma_{\text{abs}} = \sigma_T - \sigma_{\text{el}} - \sigma_q \tag{3.4}$$

where σ_{el} refers to coherent elastic scattering off a whole nucleus, and σ_q refers to quasielastic scattering from individual nucleons. In elastic and quasielastic scattering the hadron retains its identity, and its momentum is in general only slightly perturbed. We may define the absorption lengths λ of particles analogously to Eqs. 3.2 and 3.3 by replacing the total cross

Figure 3.2 Total cross section for Kn, pn, and np interactions as a function of s. (After G. Giacomelli, Phys. Rep. 23: 123, 1976.)

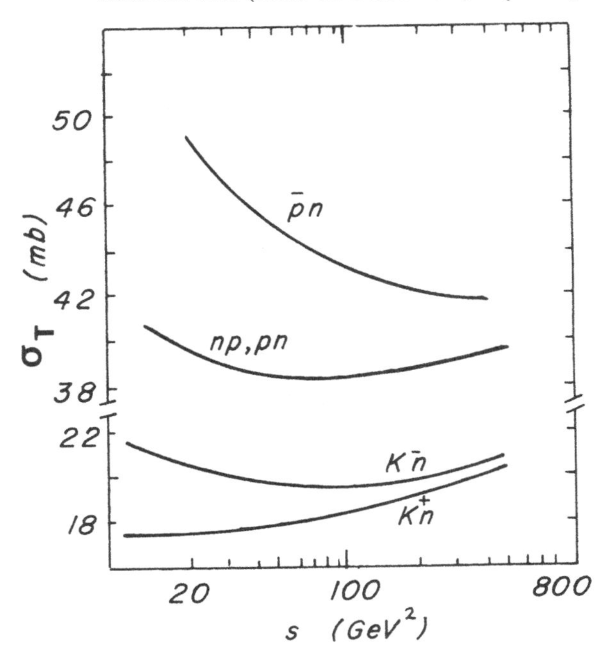

section with the absorption cross section, for example,

$$\lambda = \frac{A}{N_A \sigma_{abs}} \tag{3.5}$$

Measurements of the absorption cross section of charged hadrons from 60 to 200 GeV/c have been made by Carroll et al. [3]. The total cross section is corrected for Coulomb scattering, coherent scattering from the whole nucleus, and quasielastic scattering. The absorption cross sections are approximately independent of momentum above 20 GeV/c. Antiprotons have the largest absorption cross sections, followed in order by protons, pions, K^-, and K^+.

High energy measurements of neutron–nucleus absorption cross sections have been made by Roberts et al. [4]. Figure 3.3 shows the absorp-

Figure 3.3 Neutron–nucleus absorption (σ_I) and total (σ_T) cross sections as a function of atomic weight. (T. Roberts, H. Gustafson, L. Jones, M. Longo, and M. Whalley, Nuc. Phys. B 159: 56, 1979.)

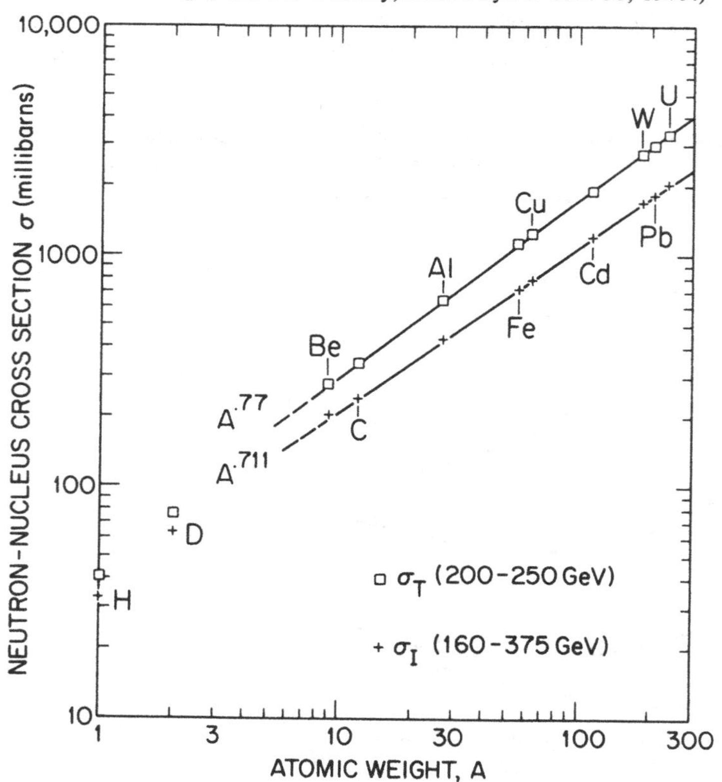

tion cross section as a function of the atomic weight of the target nucleus. It can be seen that for $A \geqslant 9$ the relation of the cross section to A is consistent with

$$\sigma_{abs}(A) = \sigma_0 A^{\alpha} \tag{3.6}$$

The fitted values of the parameters are $\sigma_0 = 41.2$ mb and $\alpha = 0.711$.

Values for nuclear collision and absorption lengths of elements are given in Table 3.1. The total cross sections refer to measurements using 80–240-GeV neutrons. Table 3.2 contains electromagnetic and nuclear properties of some common materials used in particle physics experiments.

A simple model of the nucleus indicates that $\alpha \simeq \frac{2}{3}$. Assume that the density of nuclear matter is approximately constant for all nuclei. Then the volume of the nucleus is proportional to A, the atomic weight. Assume further that the nucleons are confined in a spherical region of radius r_n. Then

$$r_n = r_0 A^{1/3} \tag{3.7}$$

where r_0 is a constant. Experimental investigations of nuclear sizes show that $r_0 = 1.38 \times 10^{-13}$ cm. Note that the experimental value for $r_0 \simeq \frac{1}{2} r_e$,

Table 3.1. *Atomic and nuclear properties of elements*

Material	Z	A	ρ^a (g/cm³)	σ_T (barns)	X_{coll} (g/cm²)	σ_{abs} (barns)	λ (g/cm²)
H_2	1	1.008	0.0708	0.0387	43.3	0.033	50.8
D_2	1	2.01	0.162	0.073	45.7	0.061	53.7
He	2	4.00	0.125	0.133	49.9	0.102	65.1
Li	3	6.94	0.534	0.211	54.6	0.157	73.4
Be	4	9.01	1.848	0.268	55.8	0.199	75.2
C	6	12.01	2.265	0.331	60.2	0.231	86.3
N_2	7	14.01	0.808	0.379	61.4	0.265	87.8
O_2	8	16.00	1.14	0.420	63.2	0.292	91.0
Ne	10	20.18	1.207	0.507	66.1	0.347	96.6
Al	13	26.98	2.70	0.634	70.6	0.421	106.4
Ar	18	39.95	1.40	0.868	76.4	0.566	117.2
Fe	26	55.85	7.87	1.120	82.8	0.703	131.9
Cu	29	63.54	8.96	1.232	85.6	0.782	134.9
Sn	50	118.69	7.31	1.967	100.2	1.21	163
W	74	183.85	19.3	2.767	110.3	1.65	185
Pb	82	207.19	11.35	2.960	116.2	1.77	194
U	92	238.03	∽18.95	3.378	117.0	1.98	199

[a] Density for solids or liquids at boiling point.
Source: Particle Data Group, Rev. Mod. Phys. 56: S1, 1984.

Table 3.2. *Properties of some common materials*

Material	ρ^a (g/cm³)	L_{rad} (cm)	X_{rad} (g/cm²)	X_{coll} (g/cm²)	λ (g/cm²)
Air	(1.29)	30420	36.66	62.0	90.0
Water	1.00	36.1	36.08	60.1	84.9
Shielding concrete	2.5	10.7	26.7	67.4	99.9
Emulsion (G5)	3.815	2.89	11.0	82.0	134
BGO	7.1	1.12	7.98	97.4	156
NaI	3.67	2.59	9.49	94.8	152
BaF_2	4.83	2.05	9.91	92.1	146
CsI	4.51	1.86	8.39	—	—
Scintillator	1.032	42.4	43.8	58.4	82.0
Lucite	1.16–1.20	∽34.4	40.55	59.2	83.6
Polyethylene	0.92–0.95	∽47.9	44.8	56.9	78.8
Mylar	1.39	28.7	39.95	60.2	85.7
Pyrex	2.23	12.7	28.3	66.2	97.6

[a] Density for solids or liquids at boiling point. Number in parentheses is for gas (gm/l; STP).
Source: Particle Data Group, Rev. Mod. Phys, 56: S1, 1984; H. Grassman, E. Lorenz, and H.-G. Moser, Nuc. Instr. Meth. 228: 323, 1985.

Figure 3.4 Total cross sections for γp and γd interactions as a function of photon energy. (After G. Giacomelli, Phys. Rep. 23: 123, 1976.)

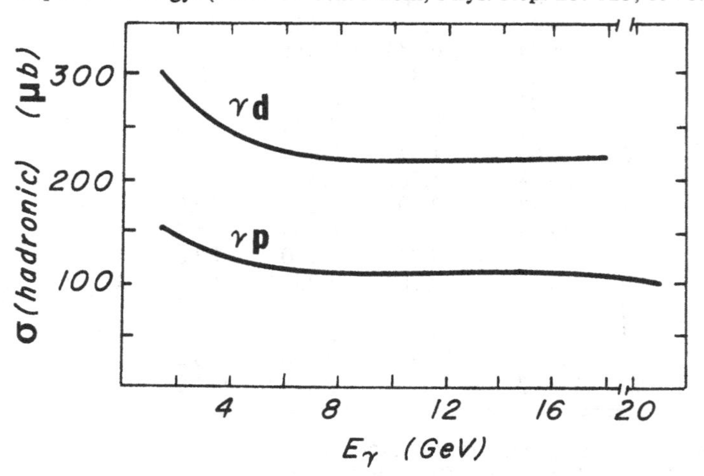

as we noted in Eq. 2.80. The geometric cross section for a point object to scatter from the nucleus is the circular area subtended by the sphere, πr_n^2, so that

$$\sigma = \pi r_0^2 A^{2/3} \tag{3.8}$$

Thus, nuclear cross sections should grow like $A^{2/3}$.

In Fig. 3.4 we show the total cross section for γp interactions. Even though the photon is not a hadron, it is believed to exist a small fraction of the time in the form of vector (spin 1) mesons. As such it exhibits behavior similar to the other hadrons. Note that the magnitude of the photoabsorption cross section is about a factor of 100 smaller than the data in Figs. 3.1 and 3.2.

The e^+e^- or e-nucleon initial state can also couple to a hadronic final state through intermediate virtual photons [5]. This is studied using the ratio R of the cross section for hadron production in e^+e^- interactions to the cross section for $\mu^+\mu^-$ production as a function of the total CM energy. Apart from particular energies, corresponding to the masses of resonance states, R is approximately 4.0 above 5 GeV. The muon pair production cross section is calculable in QED and is given in lowest order by [6]

$$\begin{aligned}\sigma_{\mu\mu} &= 4\pi\alpha^2/3s \\ &= 87.6/s \quad \text{nb}\end{aligned} \tag{3.9}$$

where s is in GeV^2. Thus, the e^+e^- hadronic production cross section falls off like s^{-1}.

The average number of particles produced in an interaction increases as the incident energy increases [7, 8]. The top curve in Fig. 3.5 shows the average number of charged particles produced in proton–proton interactions as a function of s. We see that pp interactions with s around 1000 GeV^2 typically produce 10 charged particles. The energy dependence of the multiplicity is similar for the nonannihilation reactions in $\pi^\pm$p, $K^\pm$p, and pp. The total cross section for these interactions contains a contribution from diffractive processes. Antiproton annihilation interactions tend to produce slightly more charged particles, particularly at low energies. The differences between the two classes of interactions can be reduced if the data is plotted as a function of the energy available for particle production. This is W for $\bar{\text{p}}$p annihilation and $W - M$ for nonannihilation reactions, where M is the total mass of the initial state particles. The average charged particle multiplicity is well fit with the equation [1]

$$\langle N \rangle = 1.8 \ln(s) - 2.8 \tag{3.10}$$

for pp interactions up to $s = 3000\ \text{GeV}^2$.

The composition of the produced particles for pp interactions is also shown in Fig. 3.5. Pions comprise about 90% of the produced particles at small angles and high energy. The fraction of charged kaons and protons tends to increase in large p_T events. Figure 3.6 shows the cross sections for inclusive production of π^0, K^0, and Λ. The measured π^0 cross sections are very close to those of π^+ and π^-. The cross sections for inclusive K^0 or Λ production are about a factor of 20 smaller at the same momentum.

Figure 3.5 Averaged charged multiplicity and charged particle composition in pp inclusive interactions. (H. Boggild and T. Ferbel, reproduced with permission, from the Annual Review of Nuclear Science, Vol. 24, © 1974 by Annual Reviews, Inc.)

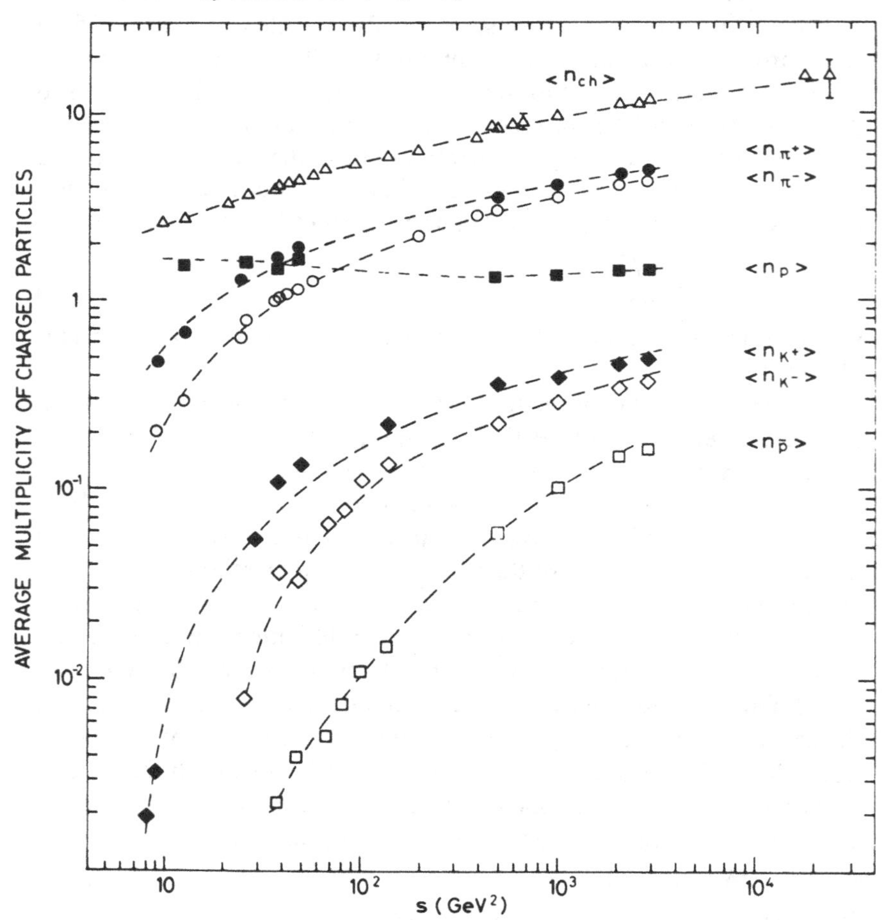

The mean charged particle multiplicity in e^+e^- interactions is similar to that observed in pp and $\bar{p}p$ interactions below $W = 5$ GeV. Above 10 GeV the muliplicity increases faster than $\ln(s)$. Particles produced with low momentum are almost entirely pions. However at $W = 30$ GeV the measured pion fraction is only around 50% for particles with momenta $\geqslant 4$ GeV/c [5].

The angular distribution of the produced particles is mostly confined to a narrow cone around the beam direction in the laboratory. The distribution of particles is nonisotropic, even when viewed in the CM coordinate system [1]. In general, there is a particle or group of particles that is

Figure 3.6 Inclusive cross sections for the production of π^0, K^0, and Λ in pp interactions. (H. Boggild and T. Ferbel, reproduced with permission, from the Annual Review of Nuclear Science, Vol. 24, © 1974 by Annual Reviews, Inc.)

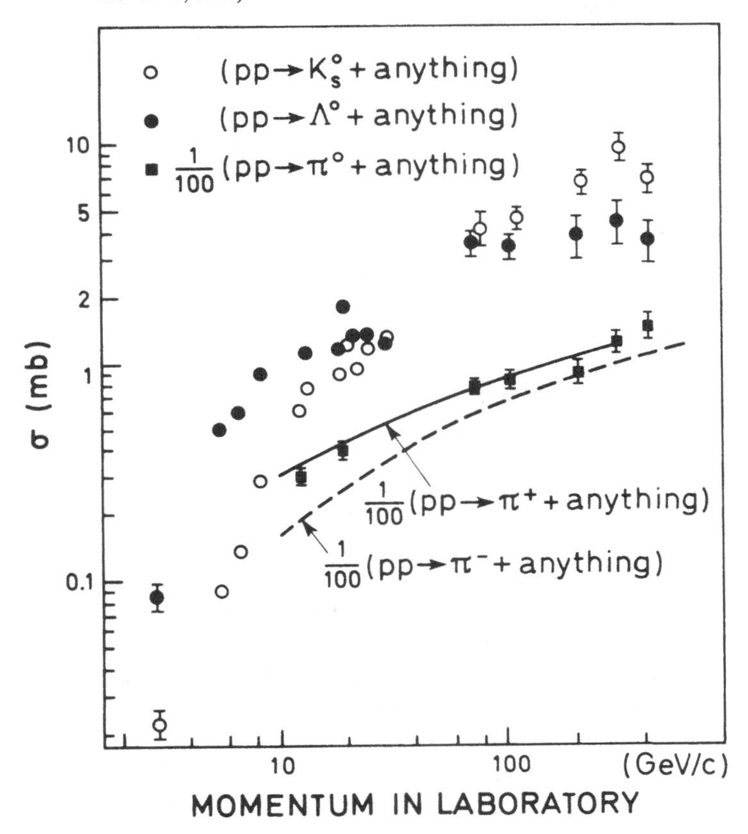

emitted at a small angle with respect to the incoming CM particles and whose momentum is larger than the other produced particles. This is referred to as the leading particle effect.

A typical pion is produced with about 300 MeV/c of transverse momentum. The transverse momentum of charged kaons or antiprotons is higher, typically $\sim$400 MeV/c. This number grows slowly with energy. The transverse momentum distribution falls exponentially for small p_T. The invariant cross section for inclusive π^+, π^-, and π^0 production show almost identical behavior as a function of p_T. The fall-off of the cross section at large p_T is much slower than the extrapolation of the low p_T exponential, particularly at high energy [9]. The production of large p_T hadrons in e^+e^- interactions is also much larger than predicted by an extrapolation of a low p_T exponential or power law dependence [5].

3.2 Weak interactions

The second fundamental nuclear interaction is the weak interaction, which is particularly important for neutrino interactions and for the decay of most quasistable particles. Since they are neutral and not hadronic, neutrinos can act as direct probes of the weak interaction [10, 11]. The price we must pay for this good fortune is, however, that the probability of a neutrino interaction in matter will therefore be very small. This makes it more difficult to detect neutrinos than other particles.

The weak interaction sometimes plays an important role in the interactions of charged leptons. Unlike the electromagnetic and strong interactions, the weak interaction does not preserve the parity (or mirror image) symmetry. Thus, the experimental signature that allows the weak interaction effects to be separated from the electromagnetic effects is a parity-violating component in the cross section. Two important examples of this are the asymmetry measured in the scattering of polarized electrons off nuclei at SLAC [12] and the asymmetry in muon pair production in e^+e^- collisions at PETRA [5].

Two major classes of weak interactions are known. Consider for example the interaction of a ν_e with a nucleon. Weak interactions conserve the value of an internal quantum number known as the electron lepton number. In a charged current neutrino interaction the final state lepton is charged, and in this example, in order to conserve electron lepton number, it must be an electron. The interaction is believed to proceed through the exchange of a virtual W charged vector boson between the neutrino and the nucleon. In a neutral current interaction, on the other hand, the final state lepton is neutral and in our example must be another ν_e. This interaction is believed to take place by the exchange of a virtual Z^0 neutral vector boson.

The total cross sections for ν and $\bar{\nu}$ on nucleons are shown in Fig. 3.7. It can be seen that at a given incident energy, the neutrino total cross section is about 2 times larger than the antineutrino total cross section. Both cross sections rise linearly with E_ν up to the highest energies yet measured. Notice, however, that the magnitude of the ν total cross section, even at 200 GeV, is only 160×10^{-38} cm^2, whereas a typical strong interaction cross section at the same energy is about 40 mb = 40×10^{-27} cm^2. The mean charged particle multiplicities for νN interactions increase like $\ln(W^2)$ above the resonance region [13].

The angular dependence is quite different for charged current ν_μN and

Figure 3.7 Total cross sections for neutrino and antineutrino interactions with nucleons as a function of the neutrino energy. (H. Fisk and F. Sciulli, reproduced with permission from the Annual Review of Nuclear and Particle Science, Vol. 32, © 1982, by Annual Reviews, Inc.)

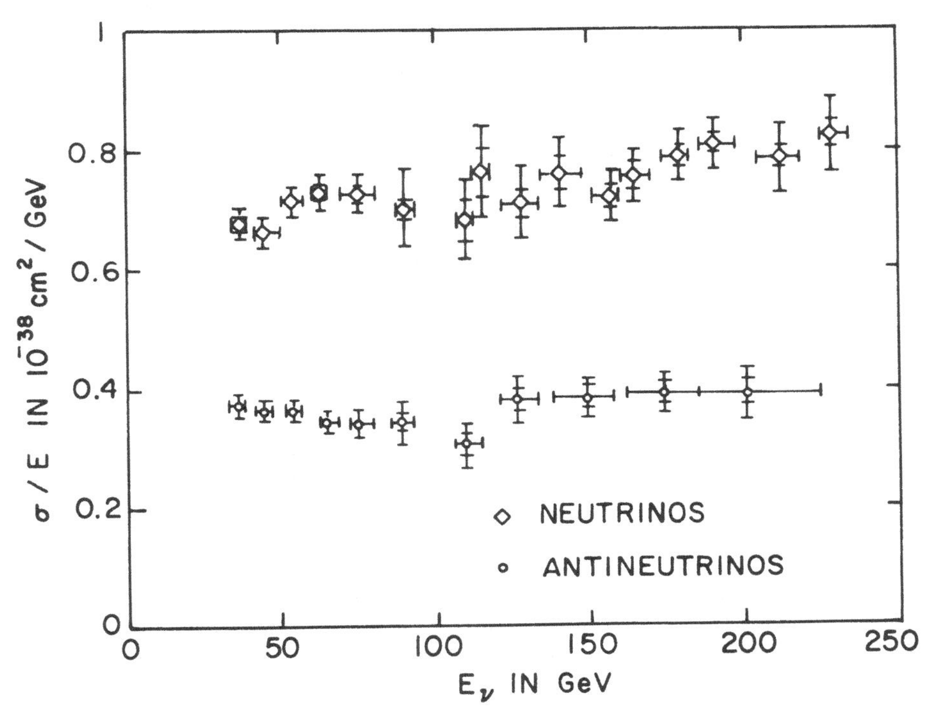

Table 3.3. *Hyperon asymmetry parameter*

Particle	Decay mode	α
Λ	$p\pi^-$	0.642 ± 0.013
	$n\pi^0$	0.646 ± 0.044
Σ^+	$p\pi^0$	-0.979 ± 0.016
	$n\pi^+$	0.068 ± 0.013
	$p\gamma$	-0.72 ± 0.29
Σ^-	$n\pi^-$	-0.068 ± 0.008
Ξ^0	$\Lambda\pi^0$	-0.413 ± 0.022
Ξ^-	$\Lambda\pi^-$	-0.434 ± 0.015
Ω^-	ΛK^-	-0.025 ± 0.028

Source: Particle Data Group, Rev. Mod. Phys. 56: S1, 1984; M. Bourquin et al., Nuc. Phys. B 241: 1, 1984.

Figure 3.8 Differential cross sections for νN and $\bar{\nu}$N scattering as a function of the variable y. (H. Fisk and F. Sciulli, reproduced with permission from the Annual Review of Nuclear and Particle Science, Vol. 32, © 1982 by Annual Reviews, Inc.)

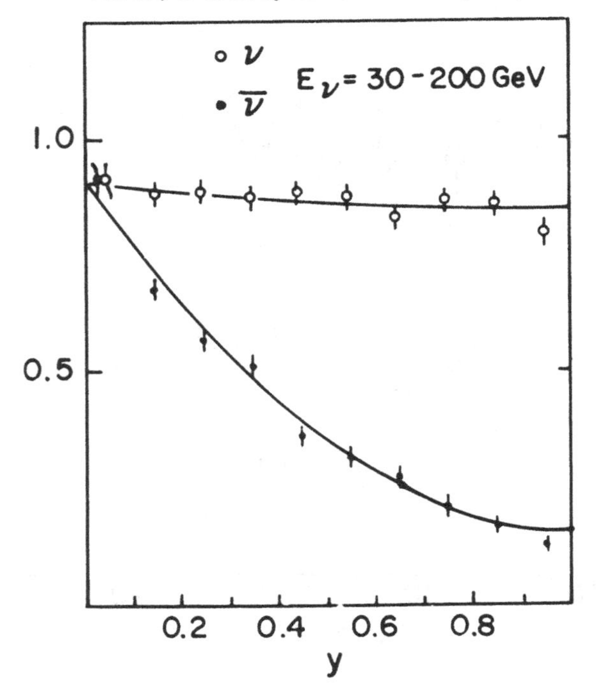

$\bar{\nu}_\mu$N scattering. Figure 3.8 shows the differential cross section plotted against the variable y, which is defined through the relation

$$1 - y = \tfrac{1}{2}(1 + \cos\theta^*) \tag{3.11}$$

where θ^* is the angle between the incident neutrino and outgoing muon in the CM system [11]. Quark–parton models predict that νN scattering should be roughly independent of y, while $\bar{\nu}$N scattering should be proportional to $(1 - y)^2$. This prediction is in good agreement with the data.

The angular distribution of the baryon in the nonleptonic decay of a spin $\frac{1}{2}$ hyperon

$$\mathrm{Y} \to \mathrm{B} + \mathrm{M}$$

is given by [14]

$$I(\theta) = I_0(1 + \alpha \mathbf{P}_\mathrm{Y} \cdot \hat{q}) \tag{3.12}$$

in the hyperon rest frame, where $\mathbf{P}_\mathrm{Y}$ is the polarization of the decaying hyperon, $\hat{q}$ is a unit vector along the direction of the decay baryon, θ is the angle between $\mathbf{P}_\mathrm{Y}$ and $\hat{q}$, and α is a constant associated with the particular type of decay. Table 3.3 shows measured values of α. Note that the decay distribution consists of a constant part plus a $\cos\theta$ part with coefficient αP_Y. The baryons from the decay of unpolarized hyperons are produced with helicity α.

References

[1] Discussions of the theory underlying strong interactions can be found in M. Perl, *High Energy Hadron Physics,* New York: Wiley, 1974.
[2] G. Giacomelli, Total cross section and elastic scattering at high energies, Phys. Rep. 23: 123–235, 1976.
[3] A. Carroll, I.-H. Chiang, T. Kycia, K. Li, M. Marx, D. Rahm, W. Baker, D. Earty, G. Giacomelli, A. Jonckheere, P. Koehler, P. Mazur, R. Rubinstein, and O. Fackler, Absorption cross sections of $\pi^\pm$, $K^\pm$, p, and $\bar{p}$ on nuclei between 60 and 280 GeV/c, Phys. Lett. 80B: 319–22, 1979.
[4] T. Roberts, H. Gustafson, L. Jones, M. Longo, and M. Whalley, Neutron-nucleus inelastic cross sections from 160 to 375 GeV/c, Nuc. Phys. B159: 56–66, 1979. These authors use the designation σ_{inel} for the quantity that we have called σ_{abs}.
[5] P. Duinker, Review of e^+e^- physics at PETRA, Rev. Mod. Phys. 54: 325–87, 1982.
[6] P. Soding and G. Wolf, Experimental evidence on QCD, Ann. Rev. Nuc. Part. Sci. 31: 231–93, 1981.
[7] H. Boggild and T. Ferbel, Inclusive interactions, Ann. Rev. Nuc. Sci. 24: 451–513, 1974.
[8] J. Whitmore, Multiparticle production in the Fermilab bubble chambers, Phys. Rep. 27: 187–273, 1976.
[9] P. Darriulat, Large transverse momentum hadronic processes, Ann. Rev. Nuc. Part. Sci. 30: 159–210, 1980.
[10] D. Cline and W. Fry, Neutrino scattering and new particle production, Ann. Rev. Nuc. Sci. 27: 209–78, 1977.

[11] H. Fisk and F. Sciulli, Charged current neutrino interactions, Ann. Rev. Nuc. Part. Sci. 32: 499–573, 1982.
[12] E. Commins and P. Bucksbaum, The parity nonconserving electron-nucleon interaction, Ann. Rev. Nuc. Part. Sci. 30: 1–52, 1980.
[13] P. Renton and W. Williams, Hadron production in lepton-nucleon scattering, Ann. Rev. Nuc. Part. Sci. 31: 193–230, 1981.
[14] W. Williams, *An Introduction to Elementary Particles,* 2nd ed., New York: Academic Press, 1971, pp. 151–6.

Exercises

1. How many centimeters of iron would be required to reduce the intensity of a 20-GeV pion beam by 90%? How many centimeters of lead would be required?

2. What is the expected QED muon pair production cross section for the LEP 50 GeV on 50 GeV e^+e^- collider?

3. What is the expected mean charged particle multiplicity at a 20 TeV on 20-TeV proton–proton collider?

4. Do a literature search to find out how the measured $\bar{p}p$ mean charged particle multiplicity at $W = 540$ GeV agrees with the extrapolation of the pp data in Fig. 3.5.

4

Particle beams

Most particle physics experiments require a beam of particles of a certain type. Usually these particles are provided by a high energy accelerator. Thus we will begin this chapter with a brief description of the characteristics of particle accelerators. These divide into two major classes, depending on whether the particle beam collides with a fixed target or with another beam of particles. We then discuss some properties of secondary beams from fixed target accelerators and the rudiments of beam transport theory. Since an important property of the beam for the experimentalist is the intensity, we will discuss flux monitoring. This is followed by a description of alternate sources of particles. The chapter concludes with a discussion of radiation protection.

4.1 Particle accelerators

A particle experimentalist is primarily concerned with four properties of the particle beam: the energy, the flux of particles, the duty cycle of the accelerator, and the fine structure in the intensity as a function of time. The duty cycle is defined to be the fraction of the time that the accelerator is delivering particles to the experiment. A detailed description of the components and acceleration process in various types of accelerators is beyond the scope of this book [1, 2]. However, we will give a brief overview in order to introduce some of the terminology.

The beam in an accelerator starts in either an electron gun or an ion source. Electrons are liberated from the filament of a high voltage triode tube. Pulses 1 – 10 μs long can be produced at a repetition rate of up to 500 Hz by triggering the grid of the tube. Electrons leave the source with $\beta \sim 0.5$. Protons originate from hydrogen molecules, which are dissociated with radio frequency (rf) energy into atoms. In a typical proton ion source, electrons oscillate back and forth inside the chamber. Collisions with neutral molecules and atoms result in the production of positive ions. These are attracted to the outlet port by an electric field and formed into a jet with a circular nozzle. Typically the output current is several milliamperes and the beam energy is several keV.

The beam from the source goes through several intermediate stages prior to injection into the main accelerating structure. For efficient injection the energy of the beam leaving an ion source must be increased. In the past this has usually been done for protons with a Cockcroft–Walton potential drop accelerator. The Cockcroft–Walton is essentially a voltage doubler circuit that can be repeated many times. The ions are accelerated by the potential differences between the stages and can reach an energy of up to 1 MeV. Electrical potential drop accelerators are ultimately limited by arcing and corona discharge. Recently the Cockcroft–Walton stage of some accelerators has been replaced with a radio frequency quadrupole (RFQ). This device can simultaneously accelerate, focus, and bunch a beam of particles.

Existing particle accelerators may be broadly classified into two groups: fixed target accelerators and storage rings. The beam in a fixed target accelerator is accelerated to its operating energy and then extracted. The beam in a storage ring, on the other hand, is accelerated to the desired energy and maintained in the ring for as long a period as possible.

4.1.1 Fixed target accelerators

Table 4.1 contains a summary of some important properties of existing fixed target accelerators with a maximum beam energy of 10 GeV or more. The chief advantages of this type of accelerator are the large number of interactions that occur when the extracted beam is directed into a liquid or solid target, and the fact that the particles emerge in the forward direction so that the required detector solid angle coverage is small.

Fixed target accelerators can be further divided into linear and circular accelerators. A linear accelerator (or linac) consists of an evacuated waveguide with a periodic array of gaps or cavities. High frequency oscillators establish an electromagnetic wave in the structure. The beam is forced into bunches and there may be rapid variations in the intensity of the extracted beam (rf structure) if it is not properly debunched. Linacs are frequently used as preaccelerators for synchrotrons. The largest linac currently in operation is the 2-mile-long electron accelerator at the Stanford Linear Accelerator Center (SLAC). It typically produces a 22-GeV

Table 4.1. *Fixed target accelerators*

Accelerator	Location	Date of first operation	Accelerated particles	Circumference (m)	Maximum beam energy (GeV)	Particles per pulse ($\times 10^{12}$)	Repetition rate (pulses/sec)
KEK	Japan	1976	p	339	12	4	0.5
SLAC	United States	1966	e^-	3050[a]	24	0.5[b]	$\leqslant 360$
CERN PS	Europe	1959	p	628	28	5	0.5
AGS	United States	1961	p	807	32	10	0.5
Serpukhov	USSR	1967	p	1484	71	3	0.2
Fermilab	United States		p	6280			
Main Ring		1972			500	40	0.1
Tevatron II		1983			1000	10	0.03
CERN SPS	Europe	1976	p	6910	500	25	0.1

[a] Linac.
[b] For 1.7-μs bursts.
Source: M. Crowley-Milling, Rep. Prog. Phys. 46: 51, 1983; R. Wilson, Sci. Amer., Jan. 1980, p. 42.

beam with a train of ~ 5000 bunches of 1.6-μs pulses and a repetition rate of up to 180 Hz.

The high energy acceleration mechanism in all existing electron and proton machines relies on forcing the particles to travel through an electric field. In electron linacs a traveling electromagnetic wave can be created that has a longitudinal component of electric field moving in phase with the particles. So long as this phase relationship can be maintained, the electrons will continue to be accelerated. Proton linacs often use the scheme illustrated in Fig. 4.1. The linac contains a series of drift tubes separated by gaps. The drift lengths are arranged so that the bunch of protons crosses the centers of the accelerating gaps at the same time that the field across the gap is approximately maximum, causing the particles to be continuously accelerated.

The synchrotron is a cyclic machine in which the particle beam is confined to a closed orbit by a series of bending magnets around a ring. The momentum p of a particle with charge q moving with a radius of curvature ρ in a magnetic field B is

$$p = qB\rho \tag{4.1}$$

For singly charged particles the momentum (in GeV/c) is given by $p = 0.3B\rho$, where B is in tesla and ρ is in meters. On each pass around the ring the particles' momenta are increased by acceleration in a synchronized rf cavity. As the momentum increases, the magnetic field in the bending magnets has to be increased to keep ρ constant. Particles in the beam

Figure 4.1 Principle of acceleration in a proton linac. A particle bunch is shown at $t = 0$ in an accelerating gap where the longitudinal component of the electric field is maximum. If the electric field phase is properly adjusted, the field will also be a maximum when the particle reaches the next gap. Between gaps the bunch must be shielded from the field.

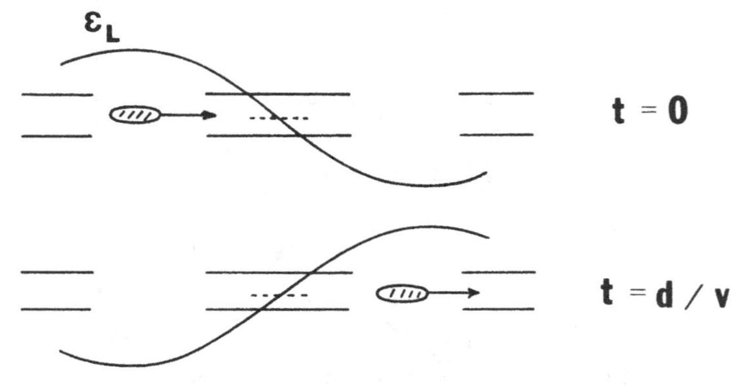

undergo transverse spatial (betatron) oscillations and energy (synchrotron) oscillations, which become smaller as the energy is increased. Quadrupole magnets are used to keep the beam focused.

Proton beams are accelerated to high energy in a synchrotron. Figure 4.2 shows the layout of the beams at the Fermilab (FNAL) accelerator [3]. This is the first large accelerator built with superconducting magnets. The linac accelerates the proton beam to 200 MeV. This is followed by a booster, which stores the linac output pulse and accelerates it to 8 GeV in $\frac{1}{15}$ sec. A series of booster pulses are injected into the main accelerator ring before the acceleration cycle begins. Each time the beam crosses the rf cavities, it gains 2.8 MeV of energy.

The acceleration cycle of a typical fixed target accelerator is shown in Fig. 4.3. Once the machine reaches peak energy, the magnetic field in the

Figure 4.2 Secondary beamlines at the Tevatron. (Courtesy of Fermi National Accelerator Laboratory.)

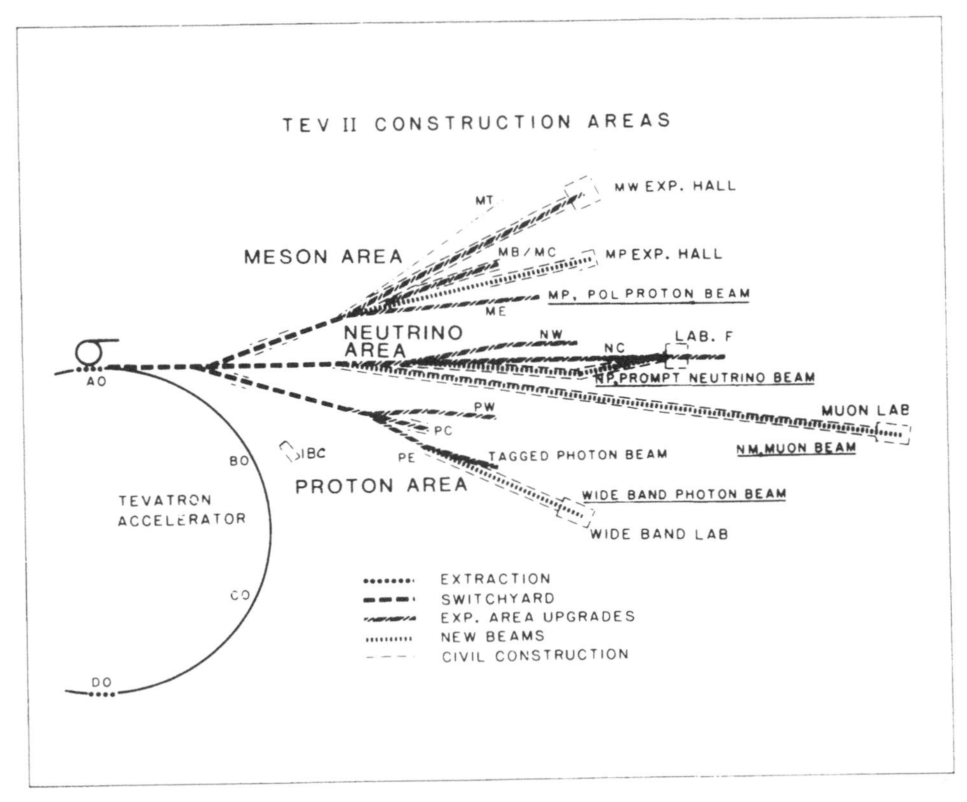

ring bending magnets can be held fixed for a period of around 1 sec (flat top), while the beam is slowly extracted from the machine for counter experiments. Alternatively, the beam can be extracted quickly for bubble chamber of neutrino experiments. The extracted FNAL beam has a peak energy of ~ 1000 GeV (Tev II) and an intensity of 2×10^{13} protons/pulse. The field is then lowered and the acceleration cycle is repeated. The beam size shrinks by a factor of about 10 between injection and extraction. The primary extracted proton beam is sent through a target "switchyard," which delivers beams to three major experimental areas: proton, meson, and neutrino.

The circular acceleration of electrons is severely hampered because of synchrotron radiation. The energy lost to radiation per revolution is [4]

$$\Delta E = (4\pi e^2/3\rho)\beta^3\gamma^4 \tag{4.2}$$

where ρ is the radius of the orbit. For electrons with $\beta \sim 1$, the energy loss in keV is $88.5E^4/\rho$, where E is in GeV and ρ is in meters. This lost energy must be replaced by the rf cavities on each revolution. The emitted radiation at high energy lies in a small cone around the particles' direction with an opening angle that goes like $1/\gamma$. The energy is deposited in a narrow strip around the circumference of the machine.

Figure 4.3 Acceleration cycle at a fixed target accelerator. (1) injection, (2) acceleration, (3) flat top and extraction, (4) deceleration.

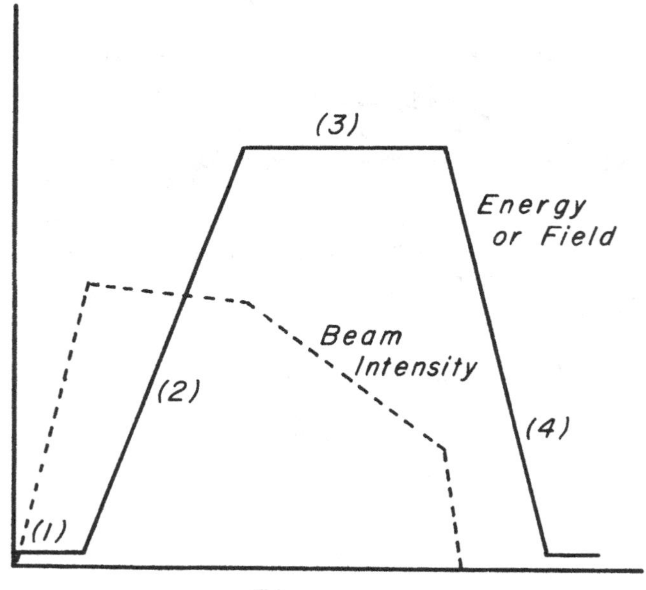

It is possible to accelerate polarized beams of particles. This is accomplished for protons by preparing and then selecting specific atomic states in the source [5]. Electron and positron beams in circular machines tend to polarize themselves in the vertical direction via the process of synchrotron radiation emission [6]. The theoretical maximum polarization arising from this (Sokolov–Ternov) effect is 92%. Electrons in linear accelerator sources can be polarized by exciting transitions in crystals using circularly polarized laser light [7]. The acceleration of polarized particles is complicated by the existence of depolarizing resonances, which occur at certain values of the particle's momentum. Most polarized sources can only produce a fraction of the intensity of an unpolarized source.

4.1.2 Storage ring accelerators

The center of mass energy available at fixed target accelerators is given by ($c = 1$)

$$E^* = (m_b^2 + m_t^2 + 2m_t E_b)^{1/2} \tag{4.3}$$

where m_b (m_t) is the beam (target) particle mass and E_b is the beam energy in the LAB frame. At high energy

$$E^* \simeq (2m_t E_b)^{1/2} \tag{4.4}$$

Thus, the available energy for the production of particles only increases as the square root of the beam energy. Energy is more readily available in a colliding beam machine, where two beams of circulating particles are made to interact head on. Here the LAB and the CM systems are the same and we find that

$$E^* = 2E_b \tag{4.5}$$

If the beams have variable energies E_1 and E_2 and cross at an angle α, the CM energy at high energy is

$$E^* \simeq 2(E_1 E_2)^{1/2} \cos(\tfrac{1}{2}\alpha) \tag{4.6}$$

Table 4.2 summarizes some characteristics of existing storage rings with a maximum energy per beam of 4 GeV or more.

A useful measure of storage ring performance is the luminosity $\mathcal{L}$. The reaction rate R is given in terms of $\mathcal{L}$ by

$$R = \mathcal{L}\sigma \tag{4.7}$$

where σ is the cross section for the reaction under consideration. Thus, luminosity has the dimensions of particles/s-cm^2. If two bunches of N particles are circling the machine with frequency f, the luminosity at an intersection point is

$$\mathcal{L} = N^2 f / A \tag{4.8}$$

Table 4.2. *Colliding beam accelerators*

Accelerator	Location	Date of operation	Stored particles	Circumference (m)	Intersection regions	Maximum energy of each beam	Maximum luminosity ($cm^{-2}\ s^{-1}$)
SPEAR	SLAC, United States	1972	e^+e^-	234	2	4.2	2×10^{31}
DORIS II	DESY, Germany	1982	e^+e^-	288	2	5.6	10^{31}
VEPP-4	Novosibirsk, USSR	1978	e^+e^-	366	3	7	2×10^{30}
CESR	Cornell, United States	1979	e^+e^-	768	2	8	2×10^{31}
PEP	SLAC, United States	1980	e^+e^-	2200	6	18	3×10^{31}
PETRA	DESY	1978	e^+e^-	2304	6	23	2×10^{31}
ISR[a]	CERN, Europe			942	8	31	
		1971	pp	—	—	—	6×10^{31}
		1980	$\bar{p}p$				10^{27}
$S\bar{p}pS$	CERN	1981	$\bar{p}p$	6910	2	315	2×10^{29}

[a] Decomissioned in 1984.

Source: M. Crowley-Milling, Rep. Prog. Phys. 46: 51, 1983; R. Wilson, Sci. Amer., Jan. 1980, p. 42; R. Kohaupt and G. Voss, Ann. Rev. Nuc. Part. Sci. 33: 67, 1983.

where A is the effective cross-sectional area of beam overlap. If two unequal Gaussian beams are bunched with k bunches per revolution in each beam, the luminosity is [8]

$$\mathcal{L} \simeq \frac{fkN_1N_2}{4\pi a_h a_v} \tag{4.9}$$

where N_i is the number of particles per bunch and a_h, a_v are the rms horizontal and vertical dimensions. For horizontal crossing

$$a_h = (a_r^2 + \alpha^2 a_b^2)^{1/2}$$

where a_b is the rms bunch length, a_r is the radial bunch dimension, and α is the crossing angle.

The electron–positron storage ring capable of reaching the highest CM energy at present is PETRA, with energies around 23 GeV per beam and luminosities of 10^{31} particles/s-cm^2. Positive and negative beams can be stored in a single ring and made to cross only at fixed intersection regions. Proton–proton storage rings require the use of two separate rings. The largest pp storage ring was the ISR with an energy of 31 GeV per beam and a luminosity of 4×10^{31} particles/s-cm^2. The beams in e^+e^- rings are tightly bunched, whereas the proton beams in the ISR were more nearly continuous. For comparison, the equivalent luminosity of a fixed target accelerator beam of $\sim 10^{13}$ particles/sec incident on 1 m of liquid hydrogen is 4×10^{37}/s-cm^2.

The design requirements for storage rings are more stringent than for fixed target machines, since the beams must be confined for periods of hours or more. The vacuum in the rings is typically 10^{-8} torr or less to prevent beam–gas interactions from limiting the lifetime of the beam. The maximum luminosity is limited by coupling of one beam with the other, space charge repulsion of the particles in each beam, and instabilities in the beam orbits [6].

The use of antiprotons in storage rings requires additional efforts to raise the luminosity to usable levels. The $\bar{p}$ are produced by proton interactions in a solid target. Those $\bar{p}$ produced in a particular solid angle and within a certain momentum interval can be accumulated in a ring. However, it is necessary to increase the density of the accumulated beam before it is accelerated and collided. This is accomplished by "cooling" the beam [9]. Two types of cooling are commonly used. In electron cooling an electron beam is injected into a straight section of the ring with the same velocity as the $\bar{p}$. Coulomb collisions between the two beams damp the transverse oscillations of the heavier $\bar{p}$. In stochastic cooling pickup electrodes on one side of the ring measure deviations of a small circumferential portion of the beam from the equilibrium orbit. A correction signal is then sent directly across the ring, so that kicker magnets can apply a correlation deflection to the beam portion when it arrives on the opposite side of the ring. Stochastic cooling can also be used to reduce the momentum spread in the beam. These cooling techniques make it possible to increase the luminosity in $\bar{p}p$ machines to useful levels.

Figure 4.4 shows part of the accelerator complex at CERN. The heart of the system is the 28-GeV proton synchrotron (PS). The extracted PS beam can be used for low energy experiments or injected into one of the higher energy machines, the SPS or ISR. The SPS can be used either as a 400-GeV proton synchrotron or as a 315 on 315-GeV $\bar{p}p$ collider. The antiprotons are produced in 26-GeV/c proton collisions with a fixed target. The collected $\bar{p}$ are cooled and stored in an accumulator ring at 3.5 GeV/c. Then they can be reaccelerated in the PS and injected into either the ISR or SPS. Experiments with high quality beams of low energy $\bar{p}$ can be performed at the LEAR storage ring.

A number of new colliding beam accelerators are presently under construction. Table 4.3 lists six major projects that should be completed within the next few years. These include circular e^+e^- colliders (LEP and TRISTAN), a linear e^+e^- collider (SLC), a new $\bar{p}p$ collider (Tevatron I), and an e^-p collider (HERA).

Figure 4.4 Layout of part of the CERN accelerator complex. (A) linac, (B) LEAR storage ring, (C) $\bar{p}$ production target, and (D) antiproton accumulator. (Assisted by CERN.)

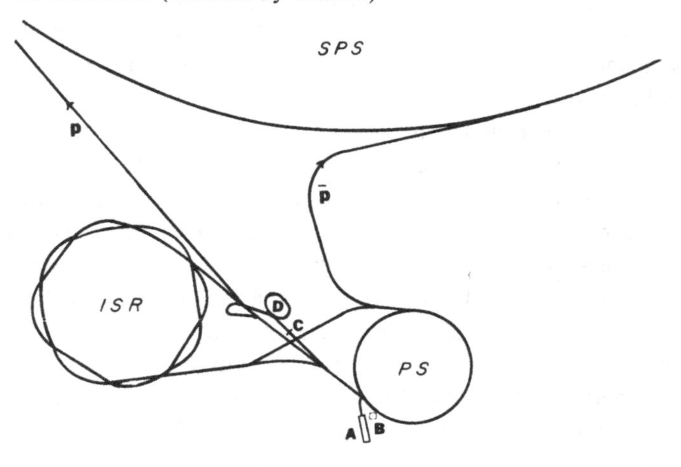

4.2 Secondary beams

Secondary beams at fixed target accelerators are created by sending a portion of the primary extracted beam into a target. Many different particles will be created, each with its own characteristic angle and momentum spectrum. A typical beamline starts with a collimator to select particles produced in some specific angular range. This is a narrow restriction built from lead bricks or other materials, which reduces the intensity or divergence of the beam or selects particles going in a specific direction.

A useful device for separating particles by mass in a low energy charged beam is the electrostatic separator. The electric portion of the Lorentz force gives a momentum kick to each particle that is proportional to the length of time the particle spends inside the separator. Since particles with different masses but the same momentum move with different velocities, they will spend different lengths of time in the separator and will spread out transversely at the end of the device. A judiciously placed slit will then permit the desired component of the beam to be passed on.

The deflection angle away from the axis is given by ($c = 1$)

$$\alpha = \frac{\Delta p_{\perp}}{p_0} = \frac{L}{\beta p_0}(e\beta B - e\mathcal{E})$$

where L is the length of the separator, β is the velocity, p_0 is the beam momentum, B is an applied magnetic field, and $\mathcal{E}$ is the applied electric field. Writing $\mathcal{E}$ in terms of the electrical potential V and the plate separation d gives

$$\alpha = \frac{eL}{p_0}\frac{V}{d}\left(\frac{1}{\beta_0} - \frac{1}{\beta}\right) \tag{4.10}$$

Table 4.3. *Major construction projects*

Project	Location	Estimated completion date	Description
LEP	CERN	1988	50 GeV × 50 GeV e^+e^- collider
SLC	SLAC	1987	50 GeV × 50 GeV e^+e^- linear collider
Tevatron I	Fermilab	1986	1 TeV × 1 Tev $\bar{p}p$ collider
TRISTAN	KEK	1986	30 GeV × 30 GeV e^+e^- collider
HERA	DESY	1990	30 GeV-electron × 820 GeV-proton collider
UNK	Serpukhov	1990	600-GeV proton Synchrotron

where β_0 is the velocity for no deflection. Note that the purpose of the magnetic field is to determine the velocity β_0 that will pass through undeflected, where

$$\beta_0 = \mathcal{E}/B \tag{4.11}$$

Using the expression $p = \beta E$, we can rewrite Eq. 4.10 as

$$\alpha = \frac{eLV}{p_0 d}\left[\left(1 + \frac{m_0^2}{p_0^2}\right)^{1/2} - \left(1 + \frac{m^2}{p^2}\right)^{1/2}\right] \tag{4.12}$$

When $p \simeq p_0 \gg m$,

$$\alpha = \frac{eLV}{p_0 d}\frac{1}{2}\frac{m_0^2 - m^2}{p_0^2} \tag{4.13}$$

Thus, at high energy the deflection angle decreases like $1/p_0^3$, and this means of separation is only practical up to about 5 GeV/c momentum. The major difficulty is the large electric field required, typically 30 kV/cm. Separation can also be accomplished using rf fields [10].

We will now consider some of the important features of various secondary beams.

4.2.1 γ

Photon beams can be created directly from an electron beam by bremsstrahlung or from a proton beam via π^0 decays. Charged particles are removed from the beam using collimators and bending magnets. A significant fraction of the neutral beam may consist of neutrons. The γ to n ratio may be improved by passing the beam through a material such as liquid deuterium, whose nuclear interaction length is significantly shorter than its radiation length. The beam contamination can be estimated by inserting lead absorbers and purposely decreasing the γ content of the beam.

The energy of the γ can be determined in a tagged photon beam. This is usually accomplished by sending an electron beam whose energy is well measured into a thin, high Z target as shown in Fig. 4.5. After the photon is created via bremsstrahlung, the electron's energy is remeasured and the photon energy determined from energy conservation. As an example, the tagged photon beam at the Omega Spectrometer at CERN uses 4×10^{12} incident 240-GeV protons to produce 2×10^7 electrons per pulse [11]. This results in $\sim 10^6$ tagged photons in the momentum range 25–70 GeV/c. The spectrum falls off with the expected p_γ^{-1} dependence. Photon beams can be polarized by production from crystals or selective absorption.

4.2.2 $\nu_\mu, \nu_e, \bar{\nu}_\mu, \bar{\nu}_e$

Most neutrino beams originate in the decays of charged pions and kaons. Muon neutrino beams are produced in the 2-body decays

$$\pi^+ \rightarrow \mu^+\nu_\mu$$
$$\pi^- \rightarrow \mu^-\bar{\nu}_\mu$$
$$K^+ \rightarrow \mu^+\nu_\mu$$
$$K^- \rightarrow \mu^-\bar{\nu}_\mu$$

Electron neutrinos are produced in the 3-body decays

$$K^+ \rightarrow e^+\pi^0\nu_e$$
$$K^- \rightarrow e^-\pi^0\bar{\nu}_e$$

The ν_e fluxes are suppressed by ~ 100 times compared to the ν_μ fluxes. A neutrino beamline typically consists of a thin production target where the π and K are produced, a decay region, and a long earth absorber where hadrons and decay muons are absorbed.

There are two general types of neutrino beams [12]. In a wide band beam the π and K are collected over a wide range of momenta and solid angle. The resultant beam has a high neutrino flux, but the energy spectrum is very broad. Sometimes a horn is used to increase the intensity. This is a thin metal sheet that surrounds the beamline after the production target. When a large current pulse is sent through the horn, the azimuthal magnetic field focuses π and K of a given charge into a parallel beam toward the detector. The oppositely charged mesons are defocused. As a result, the meson decay product beam consists primarily of neutrinos or primarily of antineutrinos. The horn-focused energy spectrum is strongly peaked at small neutrino energy. Sometimes instead of a horn a quadru-

Figure 4.5 Principle of the tagged photon beam. (e) electron beam, (C) collimators, (BM) bending magnets, (R) radiator, and (T) tagging counters.

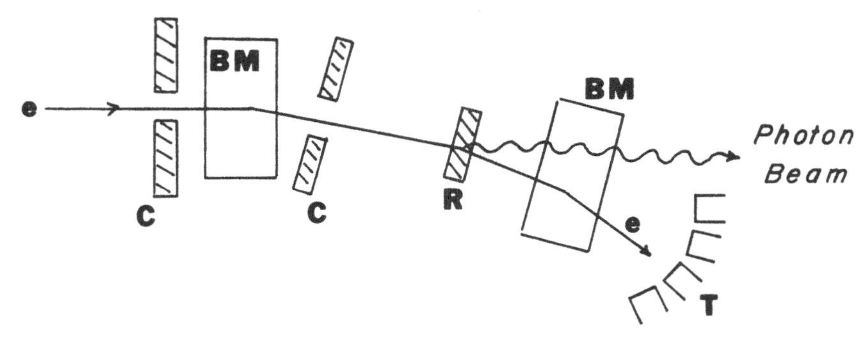

pole triplet is used to focus the mesons. The flux of 100-GeV neutrinos from the horn-focused beam at FNAL is around 2×10^6 per 10^{13} incident 400-GeV protons per m^2.

In a narrow band beam the mesons are momentum-selected before entering the decay region. The neutrinos in such a beam have a much tighter energy spectrum ($\sigma_E/E \sim 20\%$). Figure 4.6 shows the Fermilab narrow band ν spectrum. The spectrum is dichromatic with peaks corresponding to the 2-body π and K decays. The integrated intensity is much smaller than in the broad band beam. The decay angle of the neutrino is correlated with its energy.

4.2.3 $e^{\pm}$

A secondary e^+ or e^- beam can be created at a proton machine by producing π^0 in a thin production target. The decay photons can be converted in a thin radiator to yield $\pi^0 \rightarrow 2\gamma \rightarrow 2e^+ + 2e^-$.

Figure 4.6 Fermilab narrow band neutrino flux for 200-GeV π and K secondary beams. (H. Fisk and F. Sciulli, reproduced with permission from the Annual Review of Nuclear and Particle Science, Vol. 32, © 1982 by Annual Reviews, Inc.)

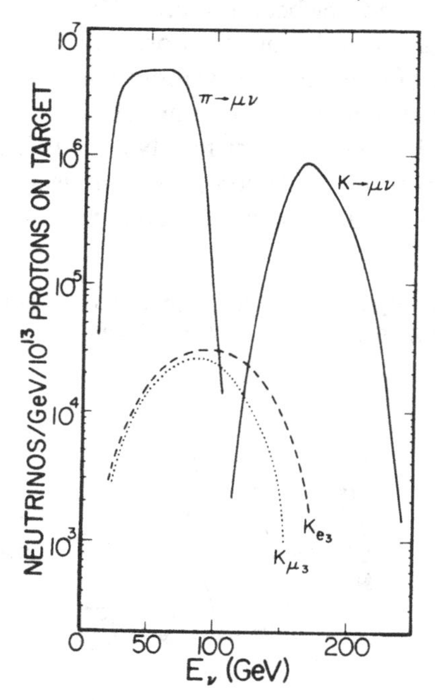

4.2.4 $\mu^{\pm}$

Muon beams are used for high energy lepton – hadron interaction studies [13]. Muon beams can be made with higher energies and intensities for these studies than electron beams can. Because of their higher mass, muon beams can be used with longer targets and have smaller, higher-order QED corrections than electrons.

A typical beam starts by first directing a secondary proton beam into a production target to create a beam of charged pions. The pions are allowed to decay until a significant fraction of the beam contains muons. Since this reaction is also used to create neutrinos, the two beams can both originate from the same decay region. The beam is then passed through a thick hadron absorber. This consists of a low Z material, such as polyethylene, in order to minimize multiple scattering of the muons. The muons readily pass through the absorber. The resulting beam has a large halo, which can, however, be significantly reduced by incorporating magnetic bending into the beamline. The 100-GeV muon beam at the SPS was created by using 400-GeV protons to produce 115-GeV pions. The μ^+ yield per proton incident on one interaction length of beryllium was 1.9×10^{-5}. A possible background consists of electrons from muon decays after the hadron absorber.

4.2.5 $\pi^{\pm}$, $K^{\pm}$, p, $\bar{p}$

High intensity proton beams are available by splitting off a portion of the primary beam at proton accelerators. Secondary beams of charged hadrons are produced by interactions of the primary beam in a nuclear target. Only particles of a given charge and produced within a certain solid angle and momentum interval will be transported down the beamline. As an example, the M1 beamline at FNAL contained 83.5% p/14.0% π^+/2.5% K^+ when the machine operated at 400 GeV [14]. The negative beam contained 95.7% π^-/3.5% K^-/0.8% $\bar{p}$. It was possible to increase the K^+ fraction of the positive beam to 15% by using a beryllium filter in the beam. At low energy the fraction of K and $\bar{p}$ in a hadron beam can be enhanced using electrostatic separators.

4.2.6 K^0, n, $\bar{n}$

Long-lived neutral hadronic beams can be made by following the production target with a thin radiator to convert photons. This in turn is followed by collimators and bending magnets to sweep all the charged particles from the beam. One obvious feature of any neutral beam is that it must point straight back to the secondary production target. Typically, the neutron content of the beam is several orders of magnitude larger than the K^0 content, unless specific absorbors are used to enhance the K^0 fraction.

4.2.7 Λ, $\Sigma^{\pm}$, Ξ^-, Ξ^0, Ω^-

Hyperons are also produced at the secondary production target. At low energy it is difficult to make beams of these particles, since their mean free path for decay is only several centimeters. However, we have seen in Eq. 1.12 that the mean free path grows with momentum. This fact has allowed useful hyperon beams to be constructed at the SPS and at FNAL [15]. Charged and neutral beams may be constructed as discussed above.

4.3 Beam transport

It is important to understand how the size and angular divergence of a beam changes as it is transported from the accelerator to an experimental area. Let us begin by briefly considering the ideal motion of a particle in a magnetic field. Consider a particle with mass m, momentum p, and charge q traveling along the z axis. If a magnetic field $\mathbf{B}$ is present, the particle is subjected to the Lorentz force $\mathbf{F} = q\mathbf{v} \times \mathbf{B}$. Assume that the field $\mathbf{B}$ has the components $(B_x, B_y, 0)$. Then the equations of motion are

$$\frac{d^2x}{dz^2} + \frac{q}{p} B_y = 0 \qquad \frac{d^2y}{dz^2} - \frac{q}{p} B_x = 0 \qquad v = \text{const} \tag{4.14}$$

where we have used $v = dz/dt$.

We now consider three important cases.

1. *Drift space.* If no field is present, the equation for x reduces to

$$x = x_0' z + x_0 \tag{4.15}$$

where x_0 and x_0' are constants. A similar equation applies to the motion in y.

2. *Bending magnet.* Now consider an ideal bending magnet or dipole with $B_x = 0$ and $B_y = B$. For the moment we neglect the presence of the fringe field at the edges of the magnet. If we consider the equations as a function of time, the coupled equations have the solution

$$\begin{aligned} x &= r\cos(\omega t) \\ z &= r\sin(\omega t) \\ \omega &= qB/m \end{aligned} \tag{4.16}$$

as can be verified by direct substitution. These are, of course, just the parametric equations for a circle of radius r.

3. *Quadrupole field.* The quadrupole field shown in Fig. 4.7 is generated by a four-pole magnet with alternating polarity. The field is of the form $\mathbf{B} = (Gy, Gx, 0)$, where the constant G is called the gradient of the quadrupole and has dimensions tesla/m. We assume in the lowest-order approximation that the particle continues to travel parallel to the z azis. The equations of motion become

$$\begin{aligned} \frac{d^2x}{dz^2} + \frac{q}{p} Gx &= 0 \\ \frac{d^2y}{dz^2} - \frac{q}{p} Gy &= 0 \end{aligned} \tag{4.17}$$

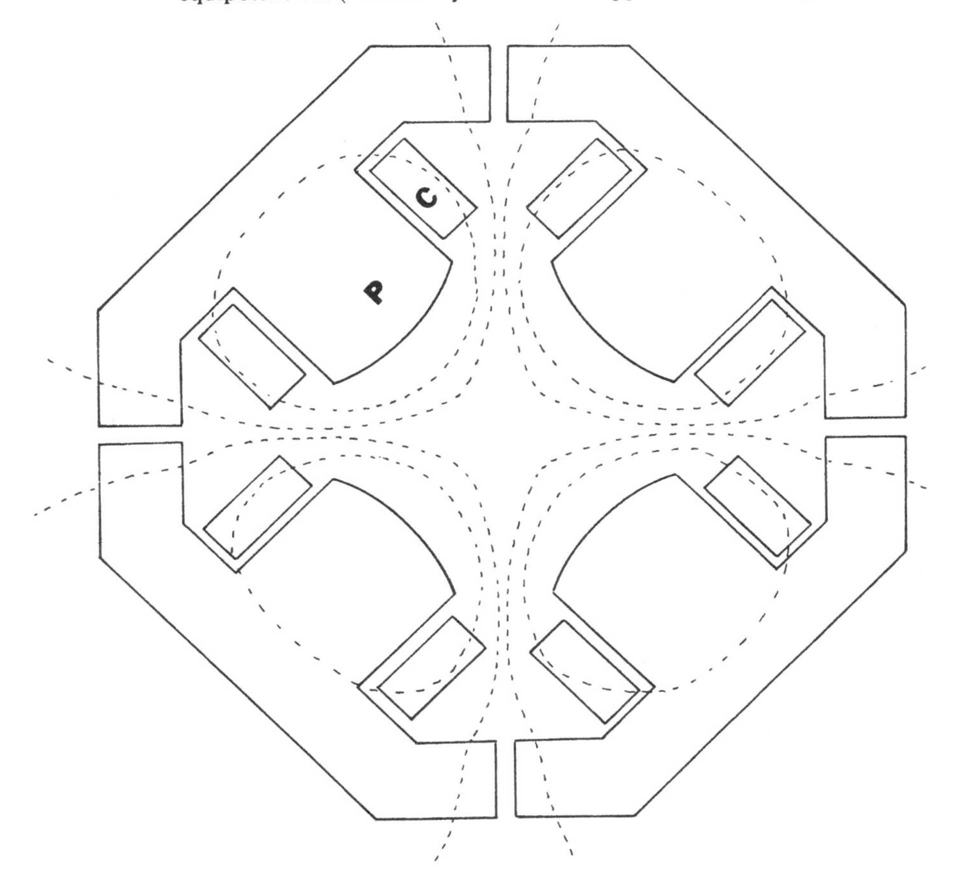

Figure 4.7 Quadrupole magnetic field. The field vanishes on the axis. (C) one of the coils and (P) one of the iron pole faces. Dotted lines show equipotentials. (Assisted by Rutherford Appleton Laboratory.)

Let

$$k^2 = \frac{q|G|}{p} \tag{4.18}$$

Then the solution for the x motion is

$$x = x_0 \cos(kz) + \frac{x_0'}{k} \sin(kz) \tag{4.19}$$

which represents oscillatory motion. The solution for y,

$$y = y_0 \cosh(kz) + \frac{y_0'}{k} \sinh(kz) \tag{4.20}$$

blows up exponentially. Thus, the quadrupole has the property of focusing a beam in one direction while defocusing it in the other. Another quadrupole with the pole polarities reversed would focus in y and defocus in x. Note that the first-order equations given here do not couple the motion in x and y.

It is convenient to combine the displacement and angular divergence of a beam into the components of a 2-dimensional vector

$$\mathbf{x} = \begin{pmatrix} x \\ x' \end{pmatrix}$$

Then the action of various devices, such as quadrupole magnets or drift spaces, can be considered to be represented by matrices M_i, which operate on an initial vector $\mathbf{x}_0$ to produce the vector $\mathbf{x}$ at the end of the device. Sometimes 3-dimensional vectors are used with the momentum dispersion as the third component.

The drift space of length L has the matrix

$$M_1 = \begin{pmatrix} 1 & L \\ 0 & 1 \end{pmatrix} \tag{4.21}$$

as can be seen from Eq. 4.15.

The displacement $\mathbf{x}$ is considered to be the distance from a central, ideal orbit. In a bending magnet with field B, the entrance angle β_i and the exit angle β_o are related by

$$\sin \beta_i + \sin \beta_0 = \frac{\int B\, dl}{3.333 p_0} \tag{4.22}$$

where p_0 is the momentum of the central orbit (or ray) and the units are tesla, meters, and GeV/c. For small β_i and β_0 this can be approximated by

$$\alpha \simeq \frac{\int B\, dl}{3.333 p_0} \tag{4.23}$$

where α is the total angle of bend. The radial deviations from the central ray are given by [16]

$$M_2 = \begin{pmatrix} \cos\alpha & \rho_0 \sin\alpha \\ -\frac{1}{\rho_0}\sin\alpha & \cos\alpha \end{pmatrix} \tag{4.24}$$

where ρ_0 is the radius of curvature of the central orbit. Any momentum dispersion in the beam produces the additional correction

$$\frac{\Delta p}{p}\begin{pmatrix} \rho_0(1-\cos\alpha) \\ \sin\alpha \end{pmatrix}$$

which must be added to **x**. A more complete treatment must also take into account the nonuniformity of the field and the focusing by the fringe fields [16].

A quadrupole magnet behaves much like a thick optical lens. We can rewrite Eq. 4.19 for the focusing direction in terms of

$$M_3 = \begin{pmatrix} \cos kL & 1/k \sin kL \\ -k\sin kL & \cos kL \end{pmatrix} \tag{4.25}$$

and Eq. 4.20 for the defocusing direction in terms of

$$M_4 = \begin{pmatrix} \cosh kL & 1/k \sinh kL \\ k\sinh kL & \cosh kL \end{pmatrix} \tag{4.26}$$

where k is given by Eq. 4.18. In the thin lens approximation the quadrupole causes a change in the divergence of the beam that is proportional to the displacement from the central orbit,

$$\Delta x' = x/f \tag{4.27}$$

but leaves the position unchanged. Here f is the focal length of the lens. In this case we have

$$M_3' \simeq \begin{pmatrix} 1 & 0 \\ -k^2L & 1 \end{pmatrix} \tag{4.28}$$

$$M_4' \simeq \begin{pmatrix} 1 & 0 \\ k^2L & 1 \end{pmatrix} \tag{4.29}$$

Comparing Eqs. 4.27, 4.28, and 4.29, we see that the focal length of the thin lens is

$$f = \pm\frac{p_0}{q|G|L} \tag{4.30}$$

where a negative focal length means focusing and a positive focal length means defocusing. Sometimes the lens power $P = 1/f$ is used instead of the focal length. A pair of quadrupoles with opposite gradients can be designed to focus in both x and y simultaneously. However, the focal lengths will be different in the horizontal and vertical directions.

Each of the transfer matrices discussed so far has unit determinant. This is a consequence of Liouville's theorem in statistical mechanics [16]. If the beamline consists of several types of devices, the overall transfer matrix can be determined by taking the product of the individual matrices.

As an example, Fig. 4.8 shows the maximum horizontal and vertical extent (beam profiles) for a medium energy separated beam. In the figure Q1 – Q8 are quadrupoles, D1 – D4 are bending dipoles, and BS1 and BS2 are electrostatic separators. The beamline accepts particles produced in a narrow solid angle near the primary beam direction. Collimator C1 can be used to adjust the particle intensity. The dipole D1 disperses the beam horizontally for subsequent momentum analysis. Quadrupoles Q1 – Q3 make the vertical beam parallel for the separators and focus the beam horizontally at the collimator C2 for momentum selection. The beam separators give a vertical divergence to the beam depending on the particle's mass. Quadrupoles Q4 and Q5 focus the beam vertically at the mass

Figure 4.8 Beam profiles for a medium energy separated beam. The beamline is shown from the production target T1 to the experimental target T2.

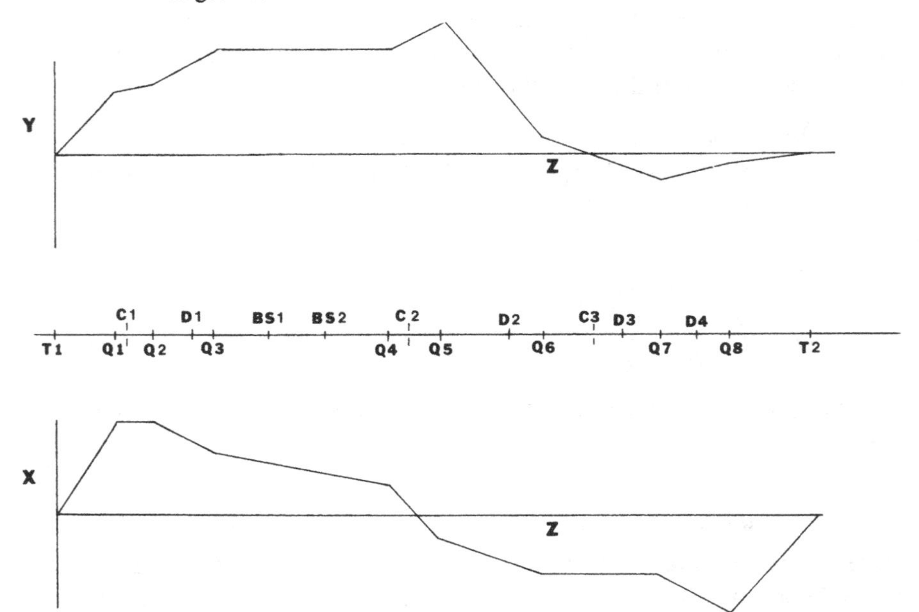

slit C3 downstream from the separators. The vertical position of the focus depends on the divergence that the particles received at the separators. The last three quadrupoles (Q6–Q8) bring the beam to a horizontal and vertical focus at the experimental target.

Now consider the profile of a beam of particles. In the 2-dimensional phase space plot the locus of the displacements and divergences at a fixed z defines an ellipse [16]. When the major axis of the ellipse is along x', the beam is said to be at a waist. This usually corresponds to a point of minimum beam size. The ratio of x to x' at a waist is defined to be the characteristic length of the beam

$$\tilde{X} = x/x' \tag{4.31}$$

The product of x and x' at a waist is known as the emittance of the beam

$$\epsilon = xx' \tag{4.32}$$

so that the area of the phase space ellipse is $\pi\epsilon$.

Let us assume that a beam that is initially at a waist, as indicated in Fig. 4.9a, enters a beamline consisting of a drift space, thin convergent lens, and another drift space. In the drift space the divergence of any particle in the beam remains constant and the phase space ellipse shears horizontally. Note that points a and b in the figure have 0 divergence initially, so their position on the figure cannot change while traversing the drift space. The ellipse at the end of the drift space is shown in Fig. 4.9b. The amount of horizontal shearing is determined by the length of the drift space. Upon crossing the thin lens the position of the particles remains the same, but all the divergences are changed by an amount given by Eq. 4.27. Thus, the phase space ellipse appears as in Fig. 4.9c. Note that points e and f, which are on the lens axis, remain fixed. Finally the second drift space shears the ellipse again horizontally and the phase space ellipse reaches another waist as shown in Fig. 4.9d (points h and i remain fixed).

One is often interested in finding the location z_2 and characteristic length $\tilde{X}_2$ of the second waist if the first drift distance z_1, the initial characteristic length $\tilde{X}_1$, and the focal length f of the lens are given. If we define the auxiliary variable

$$\rho = \frac{f^2}{\tilde{X}_1^2 + (z_1 + f)^2} \tag{4.33}$$

then the characteristic length of the outgoing beam is

$$\tilde{X}_2 = \rho \tilde{X}_1 \tag{4.34}$$

and the location of the second waist is

$$z_2 = \rho(z_1 + f) - f \tag{4.35}$$

The semiwidth or profile of the output beam is given by

$$x_2^2 = \epsilon\tilde{X}_2(1 + (z - z_2)^2/\tilde{X}_2^2) \qquad (4.36)$$

where z is measured from the center of the lens.

In practice, it is necessary to use two or more quadrupoles to achieve simultaneous focusing in both the horizontal and vertical planes. Consider a pair of quadrupoles with powers P_1 and P_2 separated by a distance d. Then if the object is a distance u before the first quadrupole and the image is required to be a distance v beyond the second quadrupole, the transfer matrix in the focusing–defocusing (FD) plane is the product

$$M = \begin{bmatrix} 1 & v \\ 0 & 1 \end{bmatrix} \begin{bmatrix} 1 & 0 \\ P_2 & 1 \end{bmatrix} \begin{bmatrix} 1 & d \\ 0 & 1 \end{bmatrix} \begin{bmatrix} 1 & 0 \\ -P_1 & 1 \end{bmatrix} \begin{bmatrix} 1 & u \\ 0 & 1 \end{bmatrix} \qquad (4.37)$$

The condition for imaging is that a given point on the object map to a given point on the image, independent of the divergence at the point on

Figure 4.9 Phase space ellipses for waist to waist beam transport.

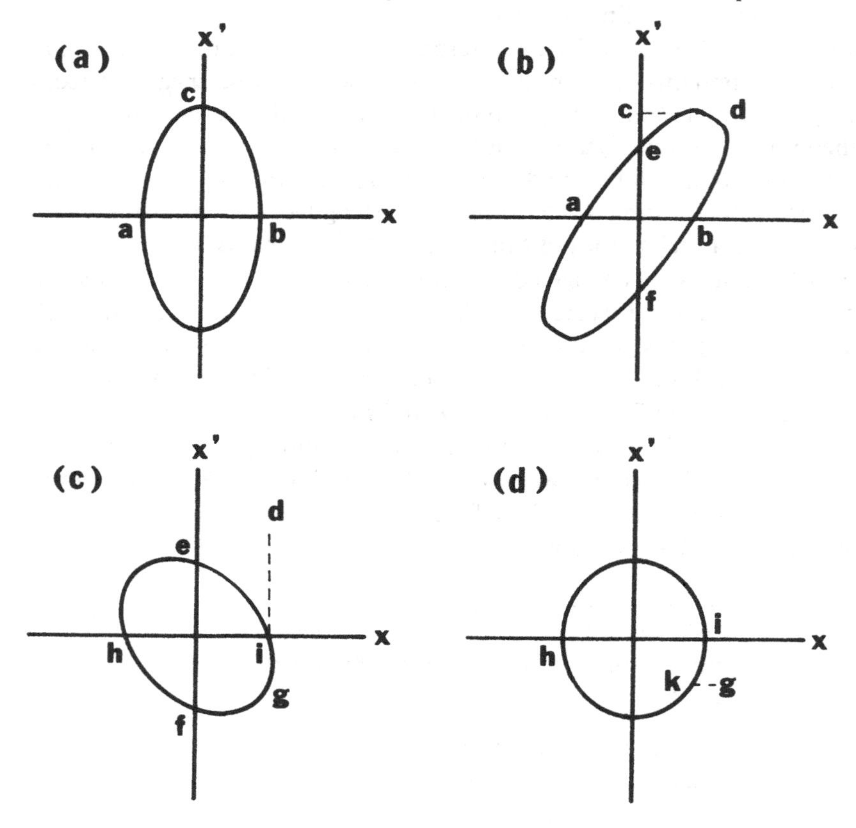

the object. This requires that $M_{12} = 0$. Developing a similar equation for the DF plane and solving for the lens power gives [16]

$$(uP_1)^2 = \frac{u+v+d}{d}\frac{u+d}{v+d}$$
$$(vP_2)^2 = \frac{u+v+d}{d}\frac{v+d}{u+d} \tag{4.38}$$

4.4 Flux monitors

Studies of total or differential cross sections require the measurement of the incident particle intensity. It is easy to make a relative monitor by constructing a small solid angle scintillation counter telescope. The telescope can be made more selective by incorporating range, energy loss, or time of flight requirements on the accepted particles.

In e^+e^- machines the incident luminosity is determined from measurements of small angle or large angle Bhabha scattering events. The differential cross section for this process may be calculated exactly, so the number of observed events together with the known acceptance of the detectors gives an absolute measurement of the luminosity. Similarly pp or $\bar{p}p$ colliders can use small angle elastic scattering to determine the luminosity.

The intensity of a photon beam can be determined if two of the quantities (total energy, maximum energy, number of photons in a known energy interval) are measured [17]. For low intensity charged beams a thin scintillation counter may be used to count the incident particles. At higher intensities a pickup electrode or calibrated ionization chamber may be used to provide a signal proportional to the beam intensity. A similar technique is to use an integrating scintillation or Cerenkov counter [17]. The charge from the photomultiplier tube is collected on a capacitor. The Cerenkov counter presents more material to the beam but offers superior background rejection. For very high currents, such as with the SLAC e^- beam, the ionization chamber output saturates, and it is necessary to use a secondary emission monitor.

Another technique that can be useful involves a measurement of the induced radioactivity in a thin metal foil [18]. Suppose we uniformly irradiate a thin foil of some material such as aluminum or gold. The number of radioactive nuclei produced in the foil per unit volume and per unit time will be

$$n = \Phi n_a \sigma_T \tag{4.39}$$

where Φ is the incident flux (the number of incident beam particles per unit area per unit time), n_a is the number of foil atoms per unit volume, and σ_T is the total cross section for producing the radioactive isotope under consideration. The number of particles produced per unit time is

$$\mathcal{N}_{sat} = \Phi\sigma_T\rho\frac{N_A}{A}\mathcal{A}d \tag{4.40}$$

where $\mathcal{A}$ is the area of the foil that is irradiated and d is the foil thickness. Now the beam intensity (number of particles per unit time) is just

$$\phi = \Phi\mathcal{A} \tag{4.41}$$

The product ρd can be conveniently measured in terms of the total mass of the foil, M_{tot}, and its total area, $\mathcal{A}_{tot}$,

$$\rho d = M_{tot}/\mathcal{A}_{tot} \tag{4.42}$$

Thus Eq. 4.40 becomes

$$\mathcal{N}_{sat} = \phi\sigma_T\frac{N_A}{A}\frac{M_{tot}}{\mathcal{A}_{tot}} \tag{4.43}$$

The activity, or number of radioactive decays per unit time, approaches $\mathcal{N}_{sat}$ exponentially since the isotope has a half-life for decaying of $t_{1/2}$. A nuclei produced at time τ has a probability $\exp[-(t_i - \tau)0.693/t_{1/2}]$ of remaining at time t_i. Let $N(t_i, t_c)$ be the number of radioactive nuclei remaining after an irradiation time t_i and a "cooling" time t_c during which it is not irradiated. The number of decaying nuclei at time t_i can be found by integrating over the contributions from all times τ

$$\begin{aligned} N(t_i, 0) &= \mathcal{N}_{sat}\int_0^{t_i}\exp[-(t_i - \tau)0.693/t_{1/2}]\,d\tau \\ &= \mathcal{N}_{sat}\frac{t_{1/2}}{0.693}(1 - e^{-0.693t_i/t_{1/2}}) \end{aligned} \tag{4.44}$$

If we allow the foil to cool for a time t_c, the number of radioactive nuclei will be

$$N(t_i, t_c) = \mathcal{N}_{sat}\frac{t_{1/2}}{0.693}(1 - e^{-0.693t_i/t_{1/2}})e^{-0.693t_c/t_{1/2}} \tag{4.45}$$

To get the activity, we take the derivative of this with respect to t_c.

$$\mathcal{N}(t_c) = \mathcal{N}_{sat}(1 - e^{-0.693t_i/t_{1/2}})e^{-0.693t_c/t_{1/2}} \tag{4.46}$$

One can measure the activity $\mathcal{N}$ of a given isotope of known half-life that has been produced in a foil irradiated for a time t_i and cooled for a time t_c. One can then determine $\mathcal{N}_{sat}$ from Eq. 4.46. Then using Eq. 4.43 and the measured properties of the foil, one can make an absolute determination of the average beam intensity ϕ.

Figure 4.10 shows the growth of activity of several isotopes produced by proton irradiation of a thin aluminum foil. Note that isotopes with short half-lives like ^{13}N saturate quickly and at a low level of total activity. Fluorine-18 is a useful isotope for intensity measurements because its half-life of 110 min is convenient for the growth of activity and because the positron emitted in its decay can easily be detected in a NaI well counter.

4.5 Other particle sources

Although most particle physics experiments utilize accelerator beams, other particle sources are sometimes used. Some experiments such as those searching for proton decays or the presence of free quarks in

Figure 4.10 Induced activities of various isotopes in aluminum as a function of the irradiation time. The curves assume the value $\rho d = 5.89$ mg/cm^2.

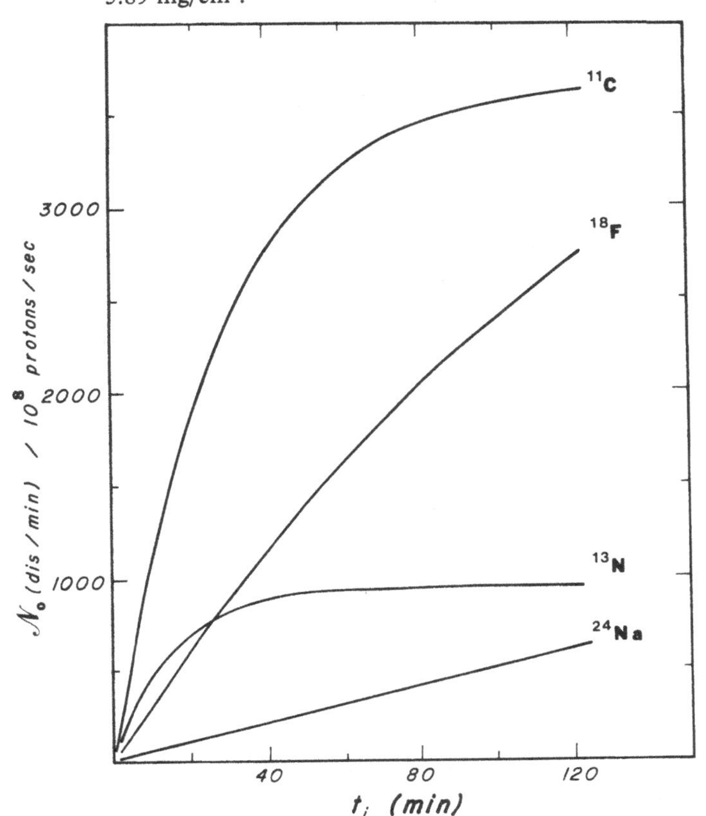

matter do not use a beam at all. Studies have been made of electron type neutrinos from the sun and $\bar{\nu}_e$ from fission reactors. Other sources of particles include cosmic rays and radioactive atoms.

In some experiments the presence of cosmic rays presents an unwanted source of background events, while for others they can be a useful tool for alignment and for checking detectors outside the beam region. The primary cosmic rays in outer space consist of protons, alpha particles, and heavier nuclei, roughly in the proportion 93 : 6 : 1. These strike particles in the upper atmosphere, producing many other elementary particles. Because of the charge imbalance of the primary particles and of the deflection of the particles in the earth's magnetic field, the intensity of various particles at sea level depends on the latitude, east–west direction, momentum interval, and angle from the vertical [19].

Cosmic ray fluxes are usually quoted in units of particles/cm^2-s, which is appropriate for flat detectors sensitive to particles from many directions or in units of particles/cm^2-s-sr for telescopic detectors with limited angu-

Figure 4.11 Flux of cosmic ray particles at sea level at 40° N geomagnetic latitude. The low energy electron and proton spectra can differ significantly from the dotted lines depending on local conditions in the atmosphere. (J. Ziegler, Nuc. Instr. Meth. 191: 419, 1981.)

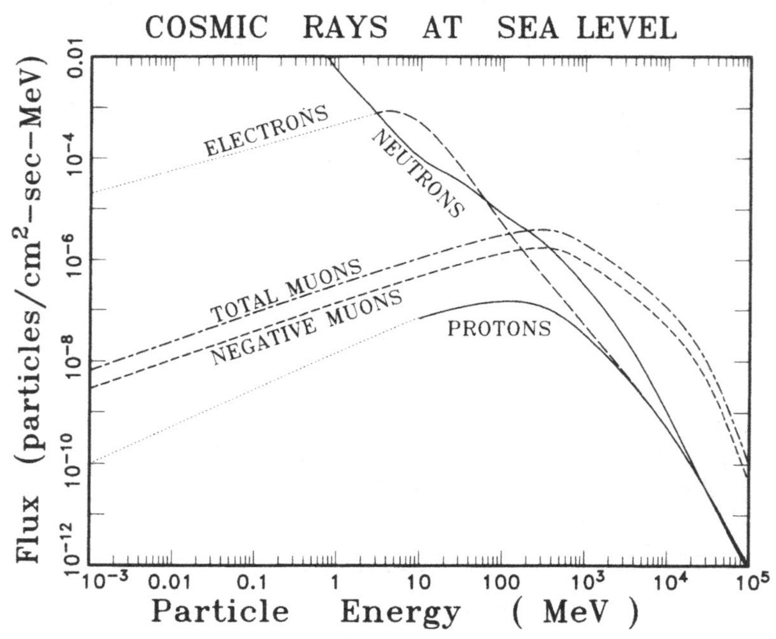

lar acceptance. The energy spectra of the main components of the cosmic ray flux at sea level are shown in Fig. 4.11. The total surface cosmic ray intensity is ~500 particles/m^2-s for midnorthern latitudes [20]. The muon momentum spectrum peaks around 800 MeV/c. The continuous spectrum of neutrons peaks around 0.1 eV.

Radioactive sources are useful for calibration and efficiency measurements. The common unit of activity is the Curie (Ci), where

$$1 \text{ Ci} = 3.7 \times 10^{10} \text{ disintegrations/sec}$$
$$= 3.7 \times 10^{10} \text{ becquerel (Bq)}$$

The unit of exposed dose is the roentgen (R). This is defined in terms of the amount of charge of either sign present in 1 cm^3 of air (STP) due to ionization or

$$1 \text{ R} = 1 \text{ esu/cm}^3$$

Table 4.4 contains a list of some radioactive sources that are useful for calibration purposes.

Table 4.4. *Radioactive sources*

Isotope	$t_{1/2}$ (yr)	Decay product	Particle energies (MeV)
^{3}H	12.26	e$^-$	0.019
^{22}Na	2.602	γ	0.511, 1.275
		e$^+$	0.54, 1.8
^{55}Fe	2.6	X-rays	0.0057
^{60}Co	5.26	β	0.315
		γ	1.173, 1.332
^{85}Kr	10.76	e$^-$	0.670
		γ	0.514
^{90}Sr	28.1	e$^-$	0.546
^{106}Ru	1.01	e$^-$	0.039
→^{106}Rh	30 sec	e$^-$	2.0, 2.4, 3.1, 3.53
		γ	0.512, 0.616
^{109}Cd	1.23	γ	0.088
^{133}Ba	7.2	γ	0.081, 0.303, 0.356
^{137}Cs	30.23	γ	0.662
		e$^-$	0.511, 1.176
^{207}Bi	30.2	γ	0.570, 1.064, 1.770
		β	0.481, 0.554, 0.976, 1.048
^{241}Am	458	γ	0.060
		α	5.486, 5.443, 5.389

Source: CRC Handbook of Chemistry and Physics, 64th ed., Boca Raton: CRC Press, 1983; Radioisotope data chart, Bicron Corp, Newbury, Ohio, 1974.

4.6 Radiation protection

High intensity beams of particles present a potential hazard to experimenters and to the accelerator's environment. For this reason, particle beams are shielded transversely and are terminated in massive beam dumps consisting of many interaction lengths of absorber. At energies less than 10 GeV neutrons and photons tend to be the chief components of stray radiation, while at energies above 100 GeV muons play a dominant role downstream of the target.

In health physics the common unit of absorbed energy dose in a given mass of material is the rad, where

$$\begin{aligned} 1 \text{ rad} &= 100 \text{ ergs/g} \\ &= 6.24 \times 10^7 \text{ MeV/g} \\ &= 10^{-2} \text{ gray (Gy)} \end{aligned}$$

It turns out, however, that 1 rad of radiation can have a different relative biological effectiveness (RBE) depending on a number of other factors. Therefore, when discussing exposure of human tissue to radiation, the preferred unit is the rem. This stands for "roentgen equivalents for man" and is given by

$$\begin{aligned} 1 \text{ rem} &= 1 \text{ rad} \times \text{RBE} \\ &= 10^{-2} \text{ sievert (Sv)} \end{aligned}$$

The RBE can depend on many factors, including the spatial distribution of the dose, dose rate, type of radiation, the type of tissue absorbing the radiation, and the energy loss per centimeter in the tissue [21]. This last quantity is particularly important and is referred to as the linear energy transfer (LET).

For radiation protection the RBE is expressed in terms of a quality factor, which is a function of the LET and takes into account how efficiently the radiation deposits energy. Quality factors vary from ~1 for photons, electrons, and minimum ionizing particles to ~20 for alpha particles and heavy ions.

Figure 4.12 shows the fluxes of several types of particles that are equivalent to an exposure of 1 mrem/hr. Note that around 1 GeV a flux of ~5 particles/cm^2-s of any type gives an exposure of 1 mrem/hr. For comparison the maximum permissible dose for the whole body in the United States is 5 rem/yr. A rough estimate of the natural background radiation due to cosmic rays is ~50 mrem/yr, while that from radioactive rocks and radioactive gases in the atmosphere is ~75 mrem/yr.

Radiation can kill cells and disturb the normal functioning of organs [22]. The most susceptible tissues and organs are the bone marrow, lymphatic system, skin, gastrointestinal tract, gonads, thyroid, and eyes. Long term, low level exposure to radiation can cause cancer, leukemia, and genetic mutations. The exact effect of this type of exposure is complicated due to the body's ability to adapt to and repair the damaged tissue. A massive short term exposure can cause radiation sickness, involving a serious decrease in the number of circulating blood cells. A dose of ~ 300 rem to the center of the body may be sufficient to cause death.

Particles entering a block of shielding material are attenuated exponentially. High energy neutrons are primarily removed through inelastic interactions. The attenuation of high energy particles is quite complicated since each interaction can produce several other particles, which are themselves capable of producing additional particles. We will discuss the resulting formation of electromagnetic and hadronic showers in more detail in Chapter 11. Accurate predictions of particle fluxes emerging from a shielding block usually require Monte Carlo calculations.

Earth and concrete are commonly used as shielding materials. The energy spectrum of the emerging particles can be strongly affected by the chemical composition of the material. For example, earthen barriers are more efficient at removing low energy neutrons than concrete blocks because of the larger fraction of water.

Figure 4.12 Particle fluxes equivalent to a dose of 1 mrem/hr. (A. Rindi and R. Thomas, adapted with permission from the Annual Review of Nuclear Science, Vol. 23, © 1973 by Annual Reviews, Inc.)

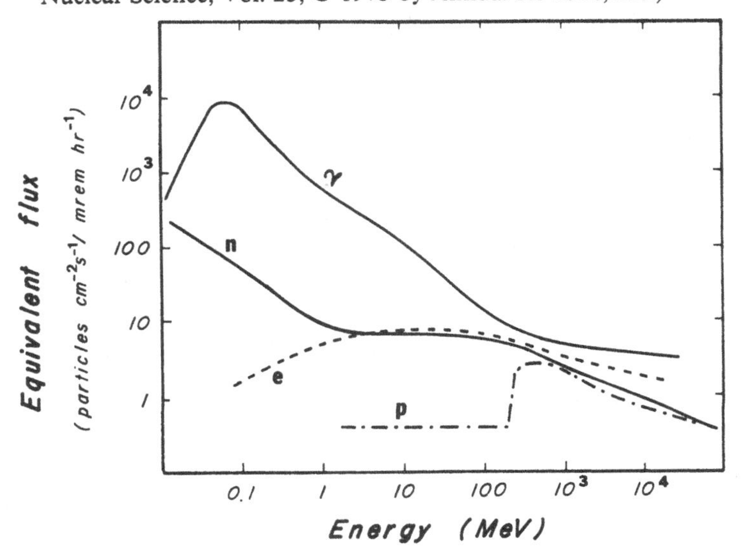

Besides the problem of stray beams there are several other environmental considerations concerning the radiation produced at an accelerator [21]. The first is radioactivity induced in accelerator components. This applies particularly to isotopes produced in plastics, oils, concrete, iron, aluminum, steel, and copper. There is a group of a few dozen isotopes whose half-lives are longer than $\frac{1}{2}$ hr, so that significant amounts may remain around long after the accelerator has been turned off. About 70% of these isotopes are γ emitters. A related nonhealth aspect of the radioactivity is radiation damage to the accelerator components. Such damage can cause failure of electronic equipment and structural weakening. Another health problem of lesser importance is the production of radioactivity in the air and in the cooling water that flows through the magnets.

References

[1] More technical details can be found in the books by E. Persico, E. Ferrari, and S. Segre, *Principles of Particle Accelerators*, Reading: Benjamin, 1968; M. Livingston and J. Blewett, *Particle Accelerators*, New York: McGraw-Hill, 1962; *Physics of High Energy Particle Accelerators*, New York: AIP, AIP Conf. Proc. Nos. 87, 105, and 127.

[2] An overview of existing and proposed high energy accelerators can be found in the review by M. Crowley-Milling, High energy particle accelerators, Rep. on Prog. Phys. 46: 51–95, 1983.

[3] J. Sanford, The Fermi National Accelerator Lab., Ann. Rev. Nuc. Sci. 26: 151–98, 1976.

[4] J. Jackson, *Classical Electrodynamics*, New York: Wiley, 1962, p. 471.

[5] R. Fernow and A. Krisch, High energy physics with polarized proton beams, Ann. Rev. Nuc. Part. Sci. 31: 107–44, 1981.

[6] R. Kohaupt and G. Voss, Progress and problems in performance of e^+e^- storage rings, Ann. Rev. Nuc. Part. Sci. 33: 67–104, 1983.

[7] E. Commins and P. Bucksbaum, The parity non-conserving e-N interaction, Ann. Rev. Nuc. Part. Sci. 30: 1–52, 1980.

[8] C. Pellegrini, Colliding beam accelerators, Ann. Rev. Nuc. Sci. 22: 1–24, 1972.

[9] F. Cole and F. Mills, Increasing the phase-space density of high energy particle beams, Ann. Rev. Nuc. Part. Sci. 31: 295–335, 1981.

[10] J. Sandweiss, Beam production at modern accelerators, in R. Shutt (ed.), *Bubble and Spark Chambers*, Vol. 2, New York: Academic, 1967, Chap. 4.

[11] D. Aston, M. Atkinson, A.H. Ball, G.R. Brookes, P.J. Bussey, B. Cake, D. Clarke, K. Connell, I.P. Duerdoth, R.J. Ellison, P.J. Flynn, W. Galbraith, P.G. Hampson, M. Ibbotson, R.E. Hughes-Jones, M.A.R. Kemp, G.D. Lafferty, J.B. Lane, J. Litt, D. Mercer, D. Newton, C. Raine, J.H.C. Roberts, K.M. Smith, K.M. Storr, R. Thompson, and A.P. Waite, The 25-70 GeV tagged photon facility at CERN, Nuc. Instr. Meth. 197: 287–96, 1982.

[12] H. Fisk and F. Sciulli, Charged current neutrino interactions, Ann. Rev. Nuc. Part. Sci. 32: 499–573, 1982.

[13] J. Drees and H. Montgomery, Muon scattering, Ann. Rev. Nuc. Part. Sci. 33: 383–452, 1983.

[14] A. Jonckheere, C. Nelson, B. Collick, S. Heppelman, Y. Makdisi, M. Marshak, E. Peterson, K. Ruddick, D. Berg, C. Chandlee, S. Cihangir, T. Ferbel, J. Huston, T.

Jensen, F. Lobkowicz, M. McLaughlin, T. Oshima, P. Slattery, and P. Thompson, Enhanced 200 GeV/c K^+ beam using a Be filter, Nuc. Instr. Meth. 180: 25–8, 1981.
[15] J. Lach and L. Pondrom, Hyperon beam physics, Ann. Rev. Nuc. Sci. 29: 203–42, 1979.
[16] A. Banford, *The Transport of Charged Particle Beams*, London: Spon, 1966.
[17] D. Caldwell and G. James, Beam monitoring methods, in D. Ritson (ed.), *Techniques of High Energy Physics*, New York: Interscience, 1961, Chap. 11.
[18] M. Babier, *Induced Radioactivity*, Amsterdam: North-Holland, 1969.
[19] S. Hayakawa, *Cosmic Ray Physics*, New York: Wiley, 1969.
[20] J. Ziegler, The background in detectors caused by sea level cosmic rays, Nuc. Instr. Meth. 191: 419–24, 1981.
[21] A. Rindi and R. Thomas, The radiation environment of high energy accelerators, Ann. Rev. Nuc. Sci. 23: 315–46, 1973.
[22] A. Upton, Effects of radiation on man, Ann. Rev. Nuc. Sci. 18: 495–528, 1968.

Exercises

1. Suppose that a proposed 20-TeV proton collider could be built with 2.5 T, 5.0 T, or 7.5 T bending magnets. Assuming that the dipoles fill 80% of the available space, calculate the circumferences of the corresponding tunnels. What are the synchrotron radiation losses for each case?

2. Find the equivalent CM energies for 400- and 1000-GeV fixed target accelerators.

3. What is the expected QED rate for two-muon events at LEP if the luminosity is 10^{31} particles/cm^2-s?

4. A 10-m-long electrostatic separator with a potential difference of 100 kV across a 2-cm gap is adjusted so that there is no deflection for a 5-GeV/c beam of pions. Find the spatial separation for kaons and antiprotons in the beam.

5. Derive Eqs. 4.34, 4.35, and 4.36.

6. A 2-m-long quadrupole magnet with a 20-kG pole tip field and a 10-cm-diameter aperture is used in a 20-GeV/c beamline. Calculate the beam transport matrices for this quadrupole. What are the corresponding matrices in the thin lens approximation? Find the focal length of this lens and the change in divergence for a particle 1 cm off the axis.

7. A beam with a maximum displacement of 1 cm and angular divergence of 10 mrad has a waist 5 m before the quadrupole described in exercise 6. Where is the beam waist following the

quadrupole? What is the characteristic length of the beam at the second waist?

8. Estimate the total rate of cosmic ray muons with energy greater than 10 MeV incident on a 500-cm^2 drift chamber oriented perpendicular to the surface of the earth.
9. Estimate the number of rads deposited by 10^{12} minimum ionizing particles passing through a cylinder of water 2 cm in diameter and 20 cm thick.

5
Targets

In this chapter we will briefly review the various types of targets used in particle physics experiments. The target normally consists of a piece of metal or liquid hydrogen. However, a number of other types of targets are sometimes used in specialized applications. These include polarized targets for spin dependent measurements, gas jets, active targets, and beam dumps.

5.1 Standard targets

Most particle physics experiments at fixed target accelerators use either a thin metal target or a cryogenic target containing liquid hydrogen or deuterium. The choice of target is dictated by considerations of the required signal rate, associated background, and necessary resolution. The finite target size and absorption, energy degradation, and multiple scattering in the target can all affect the measurement resolution.

The chief advantages of metal targets are the high interaction rate and the convenience of preparation. Solid targets such as beryllium are widely used in experiments studying muon production. A disadvantage is the amount of multiple scattering and absorption suffered by particles produced in the block. This can have a deleterious effect on experiments studying the effective mass of produced hadrons. The extraction of nucleon cross sections from scattering data with a nuclear target usually involves some sort of model-dependent or background target subtraction.

Liquid hydrogen targets offer the best resolution for the study of hadronic resonances. The multiple scattering and nuclear absorption are small. On the other hand, the density of liquid hydrogen is only 0.070 g/cm^3, so that the interaction rate per incident beam particle ($\sim 4 \times 10^{-6}/\mu$b-m) is smaller than that obtained with a solid target. Since the normal boiling point of H_2 is 20 K, the target requires an associated cryogenic system. Liquid deuterium with a density of 0.16 g/cm^3 and normal boiling point of 24 K can be used to study interactions involving neutrons. Some properties of common cryogenic fluids are given in Table 5.1.

A typical liquid hydrogen or deuterium target contains a cylinder along the beam path to hold the cryogenic fluid [1]. Figure 5.1 shows a schematic of a liquid hydrogen target. The beam enters and leaves the target through a mylar or other thin window. The cylinder is surrounded by an insulating jacket. The target gas is precooled in heat exchangers and liquified by passing the cooled gas through an expansion valve. The hydrogen density and liquid level are carefully monitored. Calibration runs with the target empty are used to determine the rate or characteristics of events coming from the target assembly.

All targets other than hydrogen have complications due to Fermi momentum and nuclear shadowing. The Fermi momentum is due to the continual motion of the nucleons inside the nucleus. It is possible to make a rough estimate of the magnitude of this momentum for various nuclei [2]. Two identical nucleons may not occupy the same energy level. Since they must obey Fermi statistics, the number of occupied states with mo-

Table 5.1. *Properties of cryogenic liquids at NBP*

	Melting point at 1 atm (K)	Normal boiling point at 1 atm (K)	Density (g/cm^3)	Heat of vaporization (J/g)	Specific heat (J/g-K)	Thermal conductivity (mW/cm-K)
^{3}He	[a]	3.19	0.059	11[b]	3.3[c]	0.20
^{4}He	[a]	4.215	0.125	21.7	4.50	0.27
H_2	14.0	20.4	0.070	447	9.5	1.19
D_2	18.7	23.6	0.163	304	5.8[d]	1.34
Ne	24.5	27.1	1.20	86.1	1.85	1.13
N_2	63.3	77.3	0.807	199	2.05	1.38
Ar	84.0	87.3	1.39	163	1.14	1.22

[a] Does not solidify at 1 atm.
[b] At 3.0 K.
[c] At 2.4 K.
[d] At 22 K.
Source: R. Scott, *Cryogenic Engineering,* Princeton: Van Nostrand, 1959; B. Colyer, Cryogenic properties of ^{3}He and ^{4}He, Rutherford Laboratory report RHEL/R138, 1966; T. Roberts and S. Sydoriak, Phys. Rev. 98: 1672, 1955; G. Haselden (ed.), *Cryogenic Fundamentals,* New York: Academic Press, 1971.

Figure 5.1 Schematic diagram of a liquid hydrogen target. (1) Liquid nitrogen supply, (2) hydrogen gas return, (3) hydrogen gas supply, (4) nitrogen gas return, (5) insulating vacuum, (6) heat exchanger, (7) liquid nitrogen reservoir, (8) expansion valve, (9) liquid hydrogen reservoir, (10) valve, (11) level indicator, and (12) thin windows.

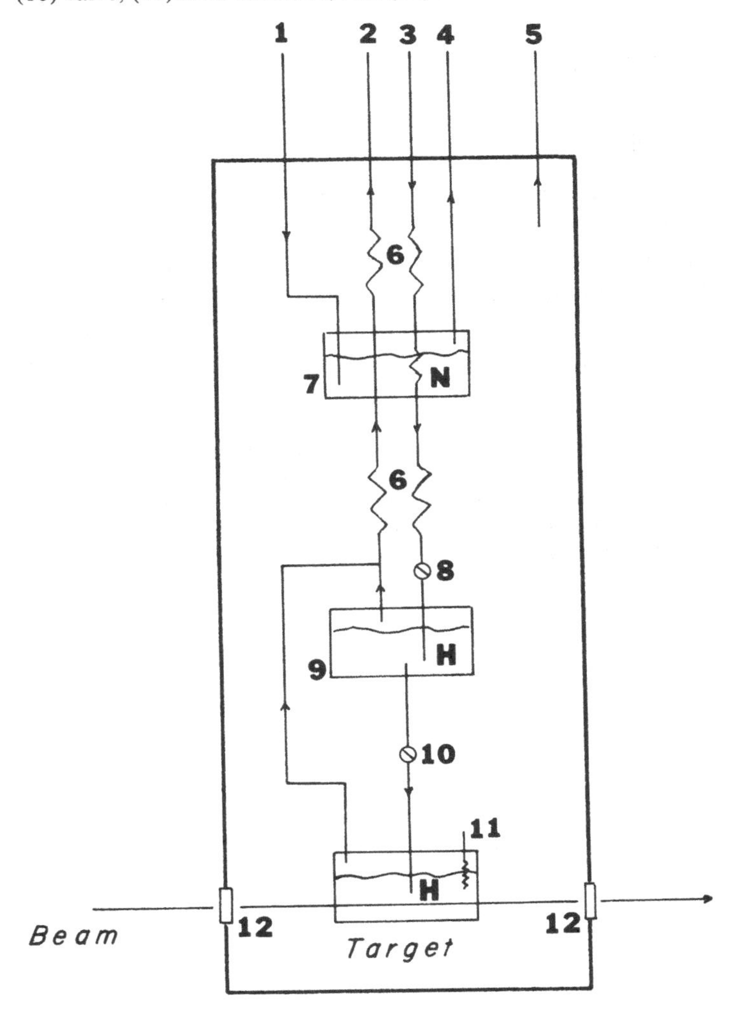

mentum less than p is

$$\frac{1}{h^3} 2 \frac{4}{3} \pi r_n^3 \frac{4}{3} \pi p^3$$

Thus, for a nucleus with Z protons and $A-Z$ neutrons we have

$$\frac{2}{h^3} \frac{4}{3} \pi r_n^3 \frac{4}{3} \pi p_p^3 = Z$$
$$\frac{2}{h} \frac{4}{3} \pi r_n^3 \frac{4}{3} \pi p_n^3 = A - Z \tag{5.1}$$

For heavy nuclei the Fermi momentum of neutrons is substantially larger than that of protons.

Figure 5.2 shows the measured Fermi momentum in deuterium. The calculated momentum peaks at approximately 50 MeV/c and has a high momentum tail extending beyond 200 MeV/c. The peak momentum in carbon is around 150 MeV/c. The Fermi momentum can have an appre-

Figure 5.2 Spectator momentum distribution for protons in deuterium. The curve follows from the potential model of T. Hamada and I. Johnston, Nuc. Phys. 34: 382, 1962. The data deviate from the curve at small momenta due to bubble chamber scanning losses. (R. Fernow, unpublished data.)

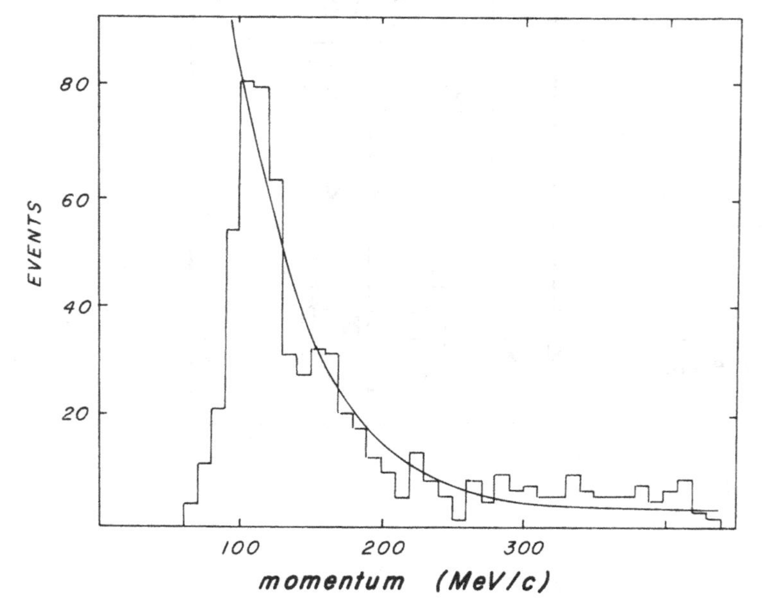

ciable effect on the kinematics of particle reactions. Interactions with a fixed beam momentum occur over a spectrum of CM energies. Uncertainty of the momentum of the target nucleon leads to a loss of resolution compared to interactions in hydrogen.

The second complication in nonhydrogen targets is nuclear shadowing [3]. Consider a high energy beam incident on a nuclear target. To the lowest-order approximation the nucleus appears as an assemblage of free nucleons. However, because of the finite probability of being absorbed in the nucleus, a beam particle is more likely to interact with the first nucleon it encounters rather than with a nucleon on the opposite side of the nucleus. Thus, the first target nucleons cast a "shadow" that reduces the probability of an interaction in the nucleus below the sum of the individual nucleon probabilities. Crudely, the shadowing effect in deuterium gives rise to the relation

$$\sigma_{\mathrm{d}} = \sigma_{\mathrm{n}} + \sigma_{\mathrm{p}} - \frac{1}{4\pi}\sigma_{\mathrm{n}}\sigma_{\mathrm{p}}\left\langle\frac{1}{r_{\mathrm{d}}^2}\right\rangle \tag{5.2}$$

where r_{d}^2 is the mean square distance between the neutron and proton in the deuteron ground state. Other effects present in nuclear targets include scattering off more than one nucleon, and combinations of scattering and absorption.

5.2 Special purpose targets

In this section we will discuss several types of targets that are used for special applications. Some experiments use an active target, meaning that the target medium is also a detector. The classic example of this is a bubble chamber. Counter neutrino experiments frequently employ liquid scintillator or segmented calorimeter targets. Experiments to measure the properties of short-lived states have used finely divided solid state detectors as targets.

Bellini et al. [4] have used a telescope made up from 40 silicon detectors, each with a thickness of 300 μm and separated from each other by 150 μm. A minimum ionizing particle deposits about 80 keV in each layer. The signal from each layer was amplified, shaped, and measured with an analog to digital converter. Figure 5.3 shows schematically the output from the layers, expressed as an equivalent number of minimum ionizing particles. The sudden jumps in the output by two equivalent particles are assumed to arise from the decays of two short-lived charged particles into three charged body decay modes.

At the opposite extreme from the active targets are beam dump experiments in which the beam is directed into a large mass of absorber [5]. Here the object is to quickly absorb normal hadrons near the interaction point. In that case muons or other particles emerging from the dump are more likely to be directly produced in the interaction and not result from the decay of ordinary hadrons.

A gas jet target delivers a narrowly collimated beam of gas. It is normally used inside the main accelerator ring. At predetermined times during the acceleration cycle a pulse of gas is directed into the circulating beam. Two advantages of a gas jet target are (1) the low gas density permits very low energy recoil particles to escape the interaction area and be detected and (2) the resolution on the measurement of the energy of the circulating incident beam is extremely good. The Fermilab gas target [6] has an equivalent thickness of only 4×10^{-7} g/cm^2. The jet has a $\pm$ 3-mm width and a 100-ms pulse length. The primary difficulty arising from the use of such targets is the background gas introduced into the

Figure 5.3 Schematic representation of the output from the active silicon target of Bellini et al. [4] for an event with two decaying particles.

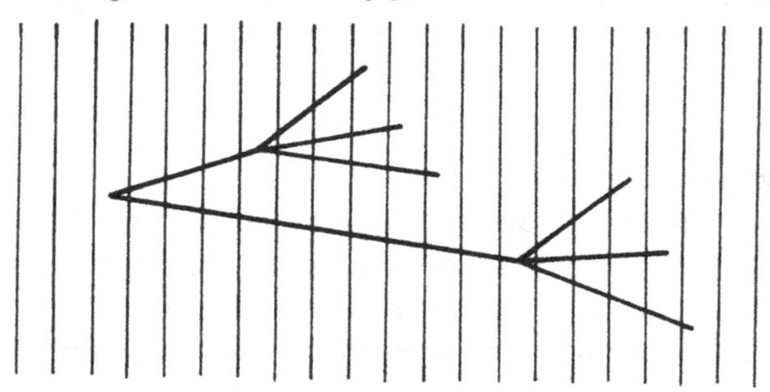

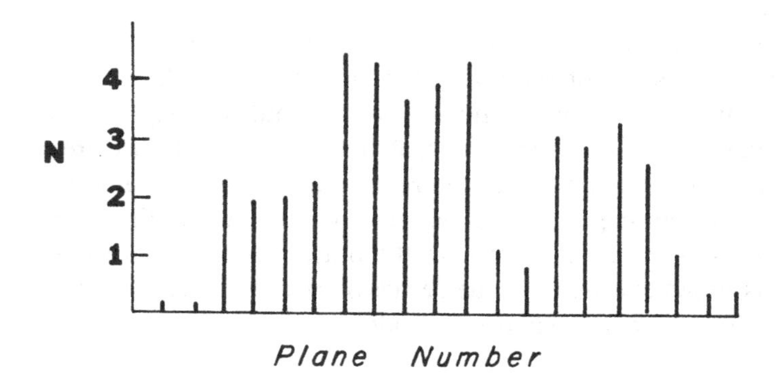

accelerator beam pipe. Large systems of diffusion pumps are necessary to reduce the gas pressure in the vicinity of the interaction point.

Polarized proton and deuteron targets are used in the study of the spin dependence of particle interactions [7]. A schematic of a polarized target system is shown in Fig. 5.4. The target contains five major subsystems: (1) the actual target material, (2) a cryogenic system, (3) a magnetic field, (4) a microwave system, and (5) a nuclear magnetic resonance (NMR) system. Polarized targets operate with magnetic fields above 2.5 T and at temperatures below 1 K. The low temperature can be achieved with ^{4}He or ^{3}He evaporation cryostats or with a dilution refrigerator.

Figure 5.4 Schematic diagram of the target region in a horizontal evaporation cryostat for a polarized proton target. (A) Magnet pole tip, (B) end of the cryostat, (C) ^{3}He fill line, (D) microwave guide, (E) Teflon seal, (F) copper cavity, (G) NMR coil, (H) NMR cable, (J) target beads, and (K) returning ^{3}He gas.

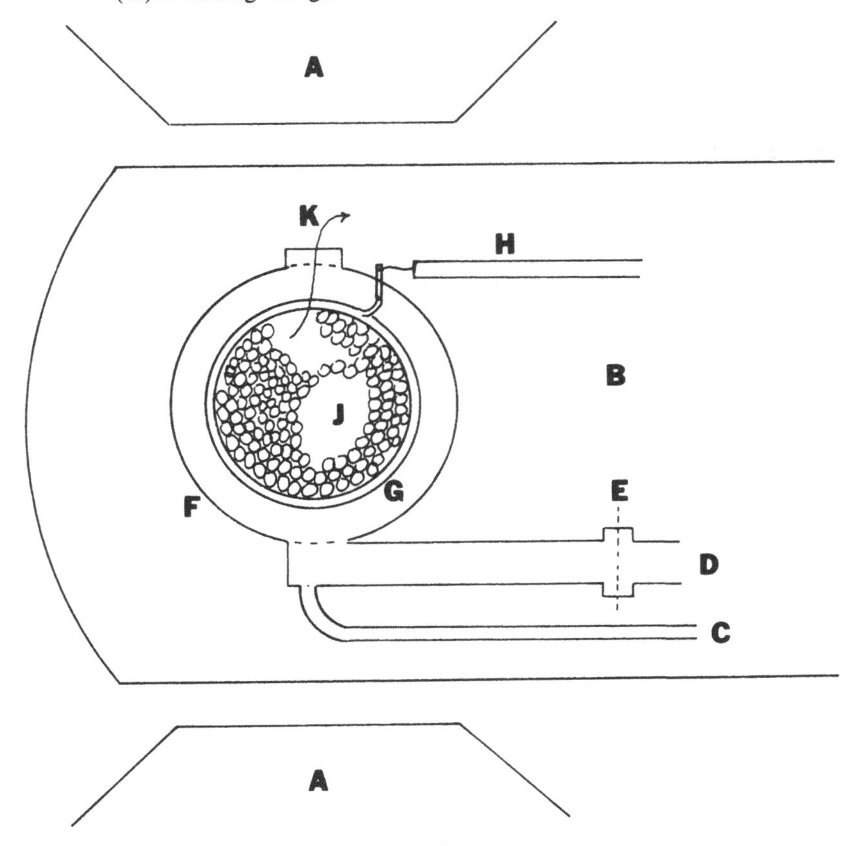

The actual target material consists of frozen beads of compounds, such as ammonia, butanol, or propanediol, which have been doped with a small proportion of a paramagnetic substance. The hydrogen atoms in the target provide the material to be polarized. Under the conditions of high magnetic field and low temperature, the free electrons in the paramagnetic additive are easily polarized to nearly 100%. Because of its much smaller magnetic moment, the proton polarization under the same conditions is $<1\%$. However, if microwaves of the proper frequency are radiated into the target mixture, the electron and proton spin systems become coupled, and it is possible to transfer substantial polarization to the protons. The magnitude of the polarization is usually determined by surrounding the target material with a coil and measuring the voltage change caused by the nuclear magnetic resonance of the polarized protons.

References

[1] G. Janes, Target preparation, in D. Ritson (ed.), *Techniques of High Energy Physics,* New York: Interscience, 1961, Chap. 10.
[2] B. Rossi, *High Energy Particles,* Englewood Cliffs: Prentice-Hall, 1952.
[3] V. Franco and R. Glauber, High energy deuteron cross sections, Phys. Rev. 142: 1195-1214, 1966.
[4] G. Bellini, P. D'Angelo, P. Manfredi, E. Meroni, L. Moroni, C. Palazzi Cerrina, F. Ragusa, and S. Sala, Active target for lifetime measurements of charmed particles and related signal processing, Nuc. Instr. Meth. 196: 351-60, 1982.
[5] R. Ball, C. Coffin, H. Gustafson, M. Crisler, J. Hoftun, T. Ling, T. Romanowski, J. Volk, S. Childress, M. Duffy, G. Fanourakis, R. Loveless, D. Reeder, D. Schumann, E. Smith, L. Jones, M. Longo, T. Roberts, B. Roe, E. Wang, C. Castoldi, and G. Conforto, Prompt muon-neutrino production in a 400 GeV proton beam dump experiment, Phys. Rev. Lett. 51: 743-6, 1983.
[6] A. Bujak, P. Devensky, A. Kuznetsov, B. Morozov, V. Nikitin, P. Nomokonov, Y. Pilipenko, V. Smirnov, E. Jenkins, E. Malamud, M. Miyajima, and R. Yamada, Proton-helium elastic scattering from 45 to 400 GeV, Phys. Rev. D 23: 1895-1910, 1981.
[7] See, for example, the reports in G. Bunce (ed.), *High Energy Spin Physics—1982,* New York: AIP, AIP Conf. Proc. No. 95, 1982, pp. 464-533.

Exercises

1. Compare the interaction rates per unit cross section, multiple scattering, and energy loss for a 5-GeV/c K^- beam incident on 20-cm-long liquid hydrogen, beryllium, and copper targets.

2. Estimate the Fermi momentum of the neutrons and protons in lead.

3. What is the maximum difference in CM energies for 20-GeV/c protons incident on deuterium due to Fermi momentum?

6

Fast electronics

Certain types of fast pulse electronics, such as discriminators and coincidence units, are used almost universally in particle physics experiments. In this chapter we review some important features of these and other electronic equipment, strictly from the point of view of a user.

6.1 Fast pulse instrumentation

An important function of fast electronics in particle physics experiments is to decide if the spatial and temporal patterns of detector signals satisfy the requirements of the event trigger. Fast in this context generally means circuits capable of processing pulses at a 100-MHz repetition rate. Most detectors produce analog signals. Discriminators are used to convert these analog signals into standardized logic levels. Logic units are available that can perform the logical operations: AND, NAND, OR, NOR, and NOT. The input and output signal amplitudes of these devices correspond to two possible states: 0 to 1 (or T or F). The logic unit signals can be joined together so that the final output is only true when a predetermined pattern of input signals is present. This output pulse can be used to signal the occurrence of a physical event of interest.

The need for certain electronic devices such as discriminators and logic units in practically every experiment lead to the establishment of the NIM standard. Devices that satisfy the NIM requirements must be housed in standard sized modules with standard rear connectors. Up to 12 units can be plugged into a NIM bin. The bin contains the power supplies for the modules and a slow gating pin that can be used, for example, to inhibit the operation of the units between beam spills. NIM modules are chiefly used in fast trigger logic. The NIM voltage levels corresponding to the states 0 and 1 are given in Table 6.1, along with levels used in the TTL and ECL families of integrated circuits.

6.2 Discriminators

A discriminator is an electronic device that converts an analog input signal into a standardized output pulse whenever the input signal amplitude exceeds some predetermined threshold voltage. Discriminators are routinely used with photomultiplier tube signals to provide uniform signals for triggering logic and for timing applications.

A block diagram for a typical discriminator is shown in Fig. 6.1. The input signal generates a pulse whenever the leading edge of the signal crosses the threshold voltage. A sharp signal corresponding to the leading edge is produced with a differentiation circuit. It is then reformed into a pulse of standard amplitude whose leading edge is related to the time of arrival of the input signal and whose width may be adjusted by the user. This circuitry introduces a delay of 10 – 30 ns between the arrival of the input signal and the leading edge of the output pulse.

Older model discriminators used shorted cable (clipping) stubs to shorten the input pulses. This was done so that the discriminator would only give one output pulse and so that the threshold would be independent of the event rate. This is unnecessary in modern discriminators designed to only give one output pulse regardless of the length of the input signal [1]. The threshold can vary slightly with temperature and the risetime and duration of the input signal. The threshold is usually adjustable from approximately 30 to 1000 mV, while the output width can be set in the range 5 – 1000 ns.

The coupling of the discriminator (or other analog signal handling device) to the input pulse has important experimental consequences. In ac coupling the pulse enters through a capacitor. In this case the time integral of a string of negative pulses must be balanced by an equal

Table 6.1. *Approximate logic levels (V)*

	State 0	State 1
NIM[a]	0.0	−0.8
TTL	0.2	2.5
ECL	−0.8	−1.6

[a] Terminated into 50 Ω.
Source: Catalog, LeCroy Research Systems Corp., Spring Valley, NY, 1983.

amount of positive signal [2]. This results in a long overshoot with small amplitude. This overshoot produces a shift in the effective signal baseline, which grows in importance as the pulse rate is increased. This rate problem is eliminated in dc coupled circuits. However, dc coupled circuits are susceptible to dc offsets in the input signal.

The rate characteristics of a discriminator are specified in terms of double pulse resolution and continuous pulse train response. The time between the leading edges of the two most closely spaced input pulses for which the discriminator gives two output pulses is called the double pulse resolution. This is 5 – 10 ns in a good discriminator. The continuous pulse train response is the maximum frequency of a continuous, equally spaced pulse train for which the discriminator will give a corresponding output train. This is 100 – 200 MHz for modern discriminators.

The input – output delay time in the discriminator is affected by a number of factors. A variation in the time delay between triggering and formation of the output pulse arising from the shape, amplitude, or rise-

Figure 6.1 Block diagram of a typical discriminator.

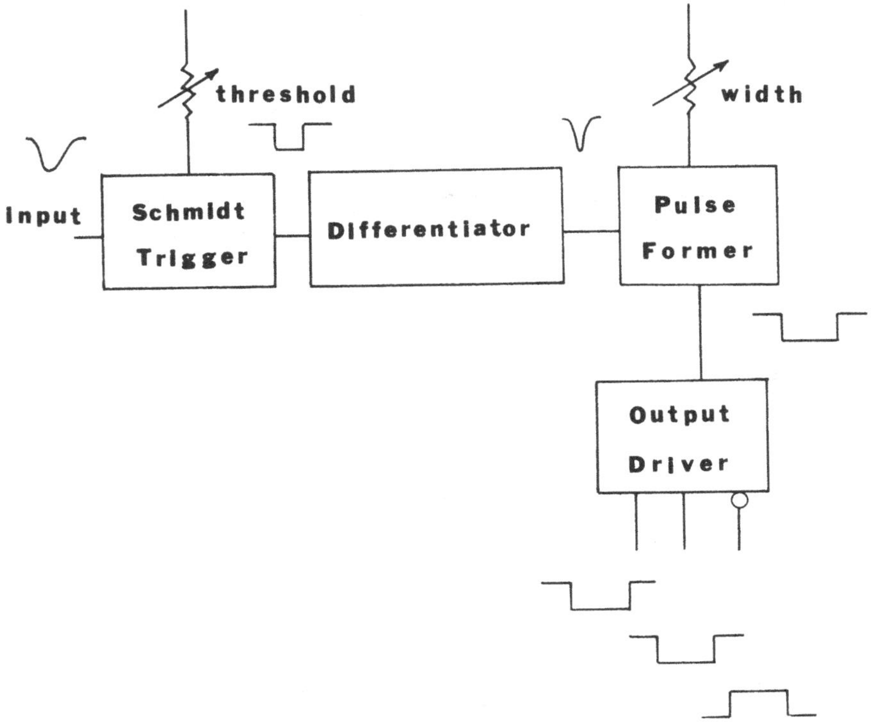

time of the input signal is referred to as time slewing or walk. This can be a large effect (~5 ns) for input signals only slightly above the threshold. The time delay also has small variations (drift) due to the aging of the components and temperature changes. A number of methods have been devised to compensate for time slewing in applications in which accurate timing information is important [3]. The most common method is to use a constant fraction discriminator. Discriminators of this type replace the leading edge method of determining the time of occurrence of an input signal with circuitry that gives a trigger signal nearly independent of the pulse height of the input signal [4]. The time walk of the output pulse can be reduced to ~200 ps for signals slightly above threshold.

Discriminators are available with a number of optional features.

1. Updating. This feature determines the action of the discriminator when a second input signal is received while an output pulse is still being generated. A nonupdating discriminator will ignore the second pulse as shown in Fig. 6.2. An updating discriminator will extend the output pulse for an additional full output width if the two pulses are separated by more than the double pulse resolution of the discriminator.

2. Burst guard. This feature allows the discriminator to respond reliably to a burst of input pulses separated by less than the double pulse resolu-

Figure 6.2 Comparison of discriminator outputs. The pulse at I is ignored in a normal discriminator. The pulses at R retrigger an updating or burst guard discriminator.

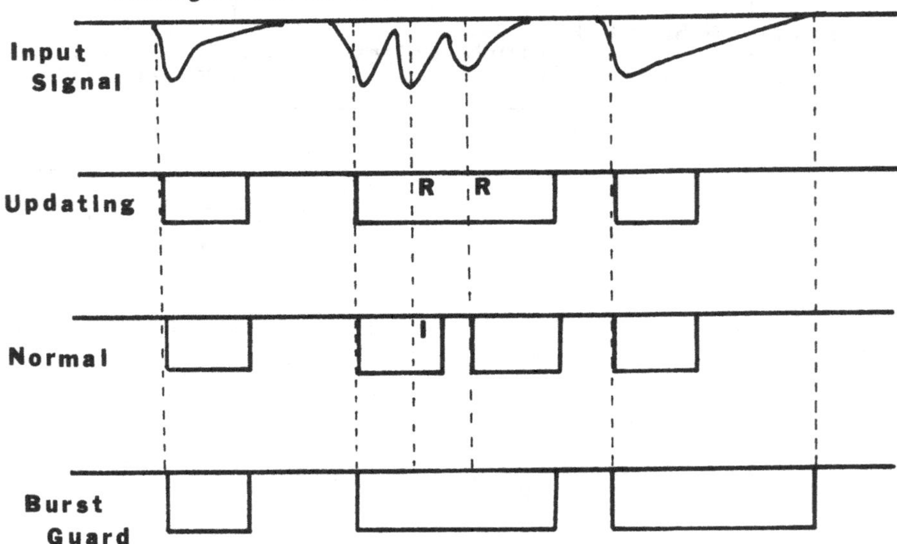

tion. The output is essentially the OR of the input signal with the output pulse width of the discriminator [1, 5]. This feature is important when considering the signals from a veto counter in a high rate environment. The burst guard feature maintains the veto signal during a rapid succession of events, thereby minimizing the possibility of a chance coincidence when the veto signal switches levels. A burst guard discriminator will maintain the output signal for the duration of an input with a long tail, as shown in Fig. 6.2. This can be important when using a proportional chamber signal as a veto in multiplicity logic.

3. *Inhibit.* This allows a fast logic signal to inhibit the operation of the discriminator. Input signals are ignored during the time the inhibit signal is present.

4. *Summing output.* Some multichannel units have a special summing output that provides a signal equal to −50 mV times the number of energized outputs. The signal could be used with another discriminator to provide an output when the input multiplicity exceeds a certain value.

5. *Inverted logic output.*

6. *Differential operation.* These discriminators have two adjustable thresholds and only provide an output pulse when the input signal is between the thresholds. It is also known as a single channel analyzer.

6.3 Coincidence units

There are many occasions when we wish to know if the signals from two or more detectors are associated in time. Consider the simple arrangement shown in Fig. 6.3. A particle passes through detector 1, travels a certain distance, and then passes through detector 2. If the signals are associated with the passage of the particle and not merely due to detector noise, we would expect that the signal from detector 2 will always come in a narrow range of times after the signal from detector 1. A coincidence unit is a device that can be used to determine if the two signals are simultaneously present.

The input signals are usually discriminated to produce standardized pulses with well-defined leading edges. One of the inputs may be passed through a variable delay before entering the discriminator. This extra delay takes into account the head start received by the signals in detector 1 while the particle traveled to detector 2. The discriminator outputs are the coincidence unit inputs.

The minimum amount of time the input signals must be simultaneously present is referred to as the coincidence overlap and is generally ~2 ns. The time interval during which the coincidence unit can produce an output is called the coincidence width or resolving time [6]. This quantity is usually determined by the widths of the input pulses as shown in Fig. 6.4. We obtain an output so long as input signals 1 and 2 are present for more than the coincidence overlap. If we consider pulse 1 to be fixed and vary pulse 2, we will obtain an overlap from the time the trailing edge of pulse 2 overlaps the leading edge of pulse 1 to the time the leading edge of pulse 2 overlaps the trailing edge of pulse 1. Thus, the time resolution of the circuit is approximately the sum of the two input pulse widths.

The time resolution can be determined experimentally from a coincidence delay curve. In Fig. 6.5 we plot the number of coincidences as a function of the variable delay in one of the input signals. The coincidence rate grows as the input signals overlap more and more. The width of the curve is a measure of the resolving time. It should be obvious from this discussion that if we desire short resolving times, the input signals should have fast risetimes and small widths.

In any experiment a random background of particles unassociated with the process of interest is usually present. It is possible that purely by

Figure 6.3 A simple coincidence circuit. (Dt1, Dt2) detectors, (V) variable delay, (D) discriminators, and (C) coincidence unit.

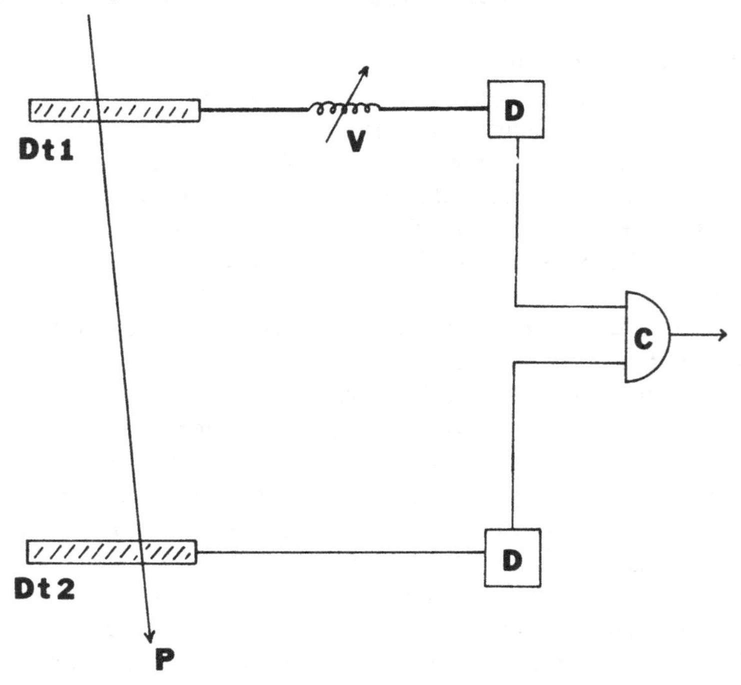

chance two of these background particles (or noise in one or both detectors) will produce signals in the detectors within the resolving time of the coincidence unit. This leads to an accidental coincidence. These "accidentals" are responsible for the tails of the delay curve.

It is possible to estimate the number of accidentals. Let us assume in our example that both discriminator output pulses have widths τ. Detector 1 will produce a certain singles rate S_1, which we assume is mostly caused by accidentals. Then for a fraction of the time $S_1\tau$ there will be a signal present at input 1 of the coincidence unit, provided that S_1 is not so large that there is significant overlap of the S_1 signals. If a random signal from detector 2 arrives during this time, an accidental coincidence will occur. Assuming the processes are independent and adding the contribution with counters S_1 and S_2 interchanged, we find that the accidental rate is

$$R_{acc} = 2S_1S_2\tau \tag{6.1}$$

Note the dependence of the accidental rate on the resolving time 2τ.

Coincidence units are available with two types of output pulses. In the

Figure 6.4 Input pulses to a coincidence unit.

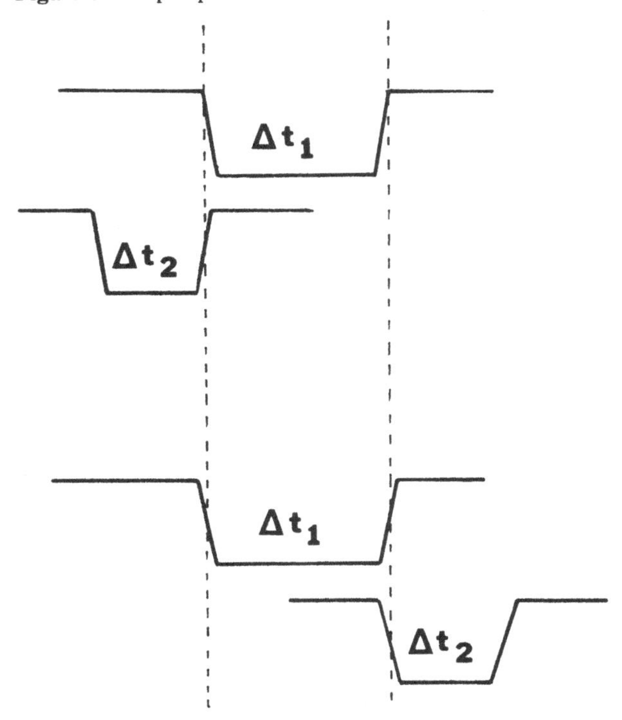

first the width of the output pulse is equal to the length of time the input signals overlap. In the second type an output pulse of fixed length is produced whenever the coincidence requirement is satisfied. The output width is usually adjustable from about 5 to 800 ns.

A large variety of coincidence units are available.

1. Logic units. These devices have three or more inputs and can perform combinations of the logical operations on the input. This permits fairly sophisticated trigger decisions to be made from the pattern of signals coming from the detectors.

2. Strobed coincidence. This device has a number of channels, all of which have a common coincidence input with the strobe signal.

3. Multiplicity logic unit. This device has a large number of input channels. When the number of simultaneously present inputs exceeds a specified threshold, it produces an output pulse. This could be used, for

Figure 6.5 Coincidence rate of a pair of counters as a function of the delay added to one of the inputs. The discriminator output widths were 17 and 6 ns. The coincidence rate was not 100% on the plateau because the counters were not geometrically matched.

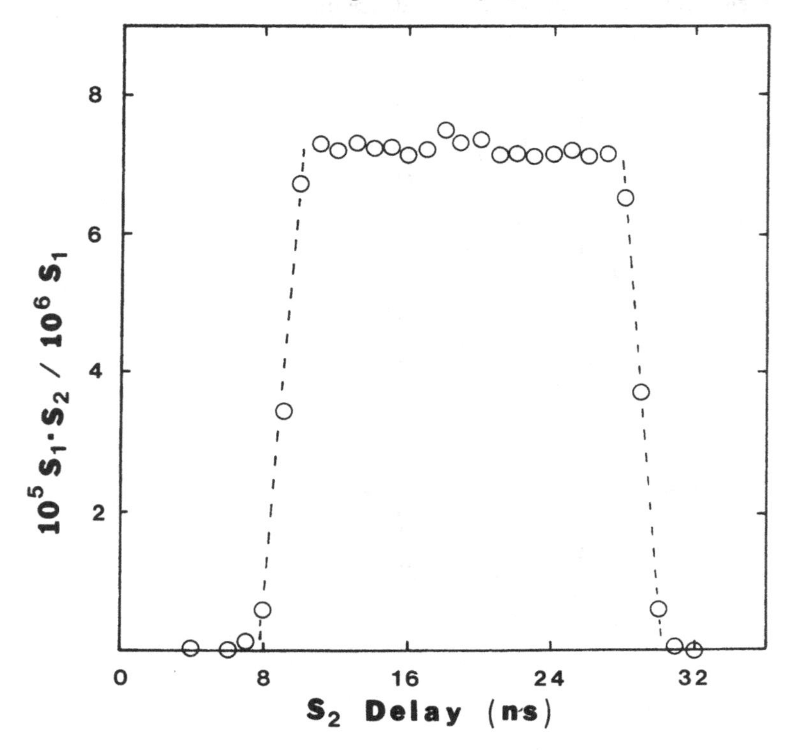

example, with a counter array to indicate the presence of a certain number of particles.

6.4 CAMAC standard

During the last decade online computer systems have played a larger and larger role in the data acquisition of particle physics experiments. CAMAC is a widely used standard with specifications for data transfer and control in addition to those for mechanical and power distribution compatibility. Mechanically, a CAMAC system consists of a crate with 25 stations, backplane connectors for power distribution, and additional lines for communication with a computer. Several crates can be connected together on a data highway. Data from the experiment enters through connectors on the front of the modules. The communication between a specific computer and the CAMAC modules is handled by a specific interface device known as a crate controller.

Each module connects to the address, data, and control lines in the crate. A particular slot in the crate has associated with it a station number (N). A device plugged into the slot may be subdivided into channels, each with its own subaddress (A). Thus, the computer can specify a particular channel of a device by giving NA. A third signal (F) is used to specify the desired operation or function. Examples of tasks include reading data into the computer or clearing the contents of the module. The computer communicates with the specified module by forming the F, N, and A information into a word and moving it into a memory location, which is in reality a register in the crate controller. The controller decodes the word back into its F, N, and A components and relays the command along to the appropriate module.

Some of the standard CAMAC function codes are given in Table 6.2. Not all CAMAC modules respond to every function. Functions F0, F2, F9, and F25 are commonly used. These commands allow the basic tasks of reading and clearing data and checking the modules' operation.

There are also dataless control commands such as CLEAR, INITIALIZE, and INHIBIT. The CLEAR function clears all registers in the crate. The INITIALIZE function sets all registers to some predetermined conditions, while INHIBIT prevents operation of the modules.

Another CAMAC signal that is important to the user is called "Look at me," or LAM. Devices that require attention can send a signal to the controller over the LAM line assigned to its station. Such a signal may indicate, for example, that a scaler has overflowed its contents or that a data run has ended. Most controllers have the ability to interrupt the computer system when a LAM is received and to transfer control to another part of the program.

As experiments have become larger, inherent limitations of the CAMAC standard have become more evident [7]. The data transfer rate is low by modern standards. The number of crates permitted on a branch is limited, and no crate to crate communications are possible. The directly addressable register space in CAMAC modules is also restricted.

As a result of these difficulties, a new data acquisition standard known as FASTBUS has been developed. The modules are physically larger than those in CAMAC, allowing a very large number (e.g., 96) of channels per module. The FASTBUS data transfer rates are at least 10 times faster than those in CAMAC. FASTBUS uses a 32-bit bus for data and address communications. Direct communications between modules is possible. The large amount of address space permits memories or look-up tables to be incorporated in the modules.

6.5 Other fast pulse devices

A number of other NIM or CAMAC devices are commonly used in particle physics experiments.

1. Fan in/out. A fan-in is a device in which many analog or logical inputs can be summed to form one output. This can be used, for example, to form the OR of the signals from a set of counters. An analog or linear

Table 6.2. *CAMAC function codes*

Function number	Function
0	read group 1 register
2	read and clear group 1 register
3	read complement of group 1 register
8	test Look at Me
9	clear group 1 register
10	clear Look at Me
16	overwrite group 1 register
18	selective set group 1 register
24	disable
25	execute
26	enable
27	test status

Source: Catalog, LeCroy Research Corporation, Spring Valley, NY, 1983.

fan-in is also useful for combining signals from detectors with a large number of elements before entering an ADC channel. The fan-out is a device with one input and many outputs, so that a signal can be used for more than one purpose. All of the outputs in a logical fan-out have the same amplitude as the input.

2. *Register.* A coincidence register or latch is a device with a large number of input channels along with a common input gate. Whenever a signal is present on an input channel simultaneously with the occurrence of the gate signal, the corresponding bit is set in a register. The pattern of bits form a word corresponding to the pattern of input signals. This word can subsequently be read by a computer.

Sometimes coincidence registers have self-contained, low threshold discriminators for every channel, which make the units useful for recording hodoscope information. The gate signal is usually generated by the trigger fast logic. The duration of the gate signal can determine the coincidence width. In addition to the register word, there may also be an output whose amplitude is proportional to the sum of the bits set in the word.

3. *Gate generator.* This device provides a relatively long variable delay. Besides the signal output, it usually provides an inverted logic output as well for a period of up to 10 sec following the input pulse. This is useful for providing deadtime at various places in the trigger logic. For example, it is desirable to inhibit a system for several hundred milliseconds after spark chambers fire because of the noise generated.

4. *Scaler.* A scaler is used to count pulses. It usually has an adjustable threshold to allow it to accept pulses from a large variety of sources. Scalers may have their own display or may be "blind" in that they can only be read out by a computer or special purpose device. Modern scalers can operate at 100 MHz continuous rate and may provide a special signal if the scaler overflows. Fast logic INHIBIT and CLEAR inputs may also be provided.

5. *Analog to digital converter.* We have seen that the presence of an analog signal above a certain threshold can be indicated by a discriminator. More detailed information about the input pulse can be obtained with an analog to digital converter (ADC). This device produces an output pulse proportional to the peak value or the integral of the input current or total charge in the pulse. During the interval in which an external gate signal is present, the charge on the input line is stored on a capacitor. The capacitor is later discharged at a constant rate while oscillator pulses are counted in a scaler. The number of counts is then proportional to the collected charge.

Modern ADCs can have a sensitivity of 0.25 pC/count with a capacity of 1024 counts. The resolution depends on the dynamic range of input signal amplitudes. This can be increased if the ADC has a bilinear range, whereby small signals have a more sensitive conversion than larger ones. The response of the ADC should be very linear in each region. In general, an ADC gives some number of counts (pedestal) even when no input signal is present.

The pedestal signal is the sum of a constant term plus a contribution that is proportional to the width of the gate signal. For example, in the LeCroy 2249A ADC the residual pedestal charge in picocoulombs is given by $1 + 0.03w$, where w is the gate width in nanoseconds [1]. The gate width dependent part of the pedestal is due to a built-in dc offset.

6. Time to digital converter. A device that converts a time interval into a digital number is called a time to digital converter (TDC). The two timing signals enter the START and STOP inputs of the unit. Timing is usually measured from the leading edges of the input signals. After the arrival of the START pulse a capacitor is charged with a constant current. The charging is terminated after the arrival of the STOP pulse, and the capacitor is discharged at a uniform rate. During the discharge period oscillator pulses are counted in a scaler and then stored in a register for readout.

Modern TDCs have a resolution as low as 50 ps/count. Errors in the timing arising from a wide range of input amplitudes or discriminator slewing may be minimized by preamplifying the input signal. The timing resolution can also be improved by running a parallel input signal into an ADC and making an amplitude dependent correction to the TDC value.

Many other special purpose CAMAC and NIM devices are available commercially, including amplifiers, digital to analog convertors, digital voltmeters, and real time clocks [1, 8].

6.6 Signal cables

It is important that detector signals and logic pulses be transferred between electronic devices with a minimum of distortion. The interconnections in logic circuits are usually made using coaxial cables. A typical cable consists of a thin copper conductor surrounded by polyethylene insulation, a copper ground braid, and a vinyl jacket. The outer grounded conductor in the cable prevents the escape of electromagnetic radiation from the cable and pickup of the radiation from other nearby cables.

A section of coaxial cable of unit length can be represented as a series resistance r_s and inductance l and a parallel resistance r_p and capacitance C. The cable has a characteristic impedance per unit length [9]

$$Z_c = (Z_s Z_p)^{1/2} \tag{6.2}$$

where Z_s is the series impedance and Z_p is the parallel impedance per unit length. In high frequency operation

$$Z_s = r_s + i\omega l \approx i\omega l$$
$$1/Z_p = 1/r_p + i\omega C \approx i\omega C$$

Thus, the characteristic impedance is:

$$Z_c \simeq (l/C)^{1/2} \tag{6.3}$$

This represents a pure resistance independent of the pulse frequency and the cable length.

Table 6.3 lists properties of some commonly used coaxial cables. The thin RG174/U cables require LEMO type connectors, while the others use the BNC type. Note that the polyethylene dielectric in the space between the conductors causes the signal propagation velocity to be considerably less than c. The highest propagation velocities ($\beta \sim 0.95$) are achieved in air core cables. If we recall that

$$1 \text{ dB} = 20 \log_{10} V_0/V_i \tag{6.4}$$

we see that an attenuation of 6 dB represents a factor of 2 loss in amplitude.

A coaxial cable should be terminated into its characteristic impedance. This is done either internally in an electronic module or by using a terminator plug, which is simply a resistance connected to ground. A cable that is not properly terminated will give rise to reflections. The ratio of the amplitude of the reflected signal to the amplitude of the in-

Table 6.3. *Properties of some common coaxial cables*

Type	Outer diameter (mm)	Ground shields	Z_c (Ω)	C (pF/m)	β	Attenuation[a] (dB/100 m)
RG8/U	10.3	1	50	85.3	0.78	5.9
RG58/U	4.95	2	53.5	93.5	0.66	13.5
RG58A/U	4.95	1	50	85.3	0.78	15.8
RG59/U	6.15	1	75	56.8	0.78	9.7
RG62A/U	6.15	1	93	44.3	0.84	10.2
RG174/U	2.54	1	50	101.0	0.66	28.9
RG214/U	10.80	2	50	101.0	0.66	6.6

[a] At 100 MHz.
Source: Belden Wire Corporation, Richmond, IN.

put signal is

$$\frac{V_r}{V_i} = \frac{R - Z_c}{R + Z_c} \tag{6.5}$$

where R is the value of the terminating resistor. We see that terminating with $R < Z_c$ gives a reflection of the opposite polarity. If the cable is not terminated at all ($R = \infty$), we get a reflection of the same amplitude and polarity as the input signal. A second characteristic of a properly terminated cable is that the input impedance looking into the cable will be Z_c, independent of the length of the cable.

Sometimes a section of cable is shorted at one end to form a clipping line. The reflected signal from the clipping line can be used to form a shortened (clipped) pulse. Figure 6.6 shows a clipping circuit. Half the signal goes down the clipping line to the grounded end. Here the pulse is reflected with the same amplitude and opposite polarity. It travels back up the clipping line and combines with the other half of the input signal, thereby canceling it. Thus, we get a pulse whose width is $\Delta t = 2L/\beta$, where L is the length of the clip line and β is the signal propagation velocity in the cable.

Figure 6.6 Signal waveforms with a clipping stub. Elements denoted Z_c are coaxial cables.

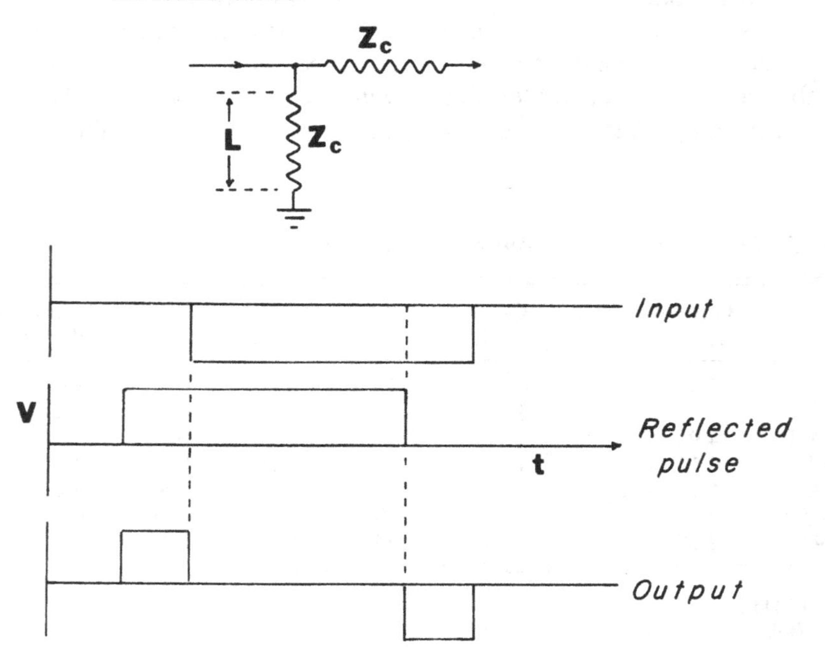

References

[1] Catalog, LeCroy Research Corporation, Spring Valley, NY 1983.
[2] R. Chase, *Nuclear Pulse Spectrometry,* New York: McGraw-Hill, 1961, pp. 27–9.
[3] B. Gottschalk, Timing discriminator using leading edge extrapolation, Nuc. Instr. Meth. 190: 67–70, 1981.
[4] Instruments for Research and Applied Science, EGG Ortec Corporation, Oak Ridge, TN.
[5] R. Mount, P. Cavaglia, W. Stockhausen, and H. Wahlen, Highly efficient scintillation counters, Nuc. Instr. Meth. 160: 23–7, 1979.
[6] A. Melissinos, *Experiments in Modern Physics,* New York: Academic, 1966, pp. 403–7.
[7] R. Dobinson, Data acquisition at LEP, Physica Scripta, 23: 487–91, 1981.
[8] A number of advertisements for such equipment is usually found in each issue of Physics Today and the CERN Courier.
[9] J. Brophy, *Basic Electronics for Scientists,* New York: McGraw-Hill, 1972, pp. 380–7.

Exercises

1. Show how a discriminator may be used to introduce a fixed deadtime after each yes signal from a coincidence unit.

2. What is the accidental rate for triple coincidence if S_1, S_2, and S_3 are the singles rate in three counters?

3. Prove Eq. 6.5.

4. Assume that the outputs from the readout electronics of two MWPCs are proportional to the number of hit wires. Show how commercial fast electronics may be combined to give a signal when there are two hits in the first chamber and four hits in the second.

5. An elastic scattering experiment has three detectors in the forwardly scattered direction (F), three detectors in the backwardly scattered direction (B), and three additional detectors (M) to monitor the overall event rate from the target. Lay out the electronics necessary to determine the elastic scattering rate. Suppose that you are suspicious that beam structure with a period of 70 ns is causing accidentals. What additional circuitry would be required to monitor the accidental rate?

7
Scintillation counters

One of the most commonly used particle detectors is the scintillation counter. A fraction of the energy lost by a charged particle can excite atoms in the scintillating medium. A small percentage of the energy released in the subsequent deexcitation can produce visible light. The technique has been used since the earliest investigations of radioactivity, when, for instance, Rutherford used scintillating ZnS crystals in his alpha particle scattering experiments.

In modern detectors light produced in the scintillator is propagated through light guides and directed onto the face of a photomultiplier tube. Photoelectrons emitted from the cathode of the tube are amplified to give a fast electronic pulse, which can be used for triggering or timing applications.

7.1 The scintillation process

We define a scintillator to be any material that produces a pulse of light shortly after the passage of a particle. The phenomenon is closely related to fluorescence, which is usually defined to be the production of a light pulse shortly following the absorption of a light quantum [1, 2]. "Shortly" here refers to time intervals on the order of 10 ns or less. Phosphorescence is a third phenomenon involving light emission, but in this case the molecules are left in a meta-stable state, and the emission may occur much later than the initiating event.

Both inorganic and organic scintillators have been discovered. The scintillation process is different for the two groups. Inorganic crystals are usually grown with a small admixture of impurity centers. The most important example is NaI doped with thallium. The positive ions and electrons created by the incoming particle diffuse through the lattice and are captured by the impurity centers [1, 3, 4]. Recombination produces an excited center, which emits light upon its return to the ground state. The light output is larger than that of the organic materials, but the electron migration through the crystal lattice results in a pulse (~ 1.5 μs) of much longer duration. The light output per unit absorbed energy of inorganic scintillators has a significant temperature dependence.

NaI crystals are hygroscopic, so the counters must be protected from water vapor. The emission spectrum of a NaI(Tl) crystal is significant over the range 340–490 nm. NaI scintillators are widely used as photon detectors because of their relatively high density and atomic number. Figure 7.1 shows the photon linear attenuation coefficient for NaI. Some properties of NaI, CsI, and bismuth germanate (BGO) are shown in Table 7.1. BGO is also finding acceptance for high resolution photon detection [5]. It has a radiation length of 1.12 cm, which is shorter than NaI, and it is nonhygroscopic. On the other hand, it is more expensive and has a smaller light output.

The organic scintillators can be further subdivided into organic crystals, liquid scintillators, and plastic scintillators. The advantages of the organic materials include transparency to their own radiation, short

Figure 7.1 Linear attenuation coefficient of NaI. (K) K absorption edge. (Data taken from J. Hubbell, National Bureau of Standards report NSRDS-NBS 29, 1969.)

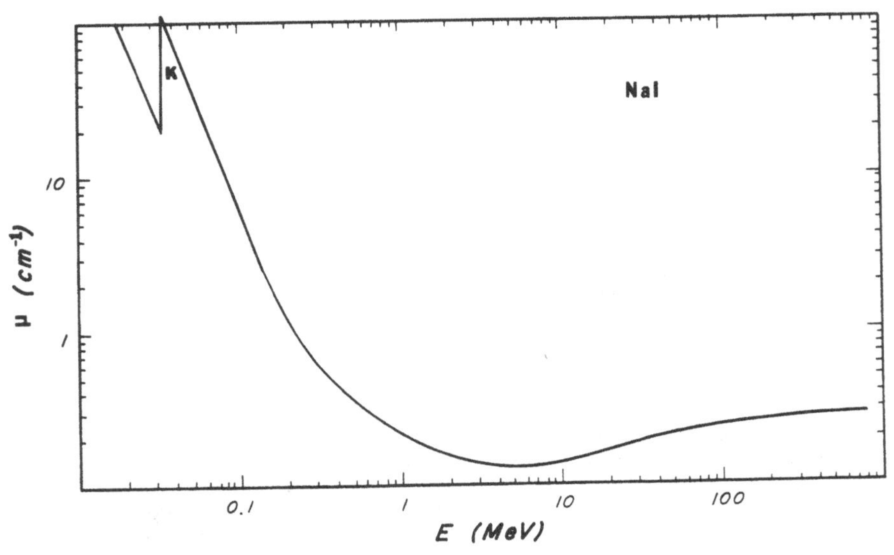

decay times, emission spectra well matched to photomultipliers, and easy adaptibility.

The scintillation mechanism in organic scintillators depends strongly on the molecular structure of the medium [1, 2, 6]. After the passage of a high energy particle many of the atoms in the scintillating medium will be excited into higher energy levels. Most of the excitation energy is given up in the form of heat and lattice vibrations. However, in a scintillator some of the excitation energy is released as radiation. The scintillation efficiency is defined to be the fraction of the deposited energy that appears as radiation. The absolute efficiency of even the best scintillators is quite low, 7% for NaI and 3.5% for anthracene, the best organic material. Deexcitation without the production of radiation is sometimes referred to as quenching. The presence of even minute quantities of impurities in the organic scintillator can lead to quenching and a consequent reduction in scintillation efficiency.

Table 7.1. *Properties of scintillators*

	Relative light output	λ_{max} emission (nm)	Decay time (ns)	Density (g/cm^3)
Inorganic crystals				
NaI(Tl)	230	415	230	3.67
CsI(Tl)	250	560	900	4.51
$Bi_4Ge_3O_{12}$ (BGO)	23–86	480	300	7.13
Organic crystals				
Anthracene	100	448	22	1.25
Trans-stilbene	75	384	4.5	1.16
Naphthalene	32	330–348	76–96	1.03
p,p'-Quarterphenyl	94	437	7.5	1.20
Primary activators				
2,5-Diphenyl-oxazole (PPO)	75	360–416	5[a]	
2-Phenyl-5-(4-biphenylyl)-1,3,4-oxadiazole (PBD)	96	360–5		
4,4''-Bis(2-butyloctyloxy)-*p*-quaterphenyl (BIBUQ)	60	365,393	1.30[b]	

[a] 4g/l in xylene.
[b] 65 g/l in toluene.
Source: E. Schram, *Organic Scintillation Detectors,* New York: Elsevier, 1963; D. Ritson (ed.), *Techniques of High Energy Physics,* New York: Interscience, 1961; P. Pavlopoulos et al., Nuc. Instr. Meth. 197: 331, 1982; Catalog, Nuclear Enterprises Inc.; M. Moszynski and B. Bengtson, Nuc. Instr. Meth. 158: 1, 1979; B. Bengtson and M. Moszynski, Nuc. Instr. Meth. 204: 129, 1982; H. Grassmann, E. Lorenz, and H.-G. Moser, Nuc. Instr. Meth. 228: 323, 1985.

The light emission is governed by the electronic transitions in the molecule [2]. The electronic levels have a typical energy spacing of ~ 4 eV. The vibrational levels of the molecule ($\Delta E \sim 0.2$ eV) also play a role. Electrons in high levels typically deexcite to the lowest excited state without emission of radiation.

The emitted light is not self-absorbed because of the differing shapes of the excited and ground state energy levels as a function of interatomic spacing. We show in Fig. 7.2 the ground state and excited vibrational energy levels of a molecule. The Frank–Condon principle states that the atoms in a molecule do not change their internuclear distances during an electronic transition [2]. Thus transition ($A \rightarrow B$) from the ground state to the excited state occurs with no change in interatomic distance. The molecule finds itself out of equilibrium with its surroundings and quickly loses vibrational energy to the lattice arriving at level *CD*. After a short period of time the molecule will decay to some vibrational level of the ground state ($D \rightarrow E$). It then falls to the ground state by interacting with the lattice. The net result is that the molecule absorbs photons corre-

Figure 7.2 Molecular potential energy versus interatomic spacing.

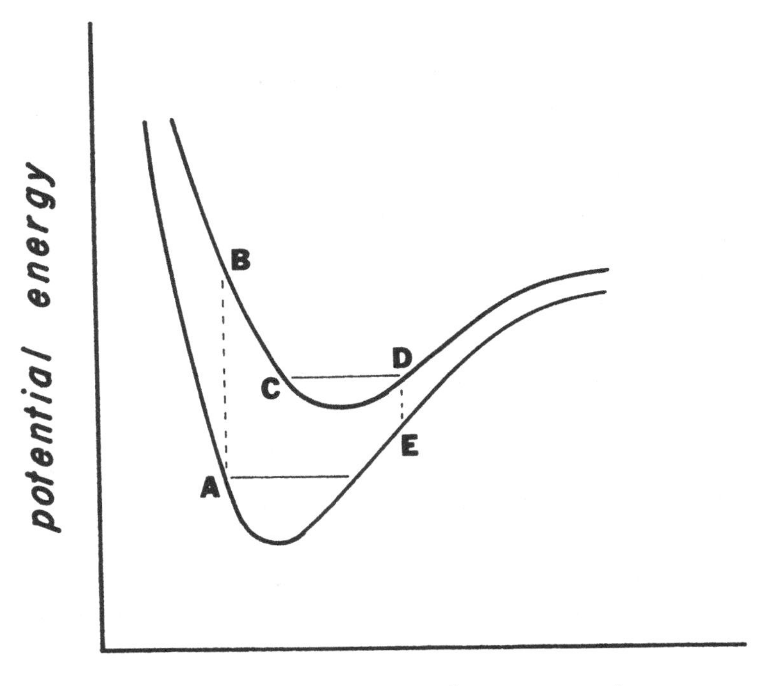

sponding to the transition $A \rightarrow B$ and later emits lower energy photons corresponding to the transition $D \rightarrow E$.

The random excitation of scintillator material leads to an exponential time dependence for photon emission.

$$n(t) = k(1 - e^{-t/\tau}) \tag{7.1}$$

where τ is a decay time characteristic of the material. Many materials have a prompt decay when most of the light is emitted, together with a smaller, long time constant decay. The density of ionization can influence the time dependence of light emission in some media.

The amount of light produced for a given energy loss is not constant but depends on the production of quenching centers. These are activated molecules raised to excited vibrational levels. The number of produced photons roughly follows the equation [4]

$$n = \frac{n_0 dE/dx}{1 + B\, dE/dx} \tag{7.2}$$

where B and n_0 are constants and dE/dx is the ionization energy loss. The parameter B is sometimes referred to as the α to β ratio. Note that the light output saturates for large energy losses.

All organic scintillators contain a benzene ring in their structure. The organic crystals have the highest scintillation efficiency. Except for special applications, they are used as solutes in liquid and plastic scintillators. The scintillation efficiency, decay times, and wavelength of maximum emission of some organic crystals are contained in Table 7.1. The most efficient crystal is anthracene, which consists of three condensed benzene rings. Scintillation efficiencies are often quoted relative to an anthracene standard.

In a liquid scintillator an organic crystal (solute) is dissolved in a solvent. A typical concentration might be 3 g/l. Fortunately, although most interactions occur with the more numerous solvent molecules, it is possible for a large fraction of the deposited energy to be transferred to the solute. The scintillation efficiency depends on the solvent used. For example, xylene and toluene give good relative pulse heights with a PPO solute [2]. The scintillator BIBUQ dissolved in toluene has the best time resolution ($\sim$ 85 ps) of any scintillator currently available [7].

Liquid scintillator can be formulated especially for photon or neutron detection. Addition of cadmium or boron can increase the efficiency for neutron detection. Liquid scintillator is sometimes useful in large volume applications, such as the active element in large calorimeters.

Usually the efficiency of a scintillator can be increased by adding a secondary solute (wavelength shifter) at about 1% the concentration of the primary solute. The purpose of the wavelength shifter is to lower the self-absorption of the light emission and to produce an output that is well matched to the spectral acceptance of the photomultiplier tube. POPOP is a commonly used wavelength shifter for this purpose. Table 7.2 lists the peaks of the emission and absorption spectra of POPOP and some other common wavelength shifters. Figure 7.3 shows measurements of the absorption spectra of wavelength shifter materials. A second important use for wavelength shifters is in the light collection systems of large scintillator calorimeters [8].

Plastic scintillators are similar in composition to liquids. Polystyrene and polyvinyl toluene are commonly used base plastics, which take the place of the solvent in liquid scintillator. Most plastic scintillators have a density around 1.03 g/cm^3 and an index of refraction of 1.58. They are rugged, easy to machine, and available in large sizes. Important properties of some commercial plastic scintillators are shown in Table 7.3. The emission spectrum for the NE-102A plastic scintillator shown in Fig. 7.4 peaks around 420 nm. Recently more economical plastic scintillators using a polymethyl methacrylate base have been developed for large volume applications, such as for calorimeters [9].

Plastic scintillators have a fast response time, making them useful for triggering. However, they have rather poor spatial resolution. With aging and mishandling they may suffer minute surface cracking (crazing),

Table 7.2. *Wavelength shifters*

		λ_{abs} (nm)	λ_{emis} (nm)
POPOP	1,4-di-(2-(5-phenyl-oxazolyl))benzene	360	415–420
BBQ		382–430	500–505
BBOT	2,5-bis(5-tert-butylbenzoxazolyl(2′))thiophene	375–398	432
Dimethyl-POPOP	1,4-di-(2-(4-methyl-5-phenyl-oxazolyl))benzene	370	425–430
bis-MSB	*p*-bis(*o*-methylstyryl)benzene	367	416–425

Source: Nuclear Enterprises, Inc.; C. Woody and R. Johnson, Brookhaven National Laboratory report OG-755, 1984; V. Eckardt et al., Nuc. Instr. Meth. 155: 389, 1978; F. Klawonn et al., Nuc. Instr. Meth. 195: 483, 1982; Catalog, National Diagnostics, Somerville, NJ.

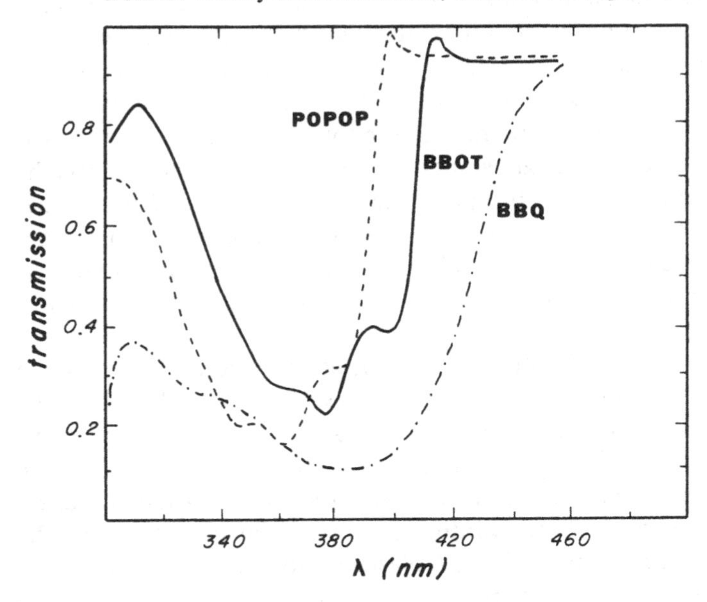

Figure 7.3 The absorption spectra of wavelength shifters. POPOP and BBOT: 20 mg/l in toluene, 3 mm; BBQ: in plastic, 2 mm. (Data taken from C. Woody and R. Johnson, Brookhaven report OG755, 1984.)

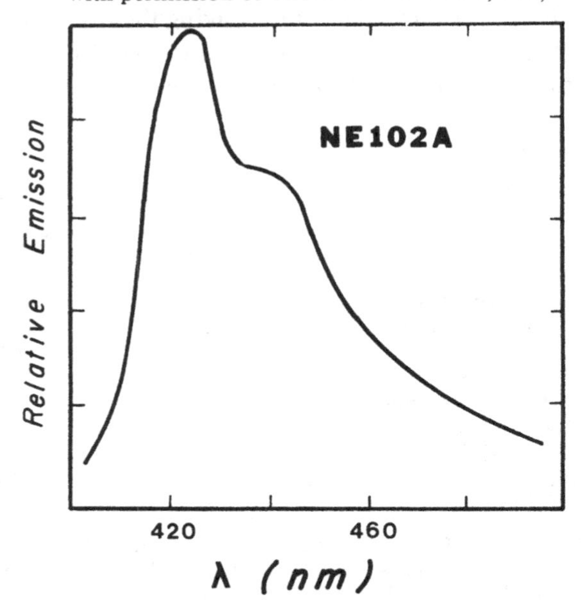

Figure 7.4 Emission spectrum of NE102A plastic scintillator. (Adapted with permission of Thorn EMI Gencom, Inc.)

which diminishes their light collection efficiency. Any large scintillator is affected by attenuation of the scintillation light traveling through the counter. The distance for the light intensity to fall to $1/e$ of its initial value is called the attenuation length and is typically on the order of 1 m.

Plastic scintillator may also be obtained in the form of thin (1–4-mm-diameter) optical fibers. Such scintillators may prove useful for a fine-grained hodoscope in a high rate environment. The scintillator is clad with a thin layer of material with a lower index of refraction than the scintillator. Encouraging results were obtained with scintillator clad with 200 μm of heat-cured silicone [10]. The major problems are in finding a suitable method for readout and in deterioration from handling.

The scintillation process also occurs in pure noble gases [11]. Atoms excited by the energy loss of a passing particle may dissipate ~ 20% of the excitation energy through the emission of ultraviolet light. The emission

Table 7.3. *Properties of common plastic scintillators*

Type	Light[a] output	λ_{max}[b] (nm)	Attenuation[c] length (cm)	Risetime (ns)	Decay[d] time (ns)	Pulse FWHM (ns)
NE 102A	58–70	423	250	0.9	2.2–2.5	2.7–3.2
NE 104	68	406	120	0.6–0.7	1.7–2.0	2.2–2.5
NE 104B	59	406	120	1	3.0	3
NE 110	60	434	400	1.0	2.9–3.3	4.2
NE 111	40–55	375	8	0.13–0.4	1.3–1.7	1.2–1.6
NE 114	42–50	434	350–400	~1.0	4.0	5.3
Pilot B	60–68	408	125	0.7	1.6–1.9	2.4–2.7
Pilot F	64	425	300	0.9	2.1	3.0–3.3
Pilot U	58–67	391	100–140	0.5	1.4–1.5	1.2–1.9
BC 404	68	408	—	0.7	1.8	2.2
BC 408	64	425	—	0.9	2.1	~2.5
BC 420	64	391	—	0.5	1.5	1.3
ND 100	60	434	400	—	3.3	3.3
ND 120	65	423	250	—	2.4	2.7
ND 160	68	408	125	—	1.8	2.7

[a] Percentage of anthracene.
[b] Wavelength of maximum emission.
[c] $1/e$ length.
[d] Main component.
Source: Catalog, Nuclear Enterprises; Catalog, Bicron Corporation; Catalog, National Diagnostics; M. Moszynski and B. Bengtson, Nuc. Instr. Meth. 158: 1, 1979; G. D'Agostini et al., Nucl. Instr. Meth. 185: 49, 1981.

spectra consist of a sharp peak near the threshold, together with a much broader continuum enhancement. Table 7.4 lists the threshold wavelength of the emission spectra and a rough estimate of the center of the enhanced portion of the spectrum. The continuum portion of the spectrum results from transitions from vibrationally excited states to the ground state. The emission spectrum is practically independent of the type of incident particle but is strongly affected by the gas pressure.

Another class of scintillating material is scintillation glass. Type SCG1-C scintillation glass contains about 43% BaO as a high Z material [12]. A small percentage of Ce_2O_3 added to the glass produces scintillation light directly and absorbs any Cerenkov light in the glass and reemits it at a longer wavelength. It has a radiation length of 4.35 cm and a fluorescent decay time of ~ 70 ns. Scintillation glass may be preferable to lead glass in certain calorimeter applications. It produces about 5 times as much light as SF5 lead glass and has better radiation resistance. The measured energy deposition was linear for 2 – 17-GeV positrons [12]. The energy resolution was $\sigma/E = 1.64\% + 1.13\%/\sqrt{E}$, compared to typical lead glass resolution $\sigma/E \simeq 1\% + 4.5\%/\sqrt{E}$. Scintillation glass can be used in more extreme environmental conditions than other scintillators, although it is sensitive to the presence of ultraviolet radiation.

7.2 Light collection

Once the scintillator light has been produced, it must be efficiently transported to the face of a photomultiplier tube. This is often one of the more difficult aspects of designing a scintillation counter. Photomultipler tubes (PMTs) cannot operate in a high magnetic field, and so it may be necessary to locate the tube a large distance from the scintillator. The scintillator often covers a much larger area than the photocathode

Table 7.4. *Scintillation of pure noble gases*

Gas	Threshold wavelength (nm)	"Center" wavelength (nm)
He	60.0	75
Ne	74.4	82
Ar	106.7	125
Kr	123.6	150
Xe	147.0	172

Source: A. Breskin et al., Nuc. Instr. Meth. 161: 19, 1979; Y. Tanaka, A. Jursa, and F. LeBlanc, J. Opt. Soc. Am. 48: 304, 1958.

surface of the tube. In some applications it may be required to collect light uniformly over the entire scintillator volume.

Light is directed to the tube by using either a light guide or a fluorescent converter. Light guides are usually constructed from plastic, although for certain applications the light is sometimes propagated through an enclosed volume of air to the PMT. The light travels down the guide via total internal reflection. On the other hand, a fluorescent converter contains a wavelength shifter, which absorbs the incident radiation and reemits it isotropically at longer wavelengths. The light guides collect light more efficiently, but the fluorescent converters are more useful for providing a compact readout in large area applications.

Let us first consider a rectanglular slab of scintillator or wavelength shifter bar. Light will be emitted isotropically from the scintillator. When the emitted light strikes a surface, total internal reflection will occur for incident angles greater than the critical angle θ_c, given by

$$\sin\theta_c = n_0/n \tag{7.3}$$

where n (n_0) is the index of refraction for the medium in (outside of) which the light is traveling. For simplicity, we take the outside medium to be air, so that $\sin\theta_c = 1/n$. The angle θ_c is measured with respect to the normal to the surface. The higher the value of n for the material, the smaller the critical angle, and the greater the range of incident angles that undergo total internal reflection.

The fraction of the emitted light transmitted via total internal reflection in one direction along the slab is

$$\begin{aligned} f &= \frac{1}{4\pi}\int_0^{\theta_c} 2\pi \sin\theta \, d\theta \\ &= \frac{1}{2}\left(1 - \frac{1}{n}\sqrt{n^2-1}\right) \end{aligned} \tag{7.4}$$

Note that this result is independent of the slab dimensions. For typical plastic scintillator with $n = 1.58$, Eq. 7.4 gives $f = 0.113$.

An improvement in light transmission in small counters can be achieved by using a reflector on the end of the scintillator opposite to the collection direction. This subject has been studied in detail by Keil [13]. The reflectors may be either specular or diffuse and may be in direct contact with the medium or separated by a small air gap. Reflectors on the other four sides of the slab will in general have a much smaller influence on the collection efficiency.

For specular reflectors, such as aluminum foil, the emitted light fraction collected (neglecting attenuation) is increased to

$$f^{\text{air}}_{\text{spec}} = (1+R)f \tag{7.5}$$

where R is the reflectivity. The superscript "air" means that the output face of the scintillator is connected to the next stage of the light collection system via an air gap. This equation holds for both direct and separated coupling to the reflector, although the case with a small reflector gap generally gives better results. If the output edge is coupled to the following stage using an optical contact medium, such as silicon oil, the emitted light fraction is increased. When using a diffuse reflector, such as TiO_2 paint, in direct contact with the scintillating medium, there is no longer any correlation between the directions of the light rays before and after the reflection.

Measurements of the effect of a specular surface on the light collection is shown in Fig. 7.5. The relative pulse height and uniformity of response is much better using total internal reflection than when the reflections take place off a vacuum deposited aluminum surface [14]. An exception to this occurs for the scintillator surface opposite from the PMT.

Plastic counters are loosely wrapped with aluminum foil to capture

Figure 7.5 Relative light collection from a test scintillator using (T) totally reflecting surfaces and (S) vacuum-deposited aluminum surface. The inset shows the test counter dimensions in centimeters. (After D. Crabb, A. Dean, J. McEwen, and R. Ott, Nuc. Instr. Meth. 45: 301, 1966.)

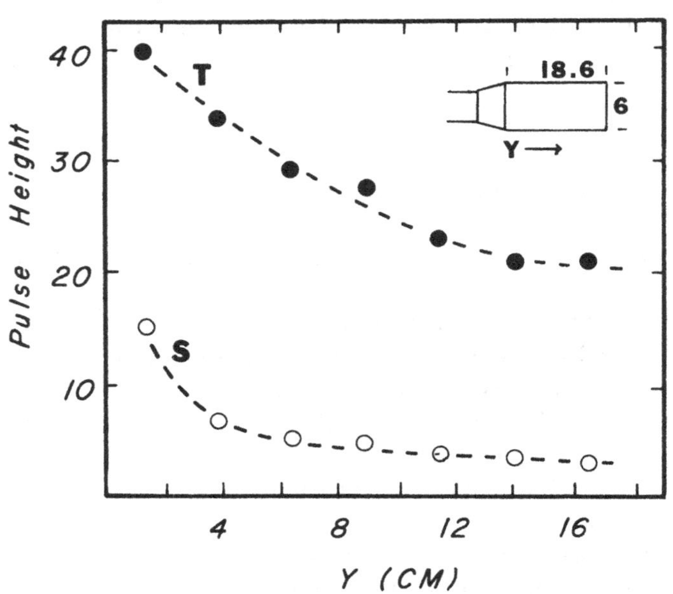

light that does not undergo total internal reflection and escapes from the surface. The counter is made light tight by wrapping black tape over the foil.

The intensity of light traveling through scintillator or plastic guides roughly follows a relation of the form

$$I(x) = I_{0s}\exp(-x/d_s) + I_{0l}\exp(-x/d_l) \tag{7.6}$$

where the characteristic distances d_s and d_l are properties of the material. The attenuation results from atomic absorption and scattering from surface imperfections, such as scratches or microscopic cracks. Short wavelength light (<360 nm) is strongly absorbed in plastics. This absorption gives rise to the short distance attenuation characterized by d_s. The attenuation of the remaining longer wavelength light is then characterized by d_l.

When the cross section of the scintillator differs significantly from that of the face of the PMT, it is necessary for most light rays to undergo a series of reflections in a light guide before arriving at the tube. The presence of the light guide improves the homogeneity of response of the scintillator [13]. Light guides are typically constructed from acrylic plastics such as lucite. Three light pipe constuctions are illustrated in Fig. 7.6. The rectangular (RT) connection is used for small pieces of scintillator [15]. In the fishtail (FT) arrangement the light pipe is gradually deformed from the

Figure 7.6 Types of light guides: (RT) rectangular, (FT) "fish tail," (TW) two twisted strips. The bottom figure shows a photomultiplier tube assembly (PM) with base (B), standard coupling piece of light guide (L), mu metal shield (dotted lines), and soft iron shield (I). (After G. D'Agostini et al., Nuc. Instr. Meth. 185: 49, 1981.)

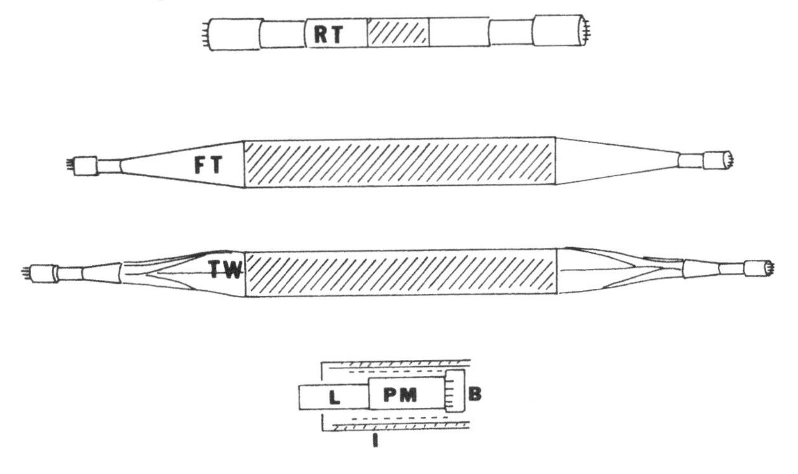

shape of the scintillator edge to the shape of the photomultiplier window. In the twisted pipe (TW) arrangement the rectangular shape of the scintillator edge is transformed into a square approximately the size of the PMT with a large number of smaller twisted pieces. The twisted arrangement is generally the most efficient.

Liouville's theorem presents a fundamental limitation on the transmission of light through the pipe [16]. The theorem requires that the light flux per unit solid angle not increase while propagating in the pipe. This has the consequence that the maximum fraction of light that can be transmitted is A_{pm}/A_{sc}, where A_{pm} (A_{sc}) is the area of the photomultiplier face (scintillator edge) and $A_{pm} < A_{sc}$.

The light collection efficiency may be improved by using a light collecting cone [17, 18]. Figure 7.7 shows the construction of a Winston cone, which is commonly used before the PMT in Cerenkov counters. The cone is designed to optically focus light at angles up to θ_{max} with respect to the optical axis onto a focus on the PMT face. The surface of the collector is a

Figure 7.7 Geometry of the Winston cone. The face of the PMT lies along *t*. Half of a section of the cone is shown with hash marks. The 3-dimensional cone is made by revolving the hashed surface around the optical axis 0*A*.

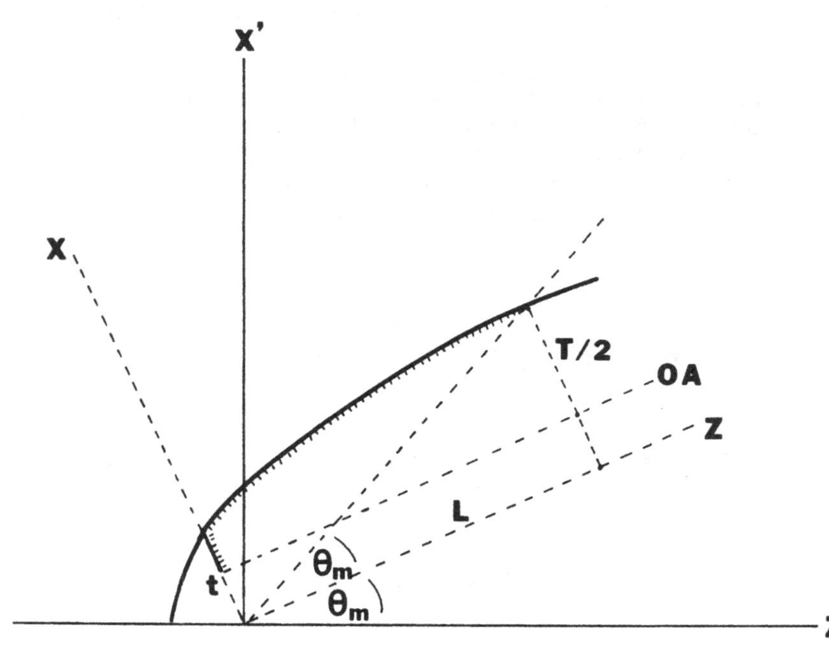

parabola in the primed coordinate system and is given by the equation

$$r(\theta) = \frac{1+\sin\theta_{max}}{1-\cos(\theta+\theta_{max})}t \tag{7.7}$$

where t is the diameter of the PMT and θ_{max} is the maximum accepted angular divergence. The length of the cone is

$$L = \frac{1+\csc\theta_{max}}{2\tan\theta_{max}}t \tag{7.8}$$

while the opening of the collector is

$$T = t/\sin\theta_{max} \tag{7.9}$$

Light guides are usually connected optically to the PMT with grease or cement with an index of refraction greater than that of the light guide, so that there is no total internal reflection from the PMT window.

7.3 Photomultiplier tubes

The great utility of scintillation counters is made possible by photomultiplier tubes (PMT), which convert the scintillator light and amplify it into an electrical pulse suitable for use with decision making, fast electronics. A schematic of a PMT is shown in Fig. 7.8. We shall discuss the characteristics of a typical tube, the RCA 8575. Properties of this and some other commercially available photomultiplier tubes are given in Table 7.5.

The incident light from the scintillator or light guide falls upon a semi-transparent photocathode. The surface is coated with a material that has a low work function in order to facilitate the emission of electrons from the photocathode by the photoelectric effect. A bialkalai material such as K-Cs-Sb is commonly used. The number of emitted electrons per incident photon, or quantum efficiency, is shown in Fig. 7.9. The quantum efficiency for this tube peaks in the near ultraviolet part of the spectrum and falls off in the visible region. Figure 7.9 also shows the absolute responsivity S (or radiant sensitivity) of the photocathode. This expresses the photoelectric current in terms of the incident radiant power. The two quantities are related by the expression QE (%) $= 123.95S\,(\text{mA/W})/\lambda(\text{nm})$. The number of photons reaching the photocathode surface also depends on the type of window material used in the tube. Fused silica can extend the accepted wavelength interval several hundred nanometers below that of normal borosilicate glass.

The emitted photoelectrons are accelerated and focused onto the first dynode of the tube. The tube has a number of dynodes made of materials with large secondary emission (e.g., Cu-Be or BeO). Typically four electrons may be emitted for each incident electron. The multiplication continues for many stages before the current is collected at the anode. The overall amplification of a 12-stage tube is then several 10^7. The exact amplification depends critically on the high voltage applied between the dynodes. The RCA 8575 pulse has a typical risetime of 3 ns and a transit time of 40 ns with 1800 V between the anode and the cathode.

Figure 7.8 Schematic diagram of the RCA 8575 photomultiplier tube. (Courtesy of RCA, New Products Division, Lancaster, PA.)

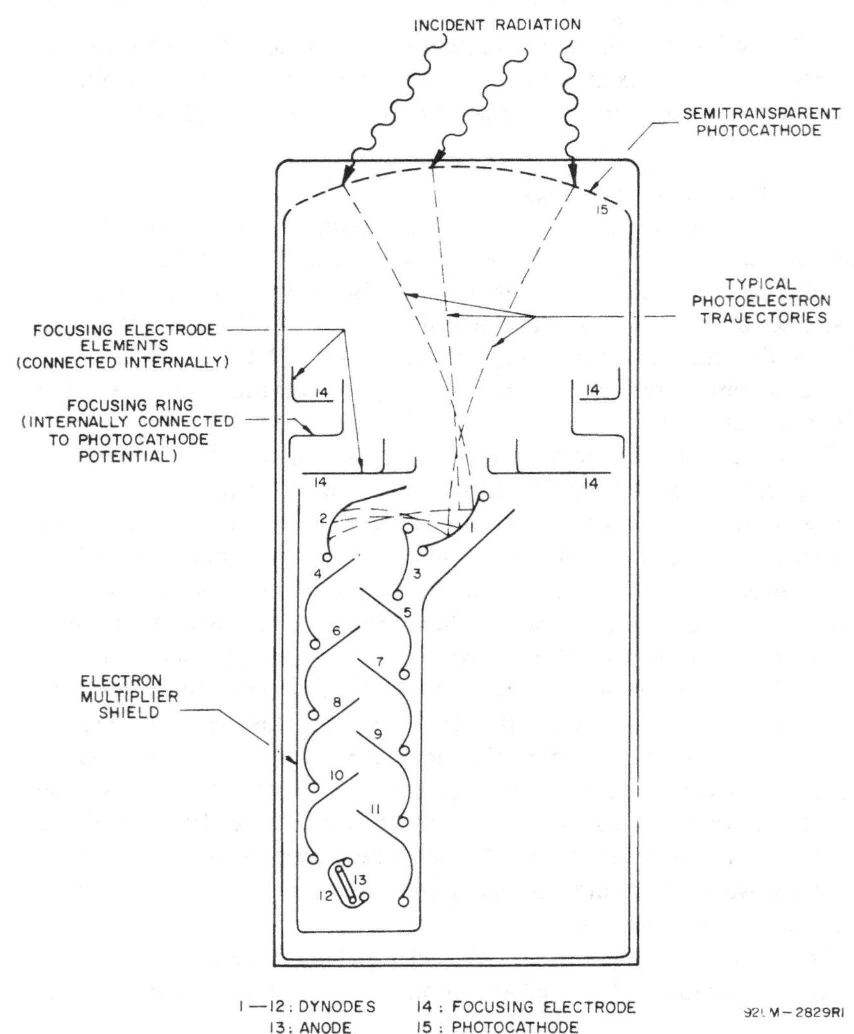

Table 7.5. *Some common photomultiplier tubes*

		Useful diameter (mm)	Gain ($\times 10^6$)	λ, peak response (nm)	Quantum efficiency at peak (%)	Risetime (ns)
EMI	9807B	46	6.7	390	27	2
	9826B	13	3.3	350	20	2.2
	9829B	46	6.7	390	29	2
	9839B	45	6.7	390	27	3
Hammamatsu	R647-01	10	1	420	25	2.3
	R1449	460	10	420	22	18
	R1635	8	1	420	24	0.8
RCA	8575	46	14	385	31	2.5
	8850	46	14	385	31	2.5
	8854	114	40	400	27	3
	C31024	46	5	400	30	1.5
Phillips (Amperex)	XP 2020	44	30	400	26	1.5
	XP 2041	110	30	400	26	2.0
	XP 2230	44	30	400	26	1.6
	56AVP	44		437	17	2.1

Source: Thorn EMI Gencom, Inc., Plainview, NY; Hammamatsu Corp., Middlesex, NJ; RCA, Electro-optics, Lancaster, PA; Amperex Electronics Corp., Hicksville, NY.

Some relevant consideration of tube performance include the photocathode size, the spectral sensitivity, uniformity of response over the photocathode surface, and pulse spread for different electron trajectories. Some tubes are specially designed to have greater response in the ultraviolet region and may use quartz entrance windows. Two other important effects are afterpulsing and dark current. Afterpulsing arises when an ion from the residual gas in the tube or from one of the dynodes makes it back

Figure 7.9 Spectral response characteristics of the RCA 8575 photomultiplier tube. (Courtesy of RCA, New Products Division, Lancaster, PA.)

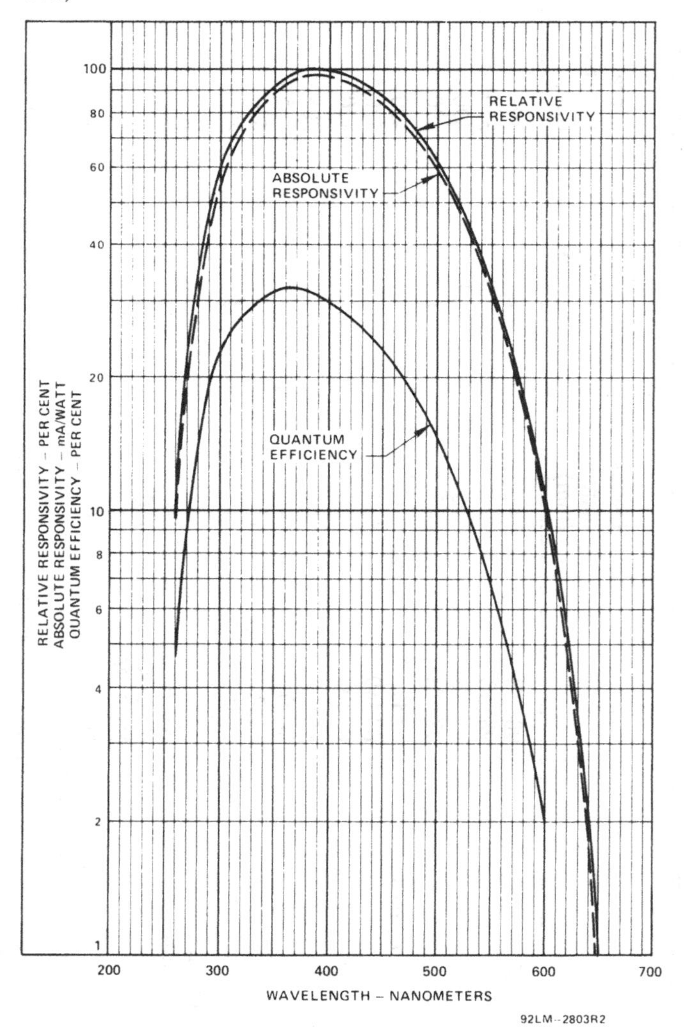

to the photocathode and initiates a second pulse. Dark current is the current in the tube when no input light is present. This arises chiefly from the thermal emission of electrons from the photocathode. Other sources include induced radioactivity, scintillation, and Cerenkov processes in the glass envelope of irradiated tubes [19] and potential differences between the glass envelope and photocathode. An electrostatic shield or mylar insulation is sometimes used to prevent current from flowing through the glass.

The high voltage distribution system is basically a resistor divider network. A typical circuit is shown in Fig. 7.10. Capacitances are used with the last dynodes to help maintain the high voltages. The current flowing through the divider to each dynode is the source of the additional electrons that enter the PMT pulse. Limitations on this current can cause the linearity of the tube to decrease for high incident light fluxes. In this case not enough current is available, and the amplification drops off. This effect can be reduced by connecting special high current sources to the last dynode stages. The high voltage supply must be very stable if the tube amplification is to remain constant. The high voltage setting may be determined with the aid of a small radioactive source by looking at the output pulses from the tube on a scope or pulse height analyzer. Often small "button" sources are wrapped inside the counter next to the scintillator. This allows convenient checking of the tube's gain.

PMTs should not be operated in a high helium environment since the gas can permeate the tube. When used continuously at a high current, the tube fatigues and loses amplification. The gain can usually be restored after a quiescent period.

Photomultiplier tubes are very sensitive to the presence of magnetic fields, which upset the delicate focusing conditions in the tube. For this reason, the tube is usually surrounded by a high permeability "mu metal" shield. Magnetic flux lines in the vicinity of the tube tend to be funneled into the shield. The shielding effect is enhanced by leaving a small gap between the tube and the shield. The shield should extend at least one shield diameter beyond the photocathode. For very strong magnetic fields multiple layers of shielding may be necessary. The outermost shield should be made of soft iron, which has a large saturation induction [19].

7.4 Performance

The most common use of scintillation counters is for triggering. The signals from the scintillators can be used by the fast electronics to decide whether to activate other apparatus, such as drift chambers, and whether to record the information from the event. Because of random

Figure 7.10 Typical photomultiplier tube base circuit. (Courtesy of RCA, New Products Division, Lancaster, PA.)

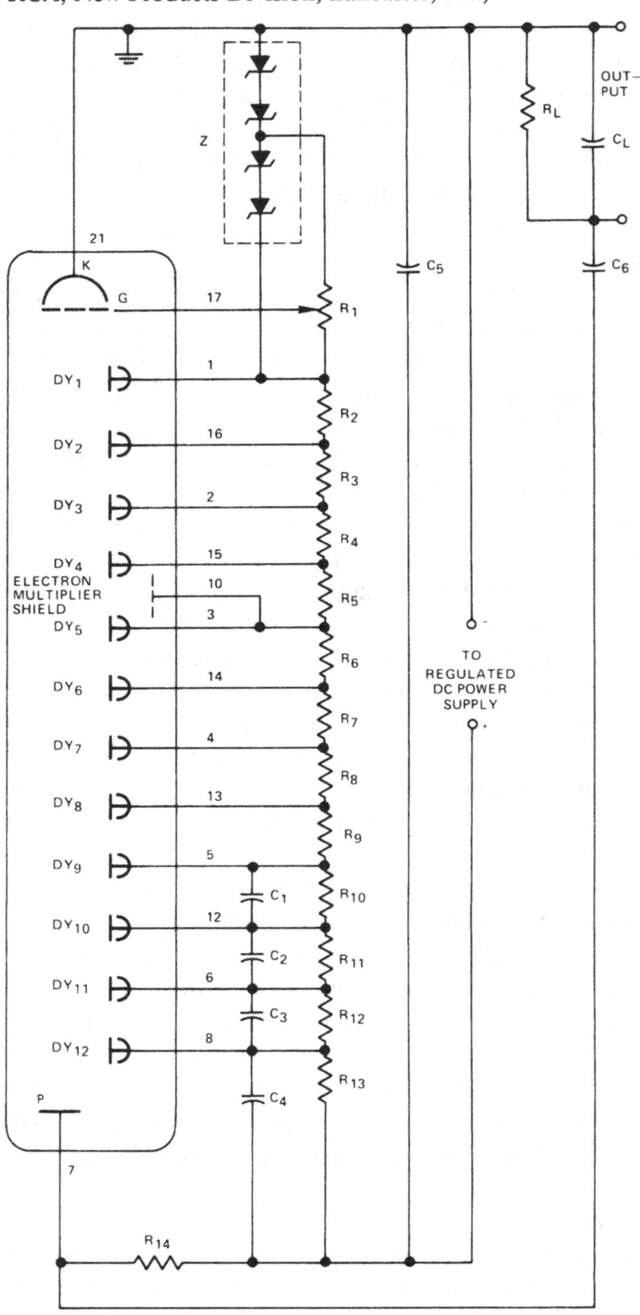

tube noise, the signal from the scintillation counter is required to arrive in coincidence with the signal from another element. Sometimes a scintillator signal may be required in anticoincidence with another signal. This is the case, for example, with a hole counter in a beamline, where we wish to eliminate particles in the beam halo. Anticounters could also be used around the target region when we wish to ensure the production of neutral particles.

A scintillator can be used to measure the differential energy loss dE/dx of particles. For small momenta the light output is proportional to the deposited energy. Gamma ray energies can be efficiently measured in NaI crystal scintillators. The high Z facilitates conversion into electron pairs that shower in the crystal. Liquid scintillators are commonly used as neutron detectors. The neutron interacts in the material, and charged particles that are produced initiate the scintillator light.

Important properties of scintillation counters include the efficiency, energy resolution, spatial resolution, and time resolution. We will discuss the first three items in this section and timing in the following section. The efficiency of a counter is ultimately determined by the number of photoelectrons emitted from the photocathode of the PMT. As an example, assume that we have a 1-cm-thick plastic scintillation counter. A minimum ionizing particle loses approximately 1.7 MeV/cm due to ionization while traversing the scintillator. Typically one photon of scintillation light will be emitted per 100 eV of deposited energy. This corresponds to a conversion of $\sim 2\%$ of the ionization energy to scintillation light. Thus, 17,000 photons should be emitted in the scintillator.

A number of losses prevent all of these photons from being detected by the phototube. Some of the light is attenuated in the scintillator and light guide. For example, NE-102 has an attenuation length of 250 cm. Much of the light strikes the edges at less than the critical angle and escapes, although some of this light is returned after reflection from the aluminum foil wrapping. Assuming a 10% light collection efficiency gives 1700 photons in our example, and assuming a quantum efficiency $\sim 25\%$ for the tube, about 400 of the photons will produce photoelectrons.

Rough design values

- Energy loss in plastic = 2 MeV/cm
- Scintillation efficiency = 1 photon/100 eV
- Collection efficiency = 0.10 (large counter)
- Quantum efficiency = 0.25

In the preceding example we calculated the expected or mean number of photoelectrons. The actual number will follow a Poisson probability distribution. The probability that no photoelectron is emitted is

$$\Pr(0) = e^{-\bar{n}} \tag{7.10}$$

In our example this is $e^{-400} \ll 1$ so that the counter is practically 100% efficient. Conversely, if we have measured the efficiency ϵ of a counter, we can make a crude estimate of the number of emitted photoelectrons from the relation

$$1 - \epsilon = e^{-\bar{n}}$$

so that

$$\bar{n} = -\ln(1-\epsilon) \tag{7.11}$$

Now let us consider several examples of the performance of actual counters. Mount et al. [20] constructed a set of highly efficient scintillation counters for use as veto counters at the CERN SPS. The inefficiency was made smaller than 10^{-6} by designing counters that could provide at least 14 photoelectrons per particle and by using burst guard discriminators. The amount of accidental vetoing was minimized by shortening the photomultiplier pulse with a resistive clip.

The energy resolution of a scintillation counter is usually determined by sampling the output pulses of the counter with a pulse height analyzer. The energy resolution $\Delta E/E$ is defined as

$$\Delta E/E \simeq \Delta M/\overline{M} \simeq (\bar{n})^{-1/2} \tag{7.12}$$

where $\overline{M}$ is the channel number of the signal peak and ΔM is the FWHM of the peak. The second equality follows from the assumption of Poisson statistics.

Scintillation counters have poor spatial resolution compared with other detectors, such as proportional or drift chambers. The best spatial resolution with scintillation counters occurs when many narrow counters are used in parallel. Such an arrangement is referred to as a scintillation hodoscope. Aubert et al. [21] have constructed a fine grained hodoscope for use in a beam at the CERN SPS. Each plane consisted of sixty 2-mm-wide counters. The spacing was chosen to match the resolution of the experimental apparatus in determining the vertex position. The length of each counter was determined by the beam size. The thickness of the counters was chosen to be 4 mm, so that the efficiency of the counter would exceed 99%. The counters were placed in two slightly overlapping planes to eliminate dead spaces due to the counter wrapping.

Long counters often use a PMT on each end. In this case measurements of the pulse height at the bottom (B) and top (T) can be used to estimate the position of the particle trajectory along the counter and in some cases the particle energy. Because of the exponential attenuation of the light traveling from the source to the ends, the position of the source above the counter midplane is

$$y \propto \ln(T/B) \tag{7.13}$$

In the region of relativistic rise, where $dE/dx \propto E$, we have

$$E \propto (BT)^{1/2} \tag{7.14}$$

Measurements of these quantities shown in Fig. 7.11 agree nicely with expectations.

As systems of scintillator hodoscopes or calorimeters become larger and larger, there are increasing problems with monitoring the pulse height and relative timing of each counter. These signals may drift or fail entirely over a period of time. For this reason, many large experiments have developed monitoring systems using LED pulsers or lasers that periodically test the response of all the counters to a known input pulse. Berglund et al. [22] have constructed a laser calibration system for use at CERN. They used an N_2 laser that produced $\sim 10^{14}$ photons in a 3-ns pulse. The light was directed into a bundle of one hundred and sixty 20-m-long optical fibers. The large number of produced photons is necessary because the solid angle for entering a given fiber is very small. The pulse to pulse light variation entering the fiber was $\pm 10\%$. The fibers were first cut to equal lengths and the delays measured. They were then cut a second time according to the measured delays. The final output of the fibers had a 150-ps (rms) time spread and 19% (rms) amplitude variation.

7.5 Timing applications

One of the important applications of scintillation counters is the measurement of time intervals at the nanosecond level. The resolution of a time interval measurement may be determined using the simple apparatus shown in Fig. 7.12. Two small counters S1 and S2 are used for triggering the system. Coincidences between S1 and S2 are used as a gate for an ADC and as a start signal for a TDC. The counter S3, which is being measured, is placed between S1 and S2. The anode pulse from the PMT is discriminated and used to stop the TDC. The dynode signal is linearly inverted and then sent to an ADC. The measured pulse height can be used to correct the TDC for slewing. The FWHM of the resulting spectrum of time intervals is a measure of the time resolution of the system.

Two groups of effects are mainly responsible for the time spread in

Figure 7.11 Position and energy information derived from a scintillator with tubes at the bottom (B) and top (T). (After C. Bromberg et al., Nuc. Phys. B 171: 1, 1980.)

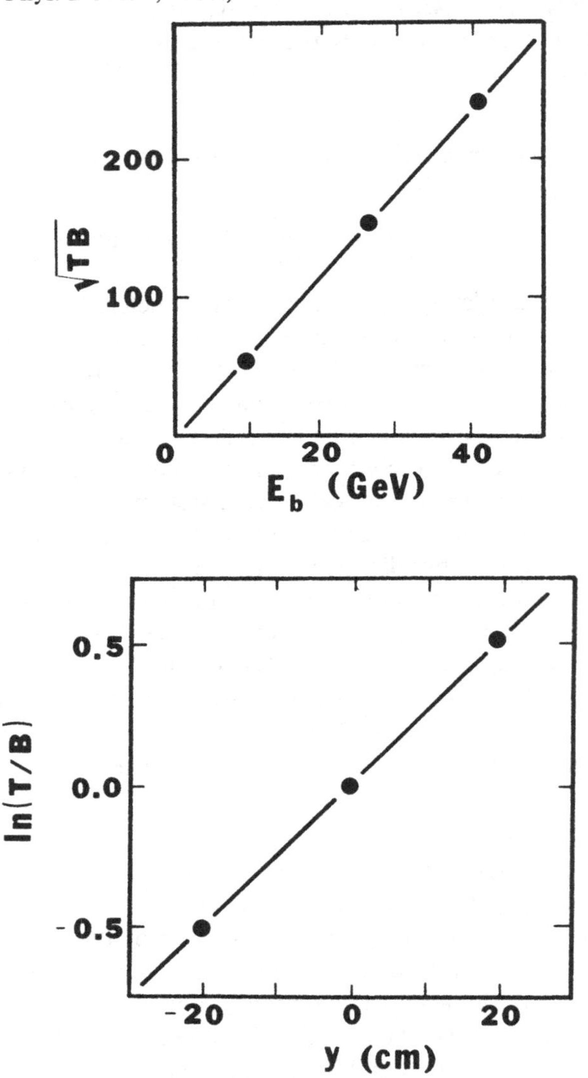

scintillation counter signals [23, 24]. The first group of limitations are concerned with variations in the time at which the detector responds to the radiation. Effects due to the scintillator include time variations in the energy transfer from the radiation to the optical levels of the scintillator and the finite decay time of the light emitting state. There are variations in transit time through the light collector. The resolution should be improved if only the light emitted straight to the PMT is utilized. This corresponds to the first part of the light pulse. This effect is one of the main reasons that small counters have significantly better time resolution (~ 50 ps) than do large ones (~ 250 ps). Effects due to the PMT include variations in the transit time from the photocathode to the first dynode due to the spread in initial photoelectron velocities and the location of the photoelectron emission on the photocathode and variations in electron amplification in the tube.

Figure 7.12 A simple arrangement for timing resolution measurements. (S1 – S3) scintillation counters, (D) discriminator, (C) coincidence unit, (LI) linear inverter, (a) anode signal, (d) dynode signal, (g) ADC gate, (s) TDC start.

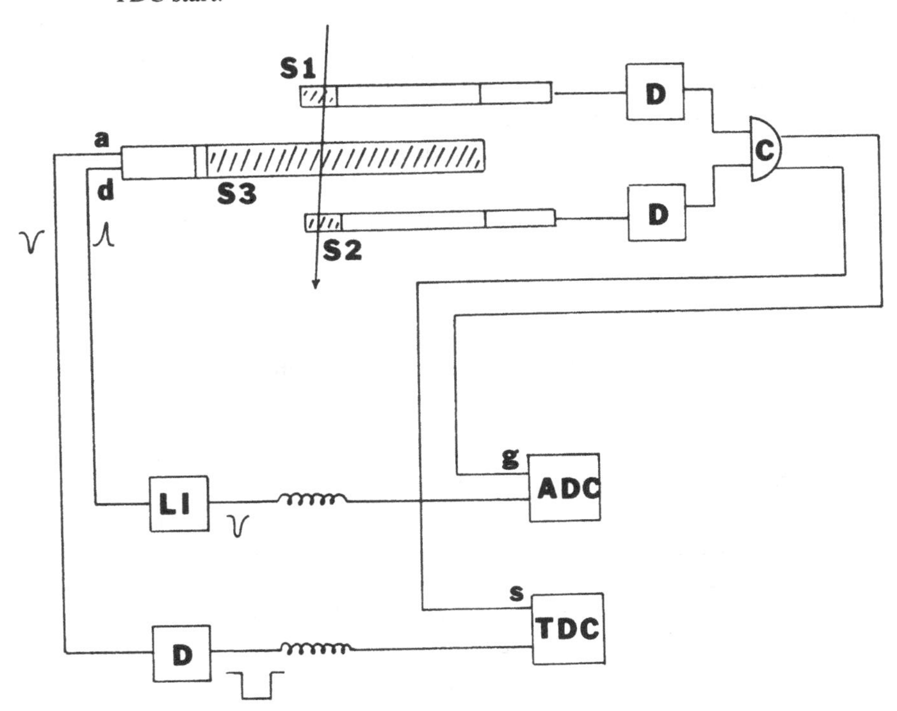

The photomultiplier transit time spread can be minimized by increasing the voltage between the photocathode and first dynode. A significant improvement in the time resolution of some tubes can also be obtained by taking the output pulse from an intermediate dynode stage instead of the anode [7]. This apparently results from the fact that electrons traveling from the next to last to the last dynode induce a signal on the anode of some tubes. The anode signal then consists of the directly collected charge plus the induced signal, resulting in a broadened pulse.

The second group of limitations are due to the electronics circuits used to process the PMT pulse. Some form of discriminator must be used to pick off a time from the pulse. The manner in which this is done can make a significant contribution to the timing resolution. Additional contributions come from electronic drift and thermal variations of the signal cable propagation velocity.

An important application of scintillation counters that makes use of their excellent timing capabilities is in time of flight (TOF) systems. This technique has been used for low momentum particles to help extract a signal from the background. For example, the TOF of the recoil proton in elastic scattering is correlated with its kinetic energy [25]. A scatter plot of the two quantities reveals a heavy band due to the protons over a scattered background due to accidentals. A cut around this band can be used to help purify the elastic scattering data sample.

A second and more common use of the TOF technique is to identify particle masses [26]. Imagine two scintillation counters separated by a distance L. A relativistic particle with momentum p will traverse the distance between the counters at a rate ($c=1$)

$$t/L = (3333/p)(p^2+m^2)^{1/2} \qquad \text{ps/m} \tag{7.15}$$

There will be a TOF difference for two equal momentum particles that have different masses. The time difference per flight path L is given by

$$(t_1-t_2)/L \simeq 1667(m_1^2-m_2^2)/p^2 \qquad \text{ps/m} \tag{7.16}$$

Figure 7.13 shows the time separation per meter between p and K, p and π, and K and π as a function of momentum. We note first of all that the time difference is very small. Even with a long 10-m flight path, the time difference between a 1-GeV/c pion and kaon is only 3.8 ns. The other important feature is that the time separation decreases like the square of the momentum. This limits the usefulness of this means of particle identification to momenta below ~4 GeV/c. Nevertheless, a large fraction of charged particles are produced with low momentum, even at the high energy accelerators. This method of particle identification is limited by the maximum permissible flight path, the timing resolution, and by systematic errors for large systems of counters.

A measured TOF spectrum for 1.4-GeV/c π and K is shown in Fig. 7.14. The arrival time of a scintillator signal is found to be related to its pulse height by [26]

$$\tau = t - K[(a_0)^{-1/2} - (a)^{-1/2}] - x/v \tag{7.17}$$

where τ is the corrected time, t is the measured time, a is the pulse height, a_0 is a reference pulse height, x/v is a position correction due to the finite

Figure 7.13 The time difference per unit flight path for πK, Kp, and πp as a function of momentum.

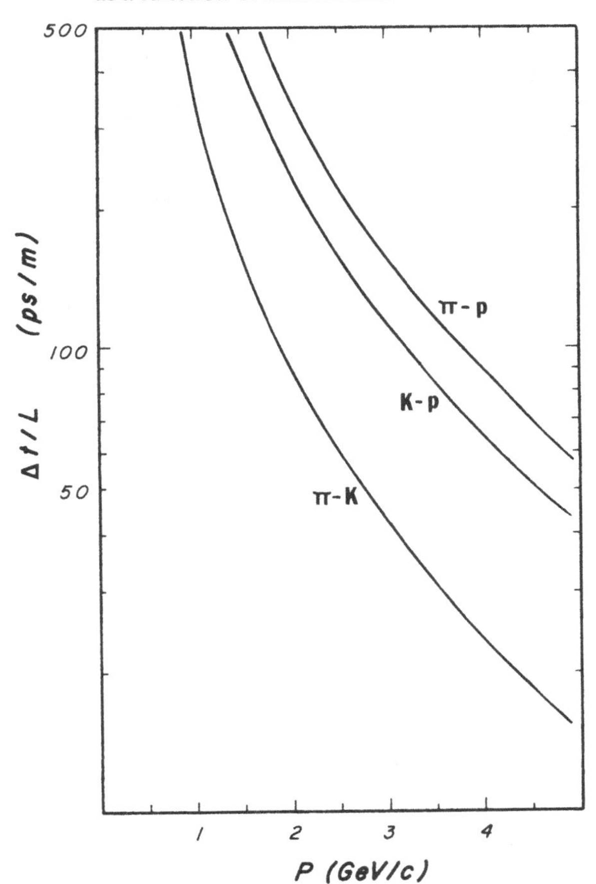

time required for light to travel a distance x with effective velocity v, and K is a free parameter. Note in Fig. 7.14 how the TOF distribution is sharpened with the application of the pulse height correction.

In timing applications attention must be paid to the risetime and FWHM of the light pulses created in the scintillator. Special scintillators such as NE-111 have been formulated especially for fast timing purposes. PMTs should be chosen to have as small a time jitter as possible. It has also been found that it is beneficial not to cover the entire counter with aluminum foil [15]. This reduces the overall efficiency, but the loss comes at the expense of slow photons, which make the pulse height correction less precise. The time resolution is found to be proportional to $(L/N_e)^{1/2}$, where L is the length of the counter and N_e is the mean number of expected photoelectrons [26].

It is fairly common for TOF systems to be incorporated in particle spectrometers. For example, the TASSO spectrometer at PETRA contains a large TOF array for low momentum particle identification [27]. There are a total of 96 separate 1.6-m-high by 33-cm-wide NE-110 scintillation counters. The start time originates from electrostatic pickups inside the beampipe, which signal a beam – beam crossing. Each counter

Figure 7.14 Time of flight spectra for 1.4-GeV/c π and K. Solid lines: raw data. Dashed lines: data corrected for signal pulse height. (After G. D'Agostini et al., Nuc. Instr. Meth. 185: 49, 1981.)

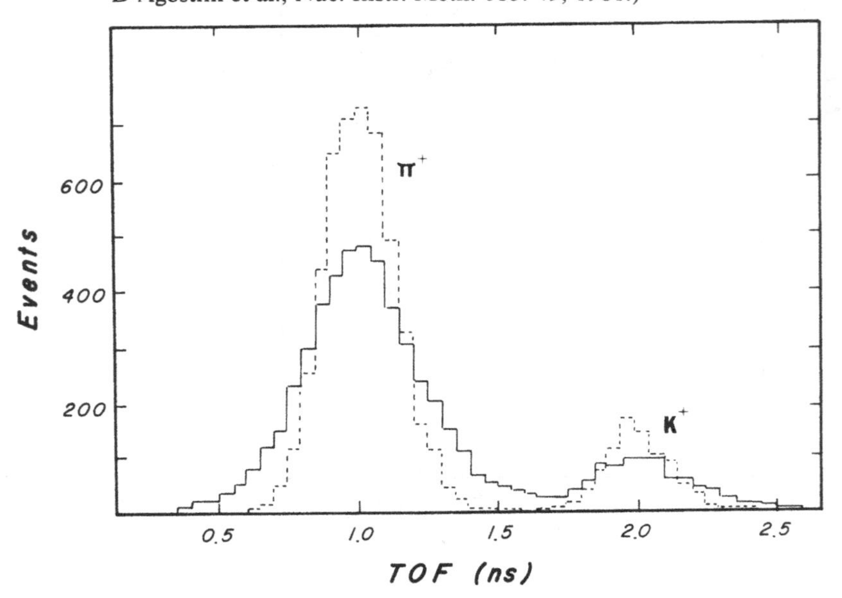

is viewed from both ends by EMI 9807B PMTs. The signal from each tube is fed into an ADC and a TDC. System operation is monitored using LED and spark gap, optical fiber arrangements. The flight path from the interaction point to the counters was 5.7 m. However, at a point approximately one-third of the way along this path, the particles had to pass through the coils of the TASSO magnet. The resolution of the system was ~450 ps, including a large contribution due to the multiple scattering of the particles passing through the magnet coil.

The MARK III detector at SPEAR used a 48-counter TOF system for particle identification up to 1.2 GeV/c particle momentum [28]. Each individual counter was 5 cm thick, 15.6 cm wide, and 318 cm long. The counters used Pilot F scintillator with XP2020 PMTs at both ends. The counters were mounted on the outer skin of the central drift chambers with a 1-cm styrofoam separator and were located in a 4-kG magnetic field. The perpendicular distance of the counters from the beamline was ~1.1 m. The light guides passed through holes in the iron end yoke. The system was monitored using an N_2 laser–optical fiber system. After all corrections were applied, the system achieved a time resolution of 170–190 ps, depending on the type of particle reaction.

References

[1] J. Birks, *The Theory and Practice of Scintillation Counting,* New York: Macmillan, 1964.
[2] E. Schram, *Organic Scintillation Detectors,* New York: Elsevier, 1963.
[3] R. Heath, R. Hofstadter, and E. Hughes, Inorganic scintillators: A review of techniques and applications, Nuc. Instr. Meth. 162: 431–76, 1979.
[4] A. Melissinos, *Experiments in Modern Physics,* New York: Academic, 1966.
[5] P. Pavlopoulos, M. Hasinoff, J. Repond, L. Tauscher, D. Troster, M. Rousseau, K. Fransson, T. Alberis, and K. Zioutas, The response of a 5 cm × 20 cm BGO crystal to electrons in the 150 to 700 MeV range, Nuc. Instr. Meth. 197: 331–4, 1982; an exhaustive study of BGO can be found in C. Holmes (ed.), International workshop on bismuth germanate, Princeton University report, 1982.
[6] F. Brooks, Development of organic scintillators, Nuc. Instr. Meth. 162: 477–505, 1979.
[7] B. Bengtson and M. Moszynski, Timing improved by the use of dynode signals studied with different scintillators and photomultipliers, Nuc. Instr. Meth. 204: 129–40, 1982.
[8] V. Eckardt, R. Kalbach, A. Manz, K. Pretzl, N. Schmitz, and D. Vranic, A novel light collection system for segmented scintillation counter calorimeters, Nuc. Instr. Meth. 155: 389–98, 1978.
[9] F. Klawonn, K. Kleinknecht, and D. Pollmann, A new type of acrylic scintillator, Nuc. Instr. Meth. 195: 483–9, 1982.
[10] S. Borenstein, R. Palmer, and R. Strand, Optical fibers and avalanche photodiodes for scintillation counters, Physica Scripta 23: 550–5, 1981.
[11] A. Policarpo, Light production and gaseous detectors, Physica Scripta 23: 539–49, 1981.

[12] B. Cox, G. Hale, P. Mazur, R. Wagner, D. Wagoner, H. Areti, S. Conetti, P. Lebrun, T. Ryan, J. Brau, and R. Gearhart, A measurement of the response of an SCG1-C scintillation glass shower detector to 2–17.5 GeV positrons, Nuc. Instr. Meth. 219: 487–90, 1984.
[13] G. Keil, Design principles of fluorescence radiation converters, Nuc. Instr. Meth. 89: 111–23, 1970.
[14] D. Crabb, A. Dean, J. McEwen, and R. Ott, Optimum design of thin, large area scintillation counters, Nuc. Instr. Meth. 45: 301–8, 1966.
[15] G. D'Agostini, G. Marini, G. Martellotti, F. Massa, A. Rambaldi, and A. Sciubba, High resolution time of flight measurements in small and large scintillation counters, Nuc. Instr. Meth. 185: 49–65, 1981.
[16] R. Garwin, The design of liquid scintillation cells, Rev. Sci. Instr. 23: 755–7, 1952.
[17] H. Hinterberger and R. Winston, Efficient light coupler for threshold Cerenkov counters, Rev. Sci. Instr. 37: 1094–5, 1966.
[18] W. Welford and R. Winston, *The Optics of Non-Imaging Concentrators,* New York: Academic, 1978.
[19] D. Ritson, Scintillation and Cerenkov Counters, in D. Ritson (ed.), *Techniques of High Energy Physics,* New York: Interscience, 1961, Chap. 7.
[20] R. Mount, P. Cavaglia, W. Stockhausen, and H. Wahlen, Highly efficient scintillation counters, Nuc. Instr. Meth. 160: 23–7, 1979.
[21] J. Aubert, G. Bassompierre, G. Coignet, J. Crespo, Y. Declais, L. Massonnet, M. Moynot, P. Payre, C. Peroni, J. Thenard, and M. Schneegans, A high resolution beam hodoscope counter for use in very intense beams, Nuc. Instr. Meth. 159: 47–51, 1979.
[22] S. Berglund, P. Carlson, and J. Jacobson, A laser based time and amplitude calibration system for a scintillation counter hodoscope, Nuc. Instr. Meth. 190: 503–9, 1981.
[23] M. Moszynski and B. Bengtson, Status of timing with plastic scintillation detectors, Nuc. Instr. Meth. 158: 1–31, 1979.
[24] P. Carlson, Prospects of scintillator time of flight systems, Physica Scripta 23: 393–6, 1981.
[25] R. Cool, K. Goulianos, S. Segler, G. Snow, H. Sticker, and S. White, Elastic scattering of $p^{\pm}$, $\pi^{\pm}$, and $K^{\pm}$ on protons at high energies and small momentum transfer, Phys. Rev. D 24: 2821–8, 1981.
[26] W. Atwood, Time of flight measurements, SLAC-PUB-2620, 1980.
[27] K. Bell, J. Blissett, B. Foster, J. Hart, A. Parham, B. Payne, J. Proudfoot, D. Saxon, and P. Woodworth, A large scintillator array for particle identification by time of flight, Nuc. Instr. Meth. 179: 27–38, 1981.
[28] J. Brown, T. Burnett, V. Cook, C. Del Papa, A. Duncan, P. Mockett, J. Sleeman, D. Wisinski, and H. Willutzki, The MARK III time of flight system, Nuc. Instr. Meth. 221: 503–22, 1984.

Exercises

1. Find the critical angle for (a) a plastic scintillator–air interface, (b) a lucite–air interface, and (c) a plastic scintillator–lucite interface.

2. Consider a rectangular slab of scintillator. Show that a fraction

$$f_t = [3(n^2 - 1)^{1/2}/n] - 2$$

of the isotropically emitted light inside the slab is reflected at all surfaces and is trapped inside the slab, provided that the index of refraction $n \geqslant 1.414$.

3. Estimate the attenuation length of the test scintillator used in Fig. 7.5.
4. Derive Eqs. 7.7, 7.8, and 7.9 for the Winston cone.
5. Find the mean number of collected photoelectrons corresponding to a detection efficiency of 99.9%. What is the probability of observing half this number of photoelectrons?
6. Estimate how many photons are produced in a 1-cm-thick plastic scintillator by (a) 1-GeV/c protons, (b) 100-MeV/c protons, and (c) 10-GeV/c protons.
7. Design a scintillation counter hodoscope for a hadron beamline with 10^6 particles/sec. and maximum extension of 10×10 cm. Specify a commercially available scintillator and photomultiplier. Make a sketch of the light guide.
8. A plastic scintillator of length L has PMTs directly coupled to both sides. If a passing particle produces a normalized pulse height of 100 mV on the left and 60 mV on the right, find the position of the particle along the counter. How is the apparent position of the particle changed if there are light guides of length $L/4$ on each side with attenuation length $\lambda_{LG} = 2\lambda_{SC}$?
9. What is the difference in flight times for 3-GeV muons and electrons over a 20-m path?
10. What path length would be required to identify 1-GeV/c pions from kaons at the 90% confidence level ($\sim 3\ \sigma$) using time of flight if the timing resolution is 250 ps?

8
Cerenkov counters

The Cerenkov (or Cherenkov) effect occurs when the velocity of a charged particle traversing a dielectric medium exceeds the velocity of light in that medium. A small number of photons will be emitted at a fixed angle, which is determined by the velocity of the particle and the index of refraction of the medium. The light may then be collected onto a photomultiplier tube to form a counter. A threshold Cerenkov counter detects the presence of a particle whose velocity exceeds some minimum amount, while a differential Cerenkov counter can measure the velocity of a particle within a certain range. Lead-glass Cerenkov counters are widely used for the detection of photons.

8.1 The Cerenkov effect

Early investigations of radioactivity had observed that radioactive substances produced a pale light when placed in certain liquids. The effect is named after the Russian physicist Cerenkov, who conducted a thorough investigation of the phenomenon in the 1930s. The Cerenkov effect occurs when the velocity of a charged particle exceeds the velocity of light in a dielectric medium (c/n), where n is the index of refraction for the medium [1, 2]. Excited atoms in the vicinity of the particle become polarized and coherently emit radiation at a characteristic angle θ, determined from the relation

$$\cos\theta = 1/\beta n \tag{8.1}$$

with $\beta > 1/n$.

The index of refraction of materials is a function of wavelength and temperature. Figure 8.1 shows the variation of n with wavelength for several materials. There is a general tendency for n to decrease with increasing λ. The variation $dn/d\lambda$ is referred to as dispersion. The dispersion is largest in the ultraviolet (*uv*) portion of the spectrum. The variation with temperature is generally small, for example, $dn/dT \simeq -15 \times 10^{-6}/$ °C for BaF_2 in the visible spectrum.

Values of $\cos\theta$ as a function of the particle's velocity β are shown in Fig. 8.2 for various values of n. Note that according to Eq. 8.1 there is a threshold velocity $\beta_t = 1/n$ below which no light is emitted. As the particle velocity increases beyond β_t, the light is given off at larger and larger angles up to a maximum $\theta_{max} = \cos^{-1}(1/n)$ which occurs for $\beta = 1$.

The threshold relation given above is strictly only true for an infinite radiator medium. In this case the angular distribution of emitted radiation is a delta function at the angle θ. For finite radiation lengths a more complete treatment shows that the angular distribution contains secondary minima and maxima [3]. However, in this discussion we will ignore any fine structure in the Cerenkov spectrum.

The Huyghen's construction derivation of Eq. 8.1 is shown in Fig. 8.3. The positions of the particle and expanding spheres of radiation are shown for four instances of time. The tangent planes to the spheres represent a plane electromagnetic wave propagating through the medium at an

Figure 8.1 Index of refraction of various materials as a function of wavelength. The vertical dashed line is the location of the sodium D line, which is often used for quoting n values. (Data from *American Institute of Physics Handbook,* 3rd ed., New York: McGraw-Hill, 1972.)

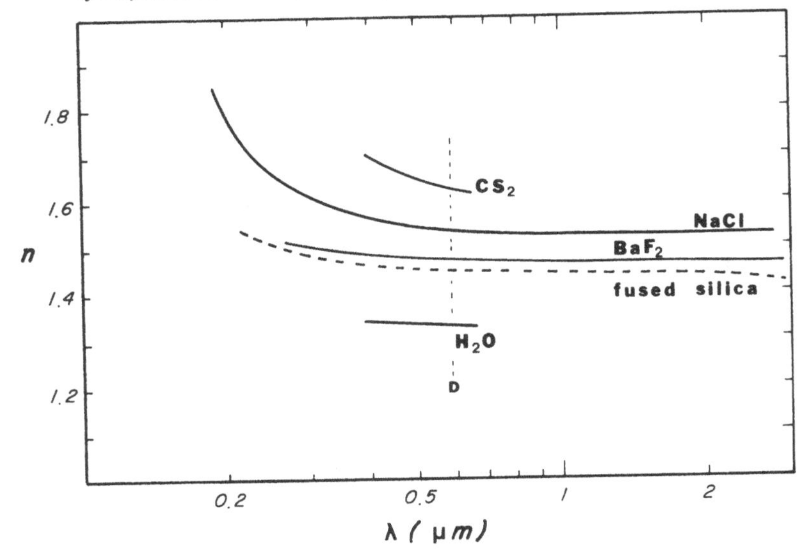

angle θ with respect to the particle's trajectory. It can be seen that a coherent wavefront can be constructed when $\beta > \beta_t$ but not when $\beta < \beta_t$.

An explanation of the effect using classical electrodynamics was given in 1937 by Frank and Tamm [1, 2]. They considered an electron moving with constant velocity through a perfect, isotropic dielectric. All the properties of the medium were encompassed in the dielectric constant or, equivalently, in the index of refraction. The solution of Maxwell's equations for the case $\beta n > 1$ has the form of a traveling electromagnetic wave.

The amount of energy emitted per unit length and per unit frequency interval $d\omega$ by a particle with charge Ze is given by

$$\frac{dE}{dx\, d\omega} = \frac{Z^2 r_e mc^2}{c^2} \left(1 - \frac{1}{\beta^2 n^2}\right) \omega \tag{8.2}$$

where r_e is the classical radius of the electron. Note that the emitted radiation increases linearly with the frequency ω and with the square of the particle's charge. For the following discussion we consider singly charged particles. It is usually more convenient to express Eq. 8.2 in terms of the wavelength λ of the radiation

$$\frac{dE}{dx\, d\lambda} = 4\pi^2 r_e mc^2 \frac{1}{\lambda^3} \left(1 - \frac{1}{\beta^2 n^2}\right) \tag{8.3}$$

Figure 8.2 Cerenkov angle θ versus particle velocity β for various values of the refractive index n.

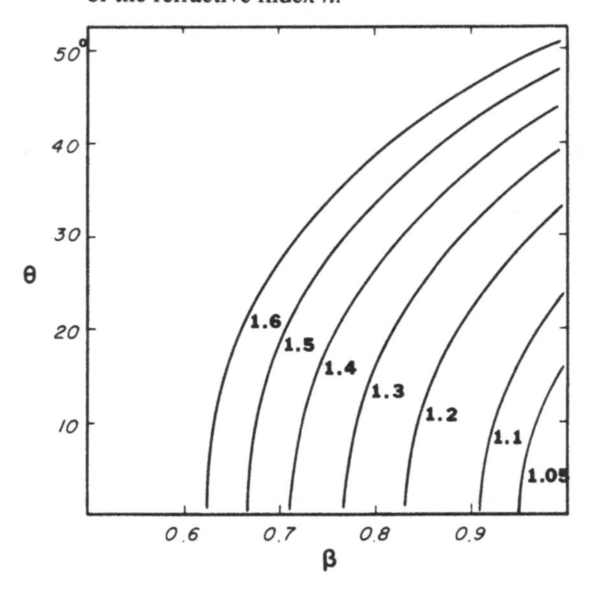

The emitted energy is strongly peaked at short wavelengths. We can rewrite Eq. 8.3 in terms of the number N of emitted photons as

$$\frac{dN}{dx\,d\lambda} = 2\pi\alpha\frac{1}{\lambda^2}\left(1 - \frac{1}{\beta^2 n^2}\right) \tag{8.4}$$

where α is the fine structure constant. The total number of photons emitted per unit path length is

$$\frac{dN}{dx} = 2\pi\alpha \int_{\beta n > 1}\left(1 - \frac{1}{\beta^2 n^2(\lambda)}\right)\frac{d\lambda}{\lambda^2} \tag{8.5}$$

Figure 8.3 Huyghen's construction of the expanding spherical wavefronts at three instances of time.

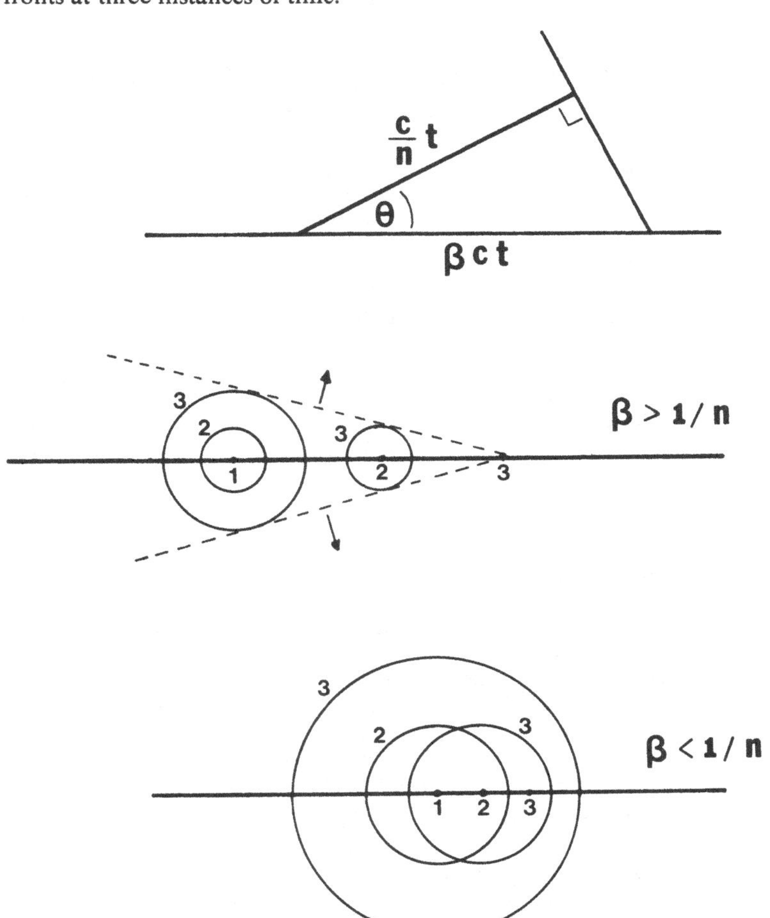

In practice, the condition $\beta n > 1$ is only satisfied for the ultraviolet to near infrared portion of the electromagnetic spectrum.

If the variation in $n(\lambda)$ is small over the wavelength region λ_1 to λ_2, the energy emitted per unit length becomes

$$\frac{dE}{dx} = 2\pi^2 r_e mc^2 \sin^2\theta \left(\frac{1}{\lambda_1^2} - \frac{1}{\lambda_2^2}\right) \tag{8.6}$$

while the photon yield is

$$\frac{dN}{dx} = 2\pi\alpha \sin^2\theta \left(\frac{1}{\lambda_1} - \frac{1}{\lambda_2}\right) \tag{8.7}$$

For example, using the wavelength interval 350–500 nm, corresponding roughly to the response range of photomultiplier tubes with Sb–Cs pho-

Table 8.1. *Properties of liquid and solid radiators*

Material	Formula	n^a	ρ (g/cm^3)
FC-75	$C_8F_{16}O$	1.276	1.76
Water	H_2O	1.333	1.00
1-Butanol	C_4H_9OH	1.397	0.810
Carbon tetrachloride	CCl_4	1.459	1.591
Glycerol	$C_3H_5(OH)_3$	1.474	1.26
Toluene	$C_6H_5CH_3$	1.494	0.867
Styrene	$C_6H_5CHCH_2$	1.545	0.910
Carbon disulfide	CS_2	1.628	1.263
Diiodomethane	CH_2I_2	1.749	3.325
Lucite	$C_5H_8O_2$	1.49	1.16–1.20
Plastic scintillator	—	1.58	1.03
Pilot 425	—	1.49	1.19
Crystal quartz	SiO_2	1.54	2.65
Borosilicate glass		1.474	2.23
Lithium fluoride	LiF	1.392	2.635
Barium fluoride	BaF_2	1.474	4.89
Sodium iodide	NaI	1.775	3.667
Cesium iodide	CsI	1.788	4.51
Bismuth germanate	$Bi_4Ge_3O_{12}$	2.15	7.13
Sodium chloride	NaCl	1.544	2.165
Silica aerogel	$n(SiO_2) + 2n(H_2O)$	$1 + 0.25\rho$	0.1–0.3

[a] For sodium light at 20–25°C.

Source: Handbook of Chemistry and Physics, 64th ed., Boca Raton: CRC Press, 1983; Particle Data Group, Rev. Mod. Phys. 56: S1, 1984; *American Institute of Physics Handbook,* 3rd ed., New York: McGraw-Hill, 1972; Catalog, Nuclear Enterprises.

tocathodes deposited on glass, we find

$$\begin{aligned} dE/dx &= 1180 \sin^2\theta \quad \text{eV/cm} \\ dN/dx &= 390 \sin^2\theta \quad \text{photons/cm} \end{aligned} \tag{8.8}$$

Indices of refraction of some common liquids and solids are given in Table 8.1. For a singly charged particle with $\beta \sim 1$ traversing water ($n = 1.33$), the Cerenkov angle is 41.2°. This implies that 513 eV/cm is given off as Cerenkov radiation. Note that this is small compared to the ~ 2 MeV/cm ionization energy loss and 100 times weaker than the light output in plastic scintillator. On the average, 170 photons will be emitted per centimeter of path length. Examining Table 8.1, note that FC-75 is a fluorocarbon with the lowest index of refraction for a room temperature liquid.

The Cerenkov light is emitted almost instantaneously. The angular distribution of the light intensity is approximately a δ function at the Cerenkov angle. The actual distribution is broadened due to dispersion, energy loss of the particle, multiple scattering, and diffraction.

8.2 Photon yield

We have seen that an important aspect of Cerenkov radiation is the small light yield compared to a scintillation counter of comparable thickness. The small photon yield means that the pulses from the counter are small with large fluctuations. Thus, every effort must be made to collect as many photons as possible. Since the bulk of the deposited Cerenkov energy is in the ultraviolet portion of the spectrum, special steps can be taken to see that these photons are accepted. If we want to ensure that fluctuations in the number of photoelectrons emitted from the photomultiplier do not play a significant role in determining the efficiency of the counter, then the counter should be designed so that at least 20 photoelectrons are emitted per particle.

Typical threshold and differential counters are shown in Fig. 8.4. The Cerenkov radiator should be transparent to the emitted radiation over the desired wavelength range. The radiator should not produce scintillation light. It is desirable to have a large index of refraction since the intensity $I \propto 1 - (\beta n)^{-2}$. In addition, the radiator should have small density and atomic number in order to minimize ionization loss and multiple scattering. The refractive indices of the radiator, optical grease, and PMT window should be as identical as possible. For example, for $n_{rad} = 1.6$ and $n_{grease} = 1.4$, a particle with $\beta \sim 1$ and incident at an angle $>30°$ to the photocathode plane, over 50% of the light is reflected [2].

Figure 8.4 Construction of typical threshold and differential Cerenkov counters. (PM) photomultiplier tube, (M) mirror, (P) particle moving along the counter axis, and (d) slit diaphragm. (J. Litt and R. Meunier, adapted with permission from the Annual Review of Nuclear Science, Vol. 23, © 1973 by Annual Reviews Inc.)

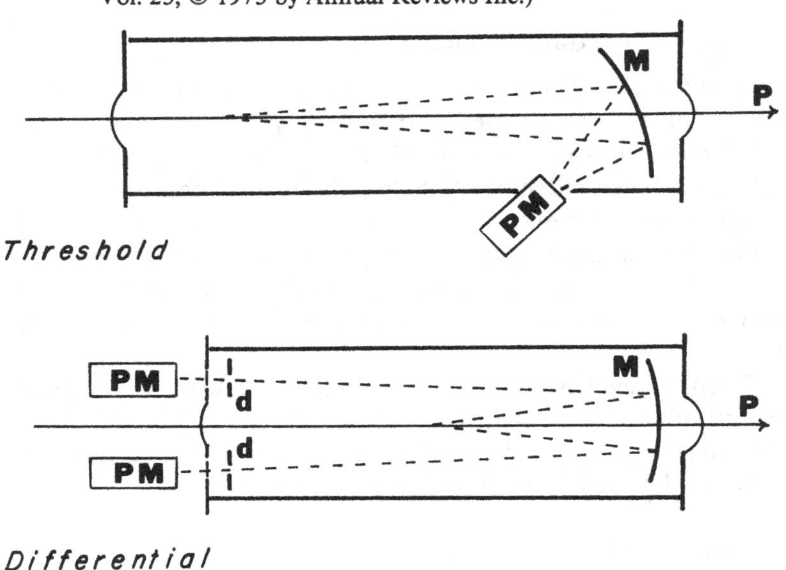

Figure 8.5 The reflectance at normal incidence from metallic surfaces. (Data from *Handbook of Chemistry and Physics,* 64th ed., Boca Raton: CRC Press, 1983.)

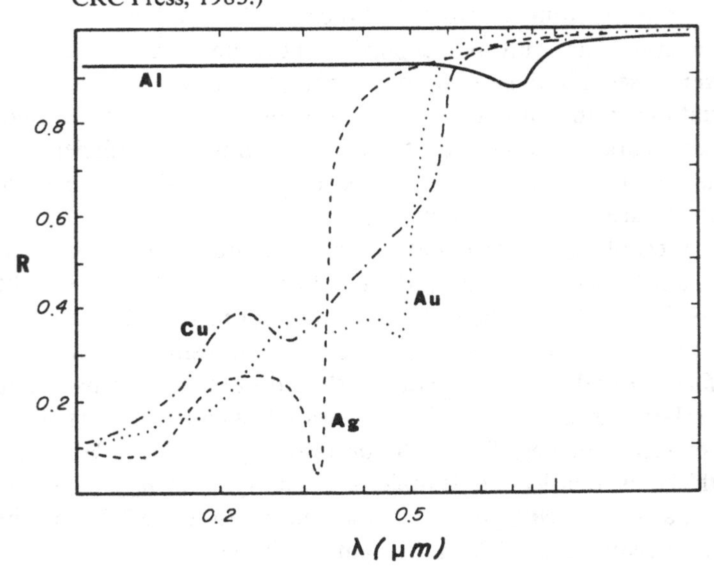

The light collection system should be as efficient as possible. This requirement is aided by the directionality exhibited by Cerenkov radiation. Simple optical arrangements can be used to gather the produced light and focus it onto the face of the PMT. The optics should be designed to minimize the light lost in reflections. The reflection coefficient for light at normal incidence on metallic surfaces is shown in Fig. 8.5. In the visible spectrum an evaporated gold layer has high reflectance. However, in the near ultraviolet region aluminum has a much higher reflectance. White paper or layers of MgO also have good reflectance in the visible. Thin layers of MgF_2 can be applied to surfaces to alter the ultraviolet reflection properties [4].

Another important consideration is transmission of the light through any windows in the optical system [5]. Transparent materials generally have absorption bands in the infrared and ultraviolet portions of the spectrum. Table 8.2 lists the regions over which various materials have a

Table 8.2. *Transmittance regions of various materials*[a]

Material	Lower limit (μm)	Upper limit (μm)
Magnesium fluoride	0.11	7.5
Lithium fluoride	0.12	9.0
Calcium fluoride	0.13	12
Sodium fluoride	0.19	15
Sodium chloride	0.21	26
Potassium chloride	0.21	30
Silver chloride	0.4	28
Potassium bromide	0.25	40
Potassium iodide	0.25	45
Cesium iodide	0.25	80
Magnesium oxide	0.25	8.5
Crystal quartz	0.12	4.5
Fused silica	0.12	4.5
Borosilicate glass	0.4	3.5
Sapphire	0.14	6.5
Diamond	0.25	80
Gallium arsenide	1.0	15
Silicon	1.2	15
Germanium	1.8	23

[a] Region between limits of 10% external transmittance for 2-mm-thick samples.
Source: American Institute of Physics Handbook, 3rd ed., New York: McGraw-Hill, 1972.

transmission greater than 10%. For efficient detection of ultraviolet light it is obviously important that the lower transmission limit be as small as possible. Ordinary borosilicate glass only transmits down to ~ 300 nm. Quartz, fused silica, and MgF_2 crystals can be used as window materials to maximize the acceptance of ultraviolet light.

High quality PMTs are necessary for single particle detection. It may be necessary to work with electronic thresholds below the single photoelectron level. The tube should have a high quantum efficiency, low dark current, good signal to noise, and no afterpulsing. The output of the tube can be used in a coincidence arrangement to minimize accidental counting. The window of the tube should have good ultraviolet transmission. A PMT window made of quartz or fused silica can extend the light transmission below 200 nm. An antireflection layer of MgF_2 can be deposited on the window. Special photocathode materials can also help extend the photomultiplier acceptance into the ultraviolet.

The photoelectron output of a given PMT is obtained by convoluting the frequency spectrum of produced Cerenkov radiation with the frequency response of the collection system and tube. Thus, using Eq. 8.4 for the number of photons produced per unit path, we find that the number of emitted photoelectrons in the tube per unit particle pathlength is

$$dN_e/dx = 2\pi\alpha \int \left(1 - \frac{1}{\beta^2 n^2}\right) \epsilon_c(\lambda) \frac{S(\lambda)}{\lambda^2}\, d\lambda \qquad (8.9)$$

where $\epsilon_c(\lambda)$ is the efficiency for collecting photons of wavelength λ at the cathode and $S(\lambda)$ is the normalized relative response of the photomultiplier. The number of photoelectrons is frequently written in the form

$$N_e = N_0 L \sin^2\theta \qquad (8.10)$$

where L is the length of the radiator and the various efficiencies and spectral responses are incorporated in the constant N_0.

An alternative method of enhancing the photon yield is to use a wavelength shifter to shift the ultraviolet region of the production spectrum to the visible region, where ordinary PMTs are most sensitive. A small concentration of photoluminescent atoms can be added to the radiator, window, or PMT face. The shifter tends to make the radiation more isotropic, and hence the collimation properties of the Cerenkov light are lost when the shifter is added to the radiator. However, this is not a problem when the shifter is applied to the PMT window [2].

8.3 Gas radiators

Cerenkov counters using a gas radiator are particularly useful for detecting particles with $\beta > 0.99$. The refractive indices of gases in the visible and ultraviolet depend critically on the presence of absorption bands. The index of refraction of the gas is related to its density ρ through the Lorenz–Lorentz law [2]

$$\frac{n^2 - 1}{n^2 + 2} \frac{M}{\rho} = R \tag{8.11}$$

where M is the molecular weight and R is the molecular refraction coefficient. The constant R approximately equals the volume occupied by 1 mol of the gas excluding empty space.

Since for gases $n \simeq 1$, we can rewrite Eq. 8.11 to a high degree of accuracy as

$$n - 1 \simeq \frac{3}{2} \frac{R}{M} \rho \tag{8.12}$$

From the ideal gas law we have

$$P = \rho R' T / M \tag{8.13}$$

where P is the pressure, T is the absolute temperature, and R' the gas constant. Substituting into Eq. 8.12, we obtain

$$n - 1 = (n_0 - 1) P / P_0 \tag{8.14}$$

where the subscript 0 indicates that the quantity is measured at atmospheric pressure. Using the notation $\eta = n - 1$, we can express this as

$$\eta = \eta_0 P / P_0 \tag{8.15}$$

The indices of refraction of a number of gases are listed for atmospheric pressure in Table 8.3. Also listed are the critical temperature and critical pressures of the gases. The critical temperature is the temperature above which a gas cannot be liquified using pressure alone. The critical pressure is the pressure at which gas and liquid exist in equilibrium when at the critical temperature.

The threshold of a gas counter can be adjusted by varying the pressure. The following relations are useful in understanding the properties of gas counters as a function of momentum ($= m\beta\gamma$).

$$\sin^2\theta = \beta_t^2 \left(\frac{1}{\beta_t^2 \gamma_t^2} - \frac{1}{\beta^2 \gamma^2} \right) \tag{8.16}$$

$$\beta_t \gamma_t = \frac{1}{\sqrt{n^2 - 1}} \approx \frac{1}{\sqrt{2\eta}} \tag{8.17}$$

Using Eq. 8.15, we find that

Table 8.3. *Properties of gas radiators, STP*

Gas	Formula	η_0[a] ($\times 10^{-4}$)	θ_{max} (degrees)	dN/dl γs/cm (350–500 nm)	T_{cr} (°C)	P_{cr} (atm)
Helium	He	0.35	0.48	0.027	−268	2.3
Neon	Ne	0.67	0.66	0.052	−229	26
Hydrogen	H_2	1.38	0.95	0.11	−240	13
Oxygen	O_2	2.72	1.33	0.21	−119	50
Argon	Ar	2.84	1.36	0.22	−122	48
Nitrogen	N_2	2.97	1.40	0.23	−147	34
Methane	CH_4	4.41	1.70	0.34	−83	46
Carbon dioxide	CO_2	4.50	1.72	0.35	31	73
Ethylene	C_2H_4	6.96	2.14	0.54	10	51
Ethane	C_2H_6	7.06	2.15	0.55	32	49
Freon 13	$CClF_3$	7.82	2.27	0.61	29	38
Sulfur hexafluoride	SF_6	7.83	2.27	0.61	46	37
Propane	C_3H_8	10.05	2.57	0.78	22	9
Freon 12	CCl_2F_2	11.27	2.72	0.88	112	41
Freon 114	$C_2Cl_2F_4$	14	3.03	1.09	146	32
Pentane	C_5H_{12}	17.1	3.3	1.3	197	33

[a] At 589 nm.

Source: J. Jelley, *Cerenkov Radiation and its Applications,* London: Pergamon, 1958; V. Zrelov, *Cherenkov Radiation in High Energy Physics,* Federal Scientific and Technical Information, Springfield, VA, 1970; W. Galbraith. Cerenkov Counters, in High Energy and Nuclear Physics Data Book, Rutherford Laboratory, 1963.

$$\beta_t \gamma_t \simeq \left(\sqrt{2\eta_0 \frac{P}{P_0}} \right)^{-1} \tag{8.18}$$

We see that increasing the pressure brings the momentum threshold down like $1/\sqrt{P}$. The threshold dependence of several gases determined from Eq. 8.18 is shown in Fig. 8.6.

For a beam of fixed velocity particles the intensity of the Cerenkov radiation varies with pressure like

$$\sin^2\theta = 1 - \frac{1}{\beta^2(1 + kP)^2} \tag{8.19}$$

Figure 8.6 Thresholds for gases versus absolute pressure.

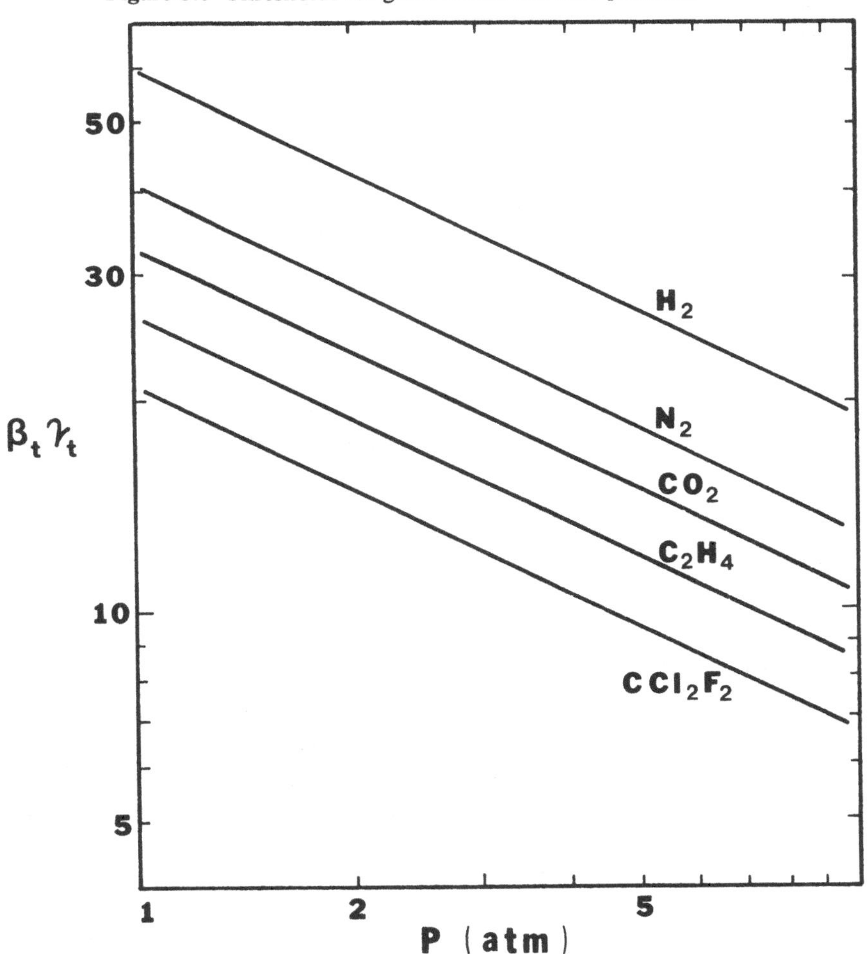

where

$$k = (n_0 - 1)/P_0 \tag{8.20}$$

The intensity is 0 until the pressure satisfies $\beta n = 1$. Then, following a threshold knee, we have ($\beta \approx 1$)

$$\sin^2\theta \simeq 2kP \tag{8.21}$$

and thus the intensity goes up linearly with pressure.

A simple pressure manometer is usually not sufficiently accurate for determining the index of refraction [2]. The pressure may be determined to greater precision using a pressure sensitive capacitance transducer, or the index of the gas may be measured directly using an interferometer.

A simple interferometer may be constructed by connecting a gas line from the counter to a small chamber containing accurately positioned Fabry–Perot plates. Light from a laser will undergo multiple reflections between the plates. A photodiode located at a fixed angle on the opposite side of the chamber from the laser can be used to measure the relative amount of transmitted light. If the gas in the chamber is slowly evacuated, the conditions for constructive interference change, resulting in a series of light and dark fringes appearing at the photodiode. The number of such fringes can be simply related to the absolute value of the refractive index.

8.4 Threshold counters

We have seen that no Cerenkov radiation is produced until the velocity of the particle exceeds a minimum value $\beta_t = 1/n$. Because of this, a Cerenkov counter can be used as a threshold device to indicate the presence of a particle whose velocity exceeds a certain minimum value. Sets of counters are commonly used this way to identify particles with different masses in a beam of fixed momentum.

Gas counters are particularly advantageous for high momentum particles since indices of refraction near 1 are required, and the indices of gases near this value can be controlled by varying the pressure of the gas. A simple threshold counter was shown in Fig. 8.4. The radiator gas is contained in a high pressure, light tight container. Cerenkov light is reflected by the mirror and transmitted through a pressure window onto the PMT. As the pressure is raised, thresholds for various mass particles are passed, and the counting rate passes through plateaus. The levels of the plateaus allow a determination of the relative particle abundances in a beam.

A class of Cerenkov counters uses total internal reflection as a filter for the collection of light [1, 2]. Consider a particle traversing normally a wide, relatively thin slab of material bounded on the exit side by air. The

Cerenkov light in the material will pass through the exit face provided that the Cerenkov angle is smaller than the critical angle. Thus, the detected particles satisfy the relation

$$\cos^{-1}(1/n\beta) \leq \sin^{-1}(1/n) \tag{8.22}$$

The counter detects particles over a velocity range starting at $\beta_t = 1/n$ and continuing up to

$$\beta_{max} = (n^2 - 1)^{-1/2} \tag{8.23}$$

instead of $\beta_{max} = 1$. Total internal reflecting counters have been constructed using lucite as a radiator.

The detection efficiency of a threshold counter increases rapidly as the velocity of the incident particles increases over the calculated threshold velocity. The detection efficiency can be found experimentally with the counter arrangement in Fig. 8.7. The Cerenkov counter efficiency is given by the ratio of coincidences $S_1 \cdot S_2 \cdot C/S_1 \cdot S_2$.

The detection efficiency may be calculated as

$$\epsilon(\beta) = 1 - \Pr(0, N_e) \tag{8.24}$$

where $\Pr(0, N_e)$ is the probability that no electrons were emitted by the photocathode of the PMT if the average number is N_e. According to Eq. 8.9, N_e depends on β, the collection efficiency, and the quantum efficiency of the tube. Since the photoelectron emission follows a Poisson distribution, we have

$$\epsilon(\beta) = 1 - \exp(-N_e) \tag{8.25}$$

Since $\Pr(0, 5) = 0.01$, we must have an average of five emitted electrons to achieve a 99% detection efficiency. The efficiency curve rises more quickly if the radiator length L or quality factor N_0 is increased, since more photons would be available on the average. In addition, the shape of the efficiency curve is affected by energy loss in the gas, dispersion, and the velocity spread of the incident beam [2].

The threshold velocity resolution of the counter can be determined

Figure 8.7 A simple arrangement for measuring Cerenkov counter efficiencies. S_1 and S_2 are scintillation counters, C is the Cerenkov counter.

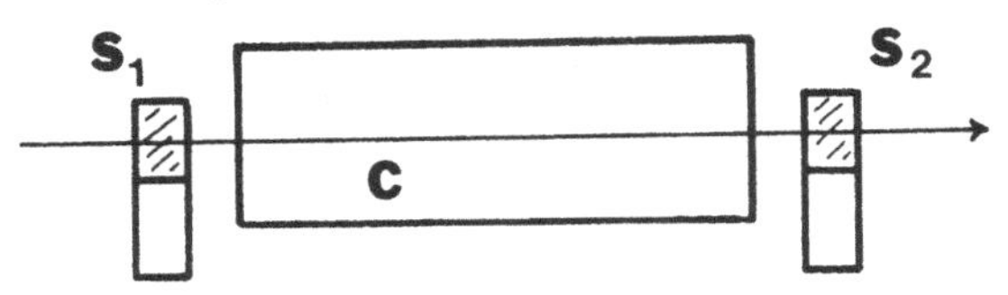

from Eq. 8.10, which we rewrite in the form

$$N_e(\beta) = N_0 L(1 - \beta_t^2/\beta^2) \tag{8.26}$$

Only keeping terms to first order in the quantity $\Delta\beta = \beta - \beta_t$, we find that

$$\frac{\Delta\beta}{\beta_t} = \frac{N_e}{2(N_0 L - N_e)} \tag{8.27}$$

It is desirable that a threshold counter achieve near 100% efficiency as quickly as possible after the particle velocity increases over threshold. For any required number of photoelectrons, the threshold velocity resolution will be improved by designing the quantity $N_0 L$ to be as large as possible.

In order to calculate the efficiency as a function of momentum, we can use Eq. 8.16 to write the mean number of photoelectrons as

$$N_e = N_0 L \beta_t^2 \left(\frac{1}{\beta_t^2 \gamma_t^2} - \frac{1}{\beta^2 \gamma^2} \right) \tag{8.28}$$

For a particle of mass m

$$N_e(p) = N_0 L \beta_t^2 m^2 (1/p_t^2 - 1/p^2) \tag{8.29}$$

where p_t is the threshold momentum. The efficiency can be found by substituting $N_e(p)$ into Eq. 8.25.

The behavior of the detection efficiency near threshold can be affected by the production of knock on electrons (delta rays). It is possible for a massive particle with $\beta < \beta_t$ to collide with an electron and knock it out with velocity greater than β_t. The number of photons $N(E)$ in the wavelength interval λ_1 to λ_2 emitted by a delta ray with initial kinetic energy E is given by [2]

$$N(E) = 2\pi\alpha \int_{\lambda_1}^{\lambda_2} \frac{d\lambda}{\lambda^2} \int_{E_t}^{E_{max}} \left(1 - \frac{(E + mc^2)^2}{n^2(E + 2mc^2)E} \right) dE \frac{dx}{dE} \tag{8.30}$$

where m is the mass of the electron, E_t is the threshold energy for Cerenkov radiation, and E_{max} is the maximum energy of the delta rays. The threshold energy depends on n through the relation

$$E_t = \frac{mc^2}{\sqrt{1 - 1/n^2}} \tag{8.31}$$

while the maximum energy of a delta ray is given by Eq. 1.24, $E_{max} = 2mc^2\beta^2\gamma^2$. Thus, the maximum electron energy increases for increasing β and equals 9.5 MeV for an incident particle with $\beta = 0.95$. Fortunately, the delta rays are emitted with an angular distribution [2]

$$\frac{dN}{d\theta} \simeq \frac{\sin\theta}{\cos^3\theta} \tag{8.32}$$

so that almost all of them come off perpendicularly to the direction of the heavy incident particle.

In gas threshold counters the choice of gas and operating pressure are determined by the desired value of γ_t. The amount of dispersion and multiple scattering may also have to be considered. The diameter of the counter must be matched to the maximum angle of radiation emission, while the length is determined by the required photon yield. Some properties of gas threshold counters currently in use are shown in Table 8.4.

Now let us consider some examples of threshold counters [6]. The TASSO group at DESY has incorporated gas and silica aerogel threshold counters into their particle spectrometer [7]. Silica aerogel consists of strings of small (~4-nm-diameter) spheres of amorphous silica surrounding spheres (~60-nm diameter) of trapped air [8]. The index of refraction of the solid material can be lowered from the value for pure silica ($n = 1.46$) by increasing the amount of trapped air. It is possible by this means to obtain a transparent solid material with a small value of n. The TASSO aerogel blocks were constructed with $n = 1.024$.

Table 8.4. *Examples of Cerenkov counter systems*

Detector	Location	Type	Medium	Pressure (atm)	Photon detection
AFS	ISR	threshold	Freon 12	4	
DELCO	PEP	threshold	isobutane	1	RCA8854
EMC	SPS	threshold	Ne, N_2	1	XP2041
Exp. 605	FNAL	RICH	He	1	PWC
HRS	PEP	ultraviolet	Ar-N_2	15	PWC
IMB	Ohio	p decay	water	1	EMI9870B
Kamiokande	Japan	p decay	water	1	R1449X
LASS	SLAC	threshold	Freon 114	1	
MD1	Novosibirsk	threshold	ethylene	25	
MPS II	BNL	threshold	Freon 12	3	RCA4501
Tagged γ	FNAL	threshold	N_2	1	
TASSO	PETRA	threshold	aerogel	1	RCA8854
		threshold	Freon 114	1	XP2041
		threshold	CO_2	1	XP2041

Source: Particle Data Group, Major detectors in elementary particle physics, Report LBL-91, UC-37, 1983; H. Gordon et al., Nuc. Instr. Meth. 196: 303, 1982; W. Slater et al., Nuc. Instr. Meth. 154: 223, 1978; Y. Declais et al., Nuc. Instr. Meth. 180: 53, 1981; G. Coutrakon et al., IEEE NS-29: 323, 1982; J. Chapman, N. Harnew, and D. Meyer, IEEE NS-29: 332, 1982; H. Burkhardt et al., Nuc. Instr. Meth. 184: 319, 1981.

The gas counters shown in Fig. 8.8 use CO_2 and Freon 114 radiators at atmospheric pressure. Concave ellipsoidal mirrors were used with one focus at the e^+e^- interaction point and the other at the PMT. The mirrors were formed from heat-treated lucite sheets with a vacuum-deposited aluminum surface. The reflectivity was around 90%. Winston cones were used to collect the light. The cathode of the XP2041 PMT was coated with wavelength shifter to improve the ultraviolet response. The efficiency of the Freon 114 counter is shown in Fig. 8.9 as a function of the pion momentum.

One difficulty with collecting the light from the aerogel counter is that visible light has a high probability for scattering. The measured scattering length is 2.4 cm at a wavelength of 436 nm [7]. In this case a light collection system using diffuse scatterings was more efficient than one using focusing mirrors. Figure 8.10 shows a plateau above 1 GeV/c in the aerogel efficiency as a function of pion momentum. The mean number of photoelectrons in the plateau is shown as a function of aerogel thickness in Fig. 8.10b. The number differs from the linear dependence predicted by Eq. 8.10 because of absorption of photons in the aerogel block. According to Eq. 8.9, the number of photoelectrons should be proportional to β^{-2}. The data plotted in Fig. 8.10c nicely confirms this.

Figure 8.8 A large Cerenkov counter system used in the TASSO spectrometer. (H. Burkhardt et al., Nuc. Instr. Meth. 184: 319, 1981.)

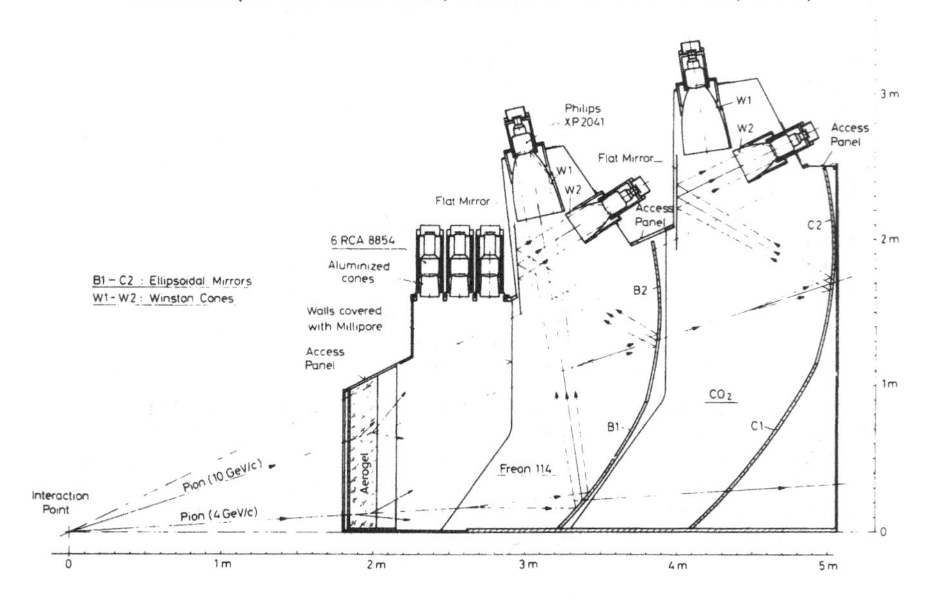

Figure 8.9 Measured Cerenkov counter efficiency as a function of particle momentum. (H. Burkhardt et al., Nuc. Instr. Meth. 184: 319, 1981.)

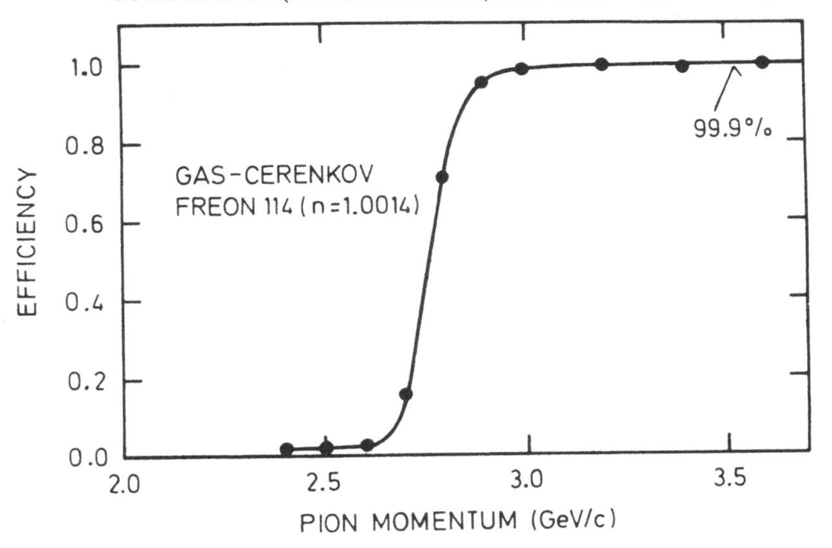

Figure 8.10 (a) Threshold curve for the TASSO aerogel counter, (b) mean number of photoelectrons as a function of the aerogel thickness, and (c) dependence of the mean number of photoelectrons on the quantity β^{-2}. (H. Burkhardt et al., Nuc. Instr. Meth. 184: 319, 1981.)

a

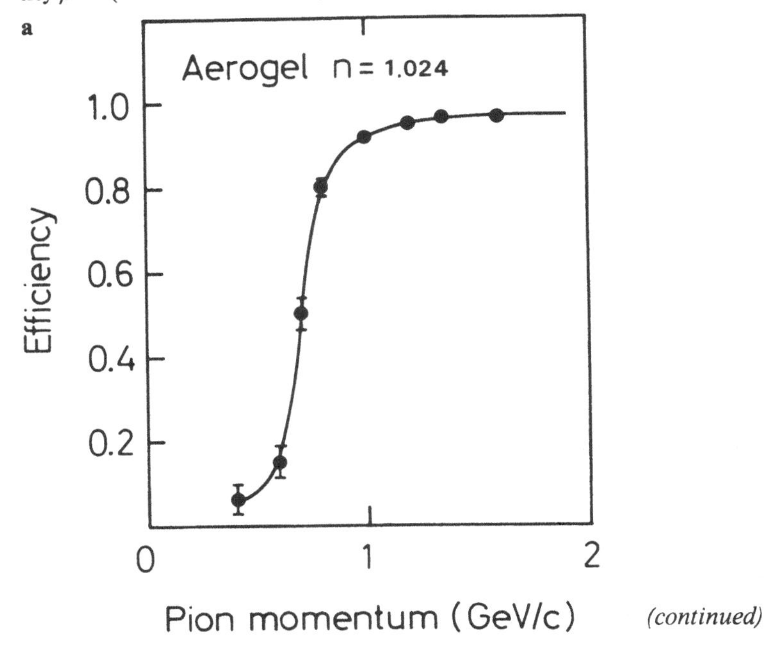

(continued)

Figure 8.10 *Continued.*

b

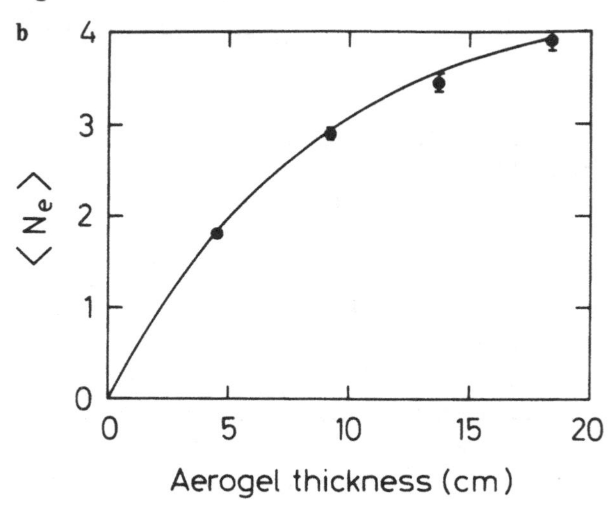

c

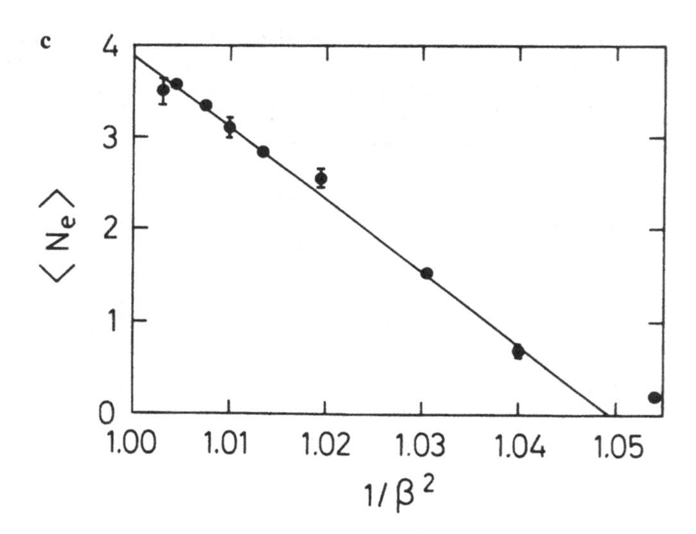

8.5 Differential counters

A differential Cerenkov counter can measure the velocity of a particle by only accepting Cerenkov light in a small annulus around some angle θ. Using such a counter, it is possible to provide a signal for the presence of a given mass particle. The primary design consideration involves the rejection of particles outside the desired mass range. Benot, Litt, and Meunier [9] have given a detailed comparison of the design criteria of threshold and differential counters. Differential counters are usually designed so that the velocity resolution is no more than one-half of the velocity separation between the closest mass particles it is desired to identify. This must be done for the highest operating momentum.

A simple differential counter was shown in Fig. 8.4. Cerenkov light in the radiator gas is reflected by a spherical mirror. The cones of light appear to the mirror as a ring source at infinity. The image consists of a ring in the focal plane of the mirror with radius

$$r = f \tan \theta \tag{8.33}$$

where f is the focal length of the mirror. The radius of curvature of a spherical mirror is $2f$.

A diaphragm containing a slit of width Δr placed in front of the PMTs will only accept light from within the angular range $\Delta\theta$ given by

$$\Delta\theta = \cos^2\theta \, \Delta r/f \tag{8.34}$$

which corresponds at a given refractive index n to a velocity resolution of

$$\Delta\beta/\beta = \tan\theta \, \Delta\theta \tag{8.35}$$

This shows that small θ and/or $\Delta\theta$ are required for good velocity resolution. On the other hand, we have seen that the photon yield increases like θ^2, so that the counter design must be carefully optimized.

The minimum obtainable angular resolution is mainly limited by dispersion. The indices of refraction of all materials are a function of the wavelength. Differentiating Eq. 8.1, we find that the dispersion limitation is

$$\Delta\theta_{\text{disp}} = \frac{\Delta n}{n \tan\theta} \tag{8.36}$$

where n is the average index of refraction and Δn is the range of n corresponding to the range of accepted wavelengths. The angular resolution is also affected by beam divergence, optical aberrations, multiple scattering in the radiator, energy loss in the radiator, and diffraction [4]. Gases with small optical dispersion include helium, neon, and SF_6, while gases with small multiple scattering include hydrogen, methane, and helium [9].

Figure 8.11 shows a differential gas counter (DISC) used in a hyperon beam at CERN. This counter uses an adjustable optical system to correct for dispersion and geometric aberrations. The accepted velocity could be changed by varying the pressure in the counter. Figure 8.12 shows the particle yield as a function of the velocity. Note the clear separation of different mass particles. A velocity resolution $\Delta\beta \sim 5 \times 10^{-5}$ was achieved.

The DISC counter is only suitable for use in a well-collimated beam of particles. However, it would clearly be desirable to use the Cerenkov effect to measure the velocities of secondary particles arising from an interaction. This information could then be combined with an independent measurement of the momenta to identify the particles' masses.

One scheme [10] for constructing such a ring imaging Cerenkov (RICH) counter is shown in Fig. 8.13. The radiating medium is contained between two spheres surrounding the target or intersection point. The Cerenkov light reflects off the mirror and is focused onto a ring at the detector surface. With this geometry the radius of the ring is directly related to the Cerenkov angle through Eq. 8.33. The resolution on the relativistic γ factor is [10]

$$\frac{\Delta\gamma}{\gamma} = \frac{\gamma^2\beta^3 n}{\sqrt{N_0 L}} \frac{2\,\Delta r}{R} \tag{8.37}$$

Figure 8.11 A DISC type differential Cerenkov counter. (M) mirror, (C) correction optics, (D) diaphragm, and (P) photomultiplier tube. (J. Litt and R. Meunier, adapted with permission from the Annual Reviews of Nuclear Science, Vol. 23, © 1973 by Annual Reviews, Inc.)

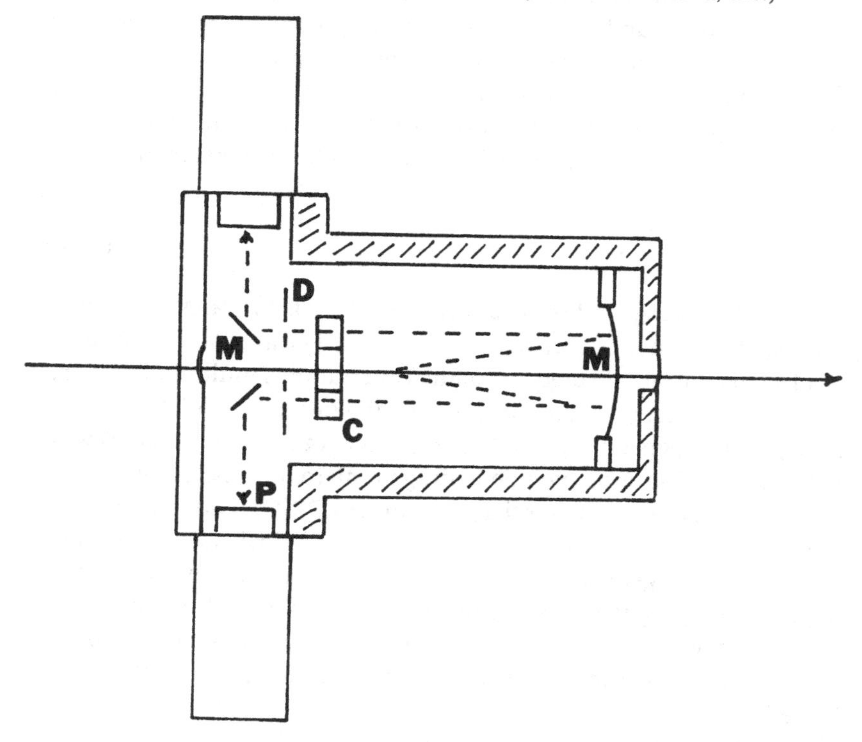

Figure 8.12 A pressure curve for the DISC counter shown in Fig. 8.11. (J. Litt and R. Meunier, adapted with permission from the Annual Reviews of Nuclear Science, Vol. 23, © 1973 by Annual Reviews, Inc.)

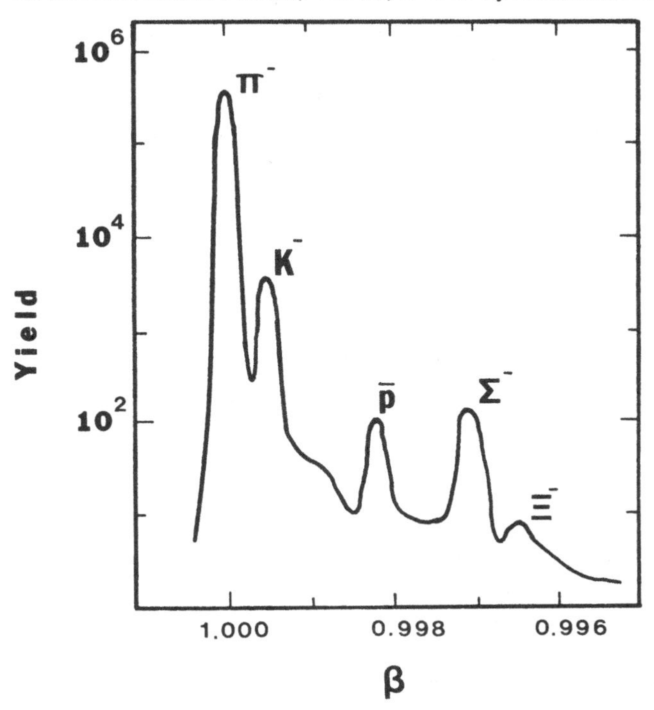

Figure 8.13 Geometrical arrangement for a ring imaging Cerenkov counter. (B) beam, (T) target, (M) spherical mirror, (*R*) radius of M, (P) produced secondary particle, (D) detector plane, and (r) Cerenkov ring on D.

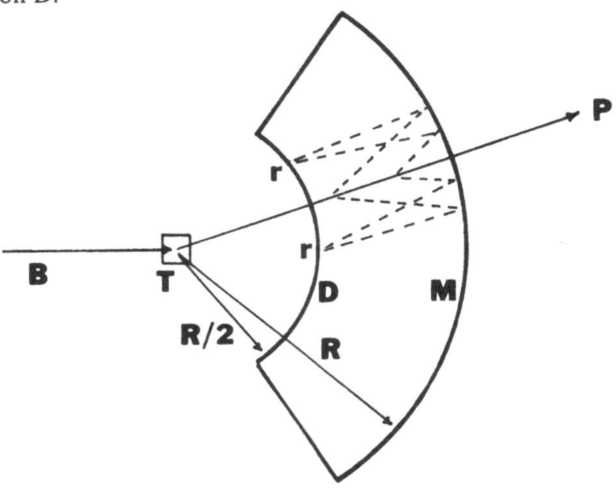

where N_0 and L are given in Eq. 8.10, R is the radius of curvature of the mirror, and Δr is the uncertainty in the ring radius at the detector plane due to all sources. In principle, if external tracking chambers are available, the center of a particle's ring is known, and the detection of only one photon is sufficient to determine the radius of the ring. However, because of complications arising from particle multiplicity and detector inefficiencies, a practical detector design would require several detected photons to determine a ring.

Since the number of photons expected per particle turns out to be very small, most RICH designs have concentrated on collecting photons from the ultraviolet portion of the spectrum. First, according to Eq. 8.4 the photon yield is much larger in the ultraviolet than in the visible. A second reason is that ultraviolet photons can be converted with practically 100% quantum efficiency using the photoelectric effect.

The Cerenkov radiating gas must be transparent in the ultraviolet. This restricts the radiator to a few gases, including the noble gases and nitrogen. The radiator must be separated from the detector by a window made from a material, such as LiF or MgF_2, that is also transparent to the ultraviolet. The detector must have a high efficiency for single photon detection. Proportional chambers, drift chambers, and Geiger needles have all been

Table 8.5. *Ultraviolet photoionization thresholds*

	Ionization threshold	
Gas	(eV)	(nm)
TMAE	5.36	231
TMPD	6.1	203
Perylene	6.95	178
Pyrathrene	7.0	177
Tri-*n*-propylamine	7.2	172
Trimethylamine	7.50	165
Dimethylamine	8.2	151
Methylamine	8.9	139
Benzene	9.24	134
Acetone	9.69	128
Ethanol	10.49	118
Ethane	11.5	108
Methane	12.6	98
Carbon dioxide	13.77	90

Source: A. Breskin et al., Nuc. Instr. Meth. 161: 19, 1979; A. Etkin et al., in Proc. of 1977 ISABELLE Summer Workshop, BNL report 50721, 1977, p. 72.

used. The gas in the detector must have a high efficiency for converting the ultraviolet photons.

Table 8.5 gives the photoionization thresholds for various gases. Only photons whose wavelengths are smaller than the threshold are able to ionize the gas. Three materials with small window thresholds are LiF, MgF_2, and CaF_2. Light is only efficiently transmitted for wavelengths greater than the window limit. Taken together only a narrow range of photon wavelengths can ionize the gas. A small percentage of benzene or other additive with a large photoionization cross section can be used to sensitize wire chambers. Ideally the detector readout should be 2-dimensional. However, some information is still obtained with a 1-dimensional readout, since the projection of the ring creates a double peak spectrum.

8.6 Total absorption counters

In a total absorption Cerenkov counter the incident particle is absorbed in the radiator medium. Two classes of reactions are commonly used with total absorption counters. The first involves incident electrons or photons. These particles initiate an electromagnetic shower through the combined processes of bremsstrahlung and pair production. The Cerenkov radiation emitted by electrons in the shower is detected by the counter. In the second type of reaction the incident particle interacts in the medium and one of the final state particles in the interaction emits the Cerenkov radiation. Examples of such secondary particles include the recoil proton in neutron elastic scatters and electrons produced in neutrino interactions.

Nemethy et al. [11] have constructed a large total absorption water Cerenkov counter for use as a neutrino detector at LAMPF. The counter was required to detect low energy electrons or positrons created in neutrino interactions on free protons or deuterons. A Cerenkov counter was chosen over a scintillation counter for this application in order to avoid background signals from recoil protons in fast neutron interactions.

The counter consisted of 6000 l of water enclosed in a cubic volume with 1.8-m sides. The walls were made of cast epoxy, strengthened by a series of struts. Wavelength shifter was mixed with the water and viewed by 96 PMTs. A fiber-optic light guide was bonded to each tube for calibration purposes. The counter surface not covered by PMTs was covered with a diffuse white reflector. The counter was surrounded by scintillation counters in order to veto cosmic ray events. The counter had a gain of 5.3 photoelectrons/MeV of energy loss. The response was uniform to within 5%. The energy resolution for electrons was $\sigma/E = 12\%$.

Arrays of lead-glass blocks are widely used detectors for photons and electrons. Lead-glass contains about 50% lead by weight. The addition of this high Z element to the glass greatly reduces the radiation length without seriously affecting the transmission of the Cerenkov radiation through the block. Typical values for the density, refractive index, and radiation length of lead-glass are $\rho = 3.9$ g/cm^3, $n = 1.7$, and $L_r = 2.6$ cm. Each block is connected to a PMT. The blocks must be long enough to contain a large fraction of the electromagnetic shower. Arrays of such blocks can cover a large solid angle. The spatial resolution depends on the cross section of the blocks. Fine granularity is usually necessary to resolve the photons originating from π^0 decays. The pulse height of the PMT is proportional to the energy of the incident particle in a properly designed counter. The energy resolution of a lead-glass counter will depend on the amount of shower leakage out of the block and on the photoelectron statistics. The energy resolution varies with energy according to the relation

$$\frac{\sigma(E)}{E} = a + \frac{b}{\sqrt{E}} \tag{8.38}$$

A typical value for b is 10 – 12%. Intense radiation causes lead-glass to darken.

Two lead-glass arrays were used as photon detectors at the European Hybrid Spectrometer at CERN [12]. One photon detector was designed to simultaneously measure the position and energy of an electromagnetic shower. It consisted of a 2-dimensional array of $5 \times 5 \times 42$ cm^3 lead-glass blocks mounted on a movable platform. The blocks were arranged longitudinally to the beam in order to totally absorb the shower. A diagram of one counter is shown in Fig. 8.14. Light is transmitted through a lead-glass light guide to the PMT. The gain of each counter was monitored using a laser – optical fiber calibration system. Spatial resolution was im-

Figure 8.14 Design of an EHS lead-glass counter. (B. Powell et al., Nuc. Instr. Meth. 198: 217, 1982.)

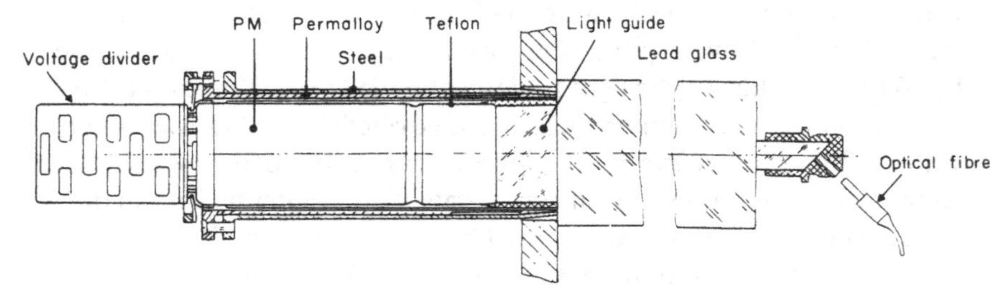

proved by accurately measuring the small amplitude tails on either side of the main shower deposition. The measured spatial resolution was 3.7 mm(FWHM), while the energy resolution was $\Delta E/E = 0.15/\sqrt{E} + 0.02$ (FWHM).

References

[1] J. Jelley, *Cerenkov Radiation and its Applications,* London: Pergamon, 1958.

[2] V. Zrelov, *Cherenkov Radiation in High Energy Physics,* Israel Program for Scientific Translations, Federal Scientific and Technical Information, Springfield, VA, 1970.

[3] V. Zrelov, M. Klimanova, V. Lupiltsev, and J. Ruzicka, Calculations of threshold characteristics of Vavilov-Cherenkov radiation emitted by ultrarelativistic particles in a gaseous Cherenkov counter, Nuc. Instr. Meth. 215: 141–6, 1983.

[4] J. Litt and R. Meunier, Cerenkov counter technique in high energy physics, Ann. Rev. Nuc. Sci. 23: 1–43, 1973.

[5] W. Galbraith, Cerenkov counters, in High Energy and Nuclear Physics Data Handbook, Sec 6, National Institute for Research in Nuclear Science, Rutherford Laboratory, 1963.

[6] The applicability of threshold counters to high energy colliders such as LEP is considered in P. Lecomte, G. Poelz, R. Riethmuller, O. Romer, and P. Schmuser, Threshold Cerenkov counters, Physica Scripta 23: 377–83, 1981.

[7] H. Burkhardt, P. Koehler, R. Riethmuller, B. Wiik, R. Fohrmann, J. Franzke, H. Krasemann, R. Maschuw, G. Poelz, J. Reichardt, J. Ringel, O. Romer, R. Rusch, P. Schmuser, R. van Staa, J. Freeman, P. Lecomte, T. Meyer, S. Wu, and G. Zobernig, The TASSO gas and aerosol Cherenkov counters, Nuc. Instr. Meth. 184: 319–31, 1981.

[8] G. Poelz and R. Riethmuller, Preparation of silica aerosol for Cherenkov counters, Nuc. Instr. Meth. 195: 491–503, 1982.

[9] M. Benot, J. Litt, and R. Meunier, Cherenkov counters for particle identification at high energy, Nuc. Instr. Meth. 105: 431–44, 1972.

[10] J. Seguinot and T. Ypsilantis, Photoionization and Cherenkov ring imaging, Nuc. Instr. Meth. 142: 377–91, 1977; T. Ypsilantis, Cerenkov ring imaging, Physica Scripta 23: 371–6, 1981.

[11] P. Nemethy, S. Willis, J. Duclos, and H. Kaspar, A water Cerenkov neutrino detector, Nuc. Instr. Meth. 173: 251–7, 1980.

[12] B. Powell, R. Heller, N. Ibold, K. Schubert, J. Stiewe, M. Vysocansky, M. Mazzucato, P. Rossi, L. Ventura, G. Zumerle, M. Boratav, J. Duboc, J. Passeneau, M. Touboul, P. Bagnaia, C. Dionisi, D. Zanello, L. Zanello, V. Bujanov, Y. Fisjak, A. Kholodenko, E. Kistenev, B. Poljakov, E. Castelli, P. Checchia, and C. Troncon, The EHS lead-glass calorimeters and their laser based monitoring system, Nuc. Instr. Meth. 198: 217–31, 1982.

Exercises

1. Find the Cerenkov angles for 2-GeV/c pions in water, lucite, and NaI.

2. Find the total Cerenkov energy emitted in visible wavelengths by a 1-GeV/c proton crossing 1 m of water.

3. Design a Cerenkov counter that gives on the average 10 collected photoelectrons for 5-GeV/c kaons.

4. Plot the convoluted frequency response of the RCA 8575 tube for Cerenkov radiation.

5. A kaon beam passes through a 2-m-long Cerenkov counter that contains CO_2 at 2 atm and has a quality factor $N_0 = 400\ \text{cm}^{-1}$. At what momentum should the efficiency of the counter reach 50%?

6. Consider a 1-GeV/c proton passing through water. Calculate numerically the number of visible photons emitted by δ rays. Use an approximation for dE/dx over the relevant energy range.

7. An atmospheric pressure, helium differential Cerenkov counter has a spherical mirror of radius 2 m and a 1-mm acceptance slit. What is the expected velocity resolution?

8. Design a total internal reflection counter that accepts *all* particles whose velocity exceeds a minimum threshold. How can you maximize the quality factor N_0? How should the radiator be coupled to the PMT?

9
Proportional chambers

Proportional chambers are particle detectors consisting essentially of a container of gas subjected to an electric field. A passing particle can leave a trail of electrons and ions in the gas. The charged particle debris are collected at the chamber electrodes and in the process provide a convenient electrical signal, indicating the passage of the particle. The detector operates as a proportional chamber when the applied electric field is large enough so that the accelerated electrons cause secondary ionization, yet small enough so that the output pulse is still proportional to the number of primary ion pairs. Multiwire proportional chambers are widely used for particle tracking and for triggering.

9.1 Elements of a proportional chamber

The essential elements of a proportional chamber can be illustrated by the apparatus shown in Fig. 9.1. The cylindrical chamber is held at ground potential. The anode wire, which runs along the axis of the cylinder, is maintained at high voltage and is electrically insulated from the cylinder. The electric field inside the chamber is

$$\mathscr{E} = \frac{V_0}{r \ln(r_c/r_w)} \tag{9.1}$$

where r_c is the radius of the cylinder and r_w is the radius of the anode wire. The interior of the cylinder contains a gas such as argon. The chamber has a capacitance per unit length of

$$C = \frac{2\pi\epsilon}{\ln(r_c/r_w)} \tag{9.2}$$

where ϵ is the dielectric constant of the chamber gas. A typical value of C is around 9 pF/m.

Figure 9.1 Basic elements of a proportional counter.

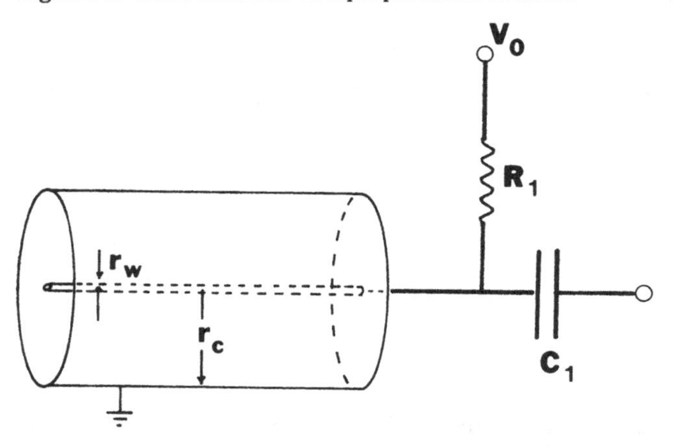

Figure 9.2 Number of collected ions in a proportional counter versus applied voltage. Regions: (R) recombination, (IC) ionization chamber, (PC) proportional chamber, (GC) Geiger counter, (D) continuous discharge. The upper (lower) curve is for a heavily (minimum) ionizing particle.

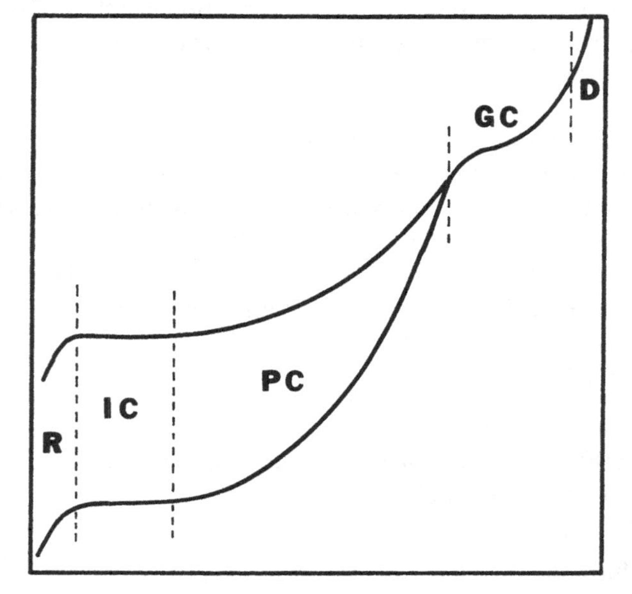

The passage of a charged particle through the gas liberates electron – ion pairs. Typically 30 eV is lost for every ion pair that is formed. As the liberated charges migrate toward the electrodes, they induce a signal in the external circuit. The magnitude of the induced voltage signal is given roughly by

$$V = \frac{MNe}{CL} \tag{9.3}$$

where N is the number of created ion pairs, M is the multiplication factor that takes into account the additional electron – ion pairs created by secondary ionization, and L is the length of the chamber. Because of the $1/r$ field dependence, most of the secondary ionization occurs within a few wire diameters of the anode.

The voltage characteristics of this type of chamber are shown in Fig. 9.2. At very low voltages some of the ions recombine before they can be collected by the electrodes. In the second region all of the ionization products are collected by the electrodes, but the accelerated charged particles do not have sufficient energy to cause secondary ionization. In this region the chamber operates as an ionization counter. Unfortunately, the signal is extremely small. Ionization counters are useful in an integrating mode in high flux environments, for example as beam monitors. As the voltage is increased, the electrons acquire sufficient energy to cause secondary ionization. Here the signal is proportional to the initial ionization, yet is amplified over the signal from an ionization chamber. At higher voltages the chamber enters the Geiger mode, where the signal is very large, but independent of the initial ionization. Finally above a certain voltage the chamber continuously discharges, independent of external conditions.

The signal induced on the anode wire receives contributions from the flow of electrons and positive ions [1 – 3]. Consider a particle with charge Q moving in the electric field. The work the charge performs in moving through a potential drop ΔV is $Q\,\Delta V$. This energy is replenished by the external circuit, which supplies a charge ΔQ_a to the anode to keep the applied voltage constant. Thus, we must have

$$Q\,\Delta V = V_0\,\Delta Q_a$$

and the voltage signal induced on the anode ΔV_a can be written

$$\Delta V_a = \frac{Q}{LCV_0}\frac{dV}{dr}\,dr \tag{9.4}$$

The induced voltage is seen to depend on the potential drop through which the particle moves. Since all the multiplication takes place very close to the anode wire, most of the electrons only move through a very small potential drop. Most of the positive ions, on the other hand, must cross almost the entire electrode gap, so they experience the full potential drop of the chamber. As a result, under typical proportional chamber conditions, the anode signal is almost entirely due to the motion of the positive ions.

9.2 Fundamental processes in gases

Before we discuss the properties of proportional chambers in more detail, it is important to understand the types of processes that can occur in a gas subjected to an electric field. A particle traversing a gaseous medium loses energy by elastic scattering, by excitation, and by ionization of the gas atoms or molecules. The energy loss of the incident particle in elastic scattering is generally so small that it does not play a significant role in the operation of gaseous detectors. The excitation process raises the gas atoms or molecules to an excited energy level, subject to the selection rules for transitions among the energy levels [4]. The atoms deexcite by photon emission. Formation of a meta-stable state is also possible if direct transition to the ground state is forbidden.

The most important process for the operation of gaseous detectors is ionization. In this case one or more electrons are liberated from the atoms of the medium, leaving positive ions and electrons. Ionization can occur when the energy imparted to the atom exceeds the ionization potential of the gas. The ionization cross section rises sharply from threshold to a peak and then decreases with increasing energy of the incident particle. Ionization cross sections for electrons incident on the noble gases peak for electron energies around 100 eV [2, 4]. Ionization potentials for some gases are given in Table 9.1. The first ionization potential is the energy necessary to remove an electron from a neutral atom or molecule. The second ionization potential is the additional energy necessary to remove a second electron, thereby leaving a doubly charged ion.

The primary ionization is defined to be the number of ionizing collisions per unit length suffered by the incident particle. Some of the ionized electrons (delta rays) may have sufficient energy to cause still more ionization. Thus, a second quantity, the total specific ionization, defined to be the total number of ions actually created per unit length, is also useful. Values of these quantities for several gases are included in Table 9.1. Notice that the average energy lost in creating an ion pair is roughly 30 eV.

Table 9.1. *Ionization properties of gases*

Gas	First ionization potential (eV)	Second ionization potential (eV)	Primary[a] ionization (cm^{-1})	Total specific[a] ionization (cm^{-1})	Average energy to create one ion pair (eV)	dE/dx[a] (keV/cm)
H_2	15.4	—	5.2	9.2	37	0.34
He	24.6	54.4	4–6	8–11	41	0.32
N_2	15.5		10–19	56	35	1.96
O_2	12.2		22	73	31	2.26
Ne	21.6	41.1	12	39	36	1.41
Ar	15.8	27.6	29	94–110	26	2.44
Kr	14.0	24.4	~22	192	24	4.60
Xe	12.1	21.2	44	307	22	6.76
CO_2	13.7		~34	91	33	3.01
CH_4	13.1		16	53	28	1.48
C_4H_{10}	10.8		~46	195	23	4.50

[a] For minimum ionizing particle at 1 atm pressure.
Source: Handbook of Chemistry and Physics, 64th ed., Boca Raton: CRC Press, 1983; P. Rice-Evans, *Spark, Streamer, Proportional, and Drift Chambers,* London: Richelieu, 1974; F. Sauli, CERN Report 77-09, 1977.

Since the number of collisions that occurs is a statistical process, the primary ionization follows a Poisson distribution. The total specific ionization, on the other hand, is influenced by the production of delta rays, which can result in large secondary ionization. Here the appropriate frequency distribution is that due to Landau or one of its later modifications, which has a significant tail for large ionization. The statistical distributions for primary ionization and total specific ionization have been measured using a streamer chamber with a variable high voltage delay [5]. The measured distribution of streamers (see Section 12.4) confirmed that the primary ionization had a Poisson distribution and that the total specific ionization had a Landau distribution.

Further ionization can occur in certain gas mixtures from the deexcitation of a meta-stable atom. In certain cases the surplus energy can ionize a second atom, a process known as the Penning effect [2]. As an example, consider a mixture of neon and argon. Two of the $2s$ states of neon are meta-stable and can deexcite via collisions with argon atoms. The exited state energy of neon exceeds the ionization potential for argon, and thus the neon can deexcite by ionizing the argon.

$$\mathrm{Ne^* + Ar \rightarrow Ne + Ar^+ + e^-}$$

The peak fraction of energy lost by an electron due to ionization can in some cases be almost doubled by the addition of 0.1% of argon to the original neon gas [2].

Once the ion pairs have been created in the gas, a number of processes may take place, including recombination, attachment, charge exchange, and absorption by the chamber walls. Positive and negative ions can recombine into neutral atoms or molecules via the process

$$\mathrm{X^+ + Y^- \rightarrow X + Y} + \gamma$$

Electrons can be removed from the gas by recombination with a positive ion

$$\mathrm{X^+ + e^- \rightarrow X} + \gamma$$

Assume that a gas contains equal concentrations n of positive and negative ions. The rate of decrease of ions due to recombination follows the relation [4]

$$dn/dt = -\beta n^2 \tag{9.5}$$

where the recombination coefficient β is a property of the gas. The concentration remaining in the gas a time t after the passage of the ionizing particle is then given by

$$n = n_0/(1 + \beta n_0 t) \tag{9.6}$$

where n_0 is the initial concentration of ions. Values of the recombination coefficient are in the range $10^{-8}-10^{-6}$ cm^3/s for various gases [2]. The values increase with Z among the noble gases.

Another means of removing electrons is by the addition of a gas with a large electron affinity. Table 9.2 lists measured values of the electron affinity for some atoms and molecules. The electron may attach itself to a neutral atom, especially if this results in a closed outer shell.

$$e^- + X \rightarrow X^- + \gamma$$

The intensity of free electrons traversing the gas falls exponentially [2],

$$I_e = I_{e0}\exp(-\eta x) \tag{9.7}$$

where η is called the attachment coefficient. This coefficient depends on the incident particle energy and is strongly affected by the presence of an electric field [1]. Some of the gases for which electron attachment is important include water vapor, oxygen, ethanol, SF_6, CCl_4, and the Freons.

Another process resulting in the elimination of the original positive ions is charge exchange. This occurs when the ionization potential of the ion is greater than that of some molecule mixed with the gas. This is usually a polyatomic gas like ethanol or methylal. The polyatomic gas quenches the ion multiplication by neutralizing the ions of the main chamber gas. It dissipates the ionization energy by dissociating into

Table 9.2. *Electron affinities*

Substance	Affinity[a] (eV)	Boiling point (°C)
Cl	3.613	−34.6
F	3.448	−188
Br	3.363	58.8
O	1.466	−183
H	0.80	−253
NO_2	3.91	21.2
UF_6	2.91	56.2
WF_6	2.74	17.5
SF_6	1.43	−63.8
BF_3	2.65	−99.9
O_2	0.45	−183

[a] For gases at 0 K.
Source: Handbook of Chemistry and Physics, 64th ed., Boca Raton: CRC Press, 1983, p. E-62.

smaller fragments and absorbs photons emitted in the radiative deexcitation process.

Proportional chambers use an electric field in the active region of the chamber. The charged ionization products in the gas have a net motion in the direction of the field. The drift velocity of an ion w^+ is related to the strength of the electric field $\mathscr{E}$ through the relation

$$w^+ = \mu^+ \mathscr{E} \tag{9.8}$$

where μ^+, the mobility, is a characteristic of the gas, which can depend on the pressure P and electric field strength. The mobility is inversely proportional to the gas pressure at constant temperature. The mobility of an ion is roughly constant for a restricted range of the variable $\mathscr{E}/P$. This result is a consequence of the fact that for moderate electric fields, the average energy of the ions is only slightly increased by the field [1]. For example, the mobility of Ar^+ ions in argon at 0 °C and 1 torr is 1200 cm/s per V/cm for $\mathscr{E}/P \lesssim 40$ V/cm-torr [4]. For larger values of $\mathscr{E}/P$ the drift velocities of many ions are roughly proportional to $(\mathscr{E}/P)^{1/2}$.

Although ions and atoms suffer a large fractional energy loss in their collisions with other atoms, electrons lose energy slowly. As a result, electrons can reach large velocities between collisions. The drift velocity for electrons is shown in Fig. 9.3 for several gases. In general, the electron mobility is a complicated function of $\mathscr{E}/P$. At high fields the drift velocity is typically 5 cm/μs, which is roughly 1000 times faster than the velocity of the heavier ions under similar conditions. The drift velocity is also influenced by the presence of a magnetic field. In the case where both an electric and a magnetic field are present, the electrons drift with a reduced velocity and at an angle with respect to the electric field (see Appendix F).

An assembly of ions will diffuse from a region of high concentration to a region of low concentration. The fraction of the original number of ions N_0 present after a time t in the interval of distance dx at x is given by [1]

$$dN/N_0 = (4\pi Dt)^{-1/2} \exp(-x^2/4Dt)\, dx \tag{9.9}$$

where D is the diffusion coefficient. This equation indicates that the ions are spread out with a Gaussian distribution, whose width increases rapidly with time. The diffusion coefficient can depend on the value of the electric field. Electrons diffuse more than ions because of their larger mean free path and higher thermal velocity. Generally the motion of ions or electrons in one dimension will be dominated by the electrical force. Motion perpendicular to the drift direction will be determined by diffusion. The radius of the distribution in two dimensions after a time t is given by [2]

$$\langle r^2 \rangle = 4Dt \tag{9.10}$$

The amount of diffusion is also affected by the presence of a magnetic field. The electrons spiral around the magnetic field lines with the consequence that if the field is sufficiently high, the diffusion perpendicular to B is reduced.

If some region contains a nonuniform concentration of ions of either charge, there will be an additional contribution to the motion due to the electrostatic space charge repulsion. The charge density ρ varies with time according to the relation [4]

$$-\partial\rho/\partial t = \nabla \cdot (\rho w) \tag{9.11}$$

where w is the drift velocity.

Now let us consider the processes leading to an electric current through the gas. The passage of the ionizing particle frees many pairs of electrons and ions. If no electric field were present, the ions and electrons would diffuse and in the process collide with other atoms and molecules in the gas. After traveling several mean free paths, the electrons can reach a thermal velocity of $\sim 10^7$ cm/s under typical conditions. The thermal velocity of positive ions, on the other hand, is only $\sim 10^5$ cm/s on account of their much larger mass. If an electric field is present in the gas, a net drift velocity toward the electrodes is superimposed on the thermal motions. If

Figure 9.3 Electron drift speeds in methane, ethane, and ethylene. (After B. Jean-Marie, V. Lepeltier, and D. L'Hote, Nuc. Instr. Meth. 159: 213, 1979.)

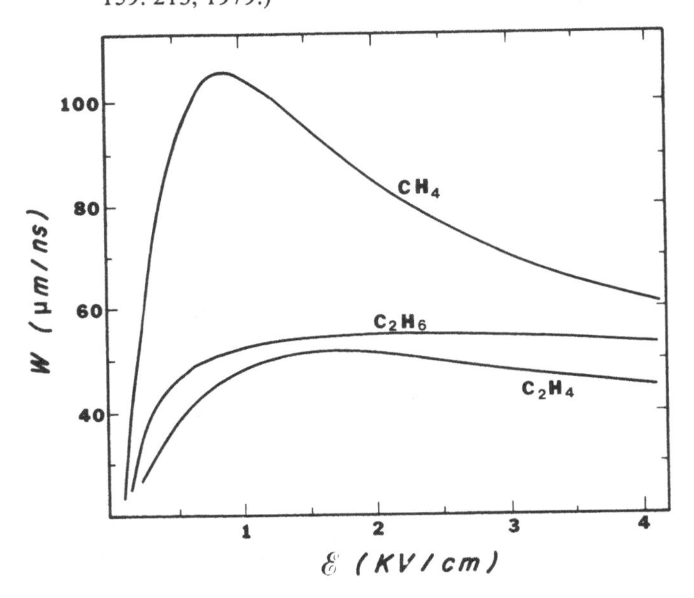

the field is high enough, the electrons can reach sufficient energies to cause secondary ionization. The number of electrons can continue to multiply until an avalanche of $\sim 10^8$ electrons is formed. At this point space charge repulsion may prevent further growth.

The number of electrons produced by a single electron traveling 1 cm along the field is known as Townsend's first ionization coefficient α. Note that α^{-1} is the electron mean free path for ionization in the gas. Values of α are given approximately by

$$\alpha/P = A \exp(-BP/\mathscr{E}) \tag{9.12}$$

where P is the pressure, $\mathscr{E}$ is the applied electric field, and A and B are constants. Table 9.3 lists values of A and B for various gases. Many gases have a peak value of α/P of about 8 ion pairs/cm-torr.

Consider a gas confined in the space between two parallel electrodes a distance d apart, as shown in Fig. 9.4. Suppose that an ionization event occurs at a distance x from the anode. Under proportional chamber conditions the total number N of free electrons grows as they are swept onto the anode. The change in the number of electrons after traveling a distance between y and $y + dy$ is

$$dN = \alpha N\, dy$$

and the total electron multiplication over the distance x is

$$M(x) = e^{\alpha x} \tag{9.13}$$

If the electric field is not uniform, α will be a function of x, and the total

Table 9.3. *Values of the* A *and* B *coefficients for gases*

Gas	A (cm^{-1} $torr^{-1}$)	B (V/cm torr)	Range of validity for $\mathscr{E}$/P (V/cm torr)
H_2	5	130	150–600
N_2	12	342	100–600
CO_2	20	466	500–1000
H_2O	13	290	150–1000
He	3	34	20–150
Ne	4	100	100–400
Ar	12–14	180	100–600
Kr	17	240	100–1000
Xe	26	350	200–800

Source: P. Rice-Evans, *Spark, Streamer, Proportional and Drift Chambers,* London: Richelieu, 1974; A. von Engel, *Ionized Gases,* London: Oxford, 1955.

multiplication will be given by

$$M(x) = \exp\left[\int_x^0 \alpha(y)\,dy\right] \tag{9.14}$$

An actual electron avalanche resembles a teardrop with the electrons highly concentrated at the leading edge [1]. The positive ions are present throughout the entire region occupied by the avalanche. Since their drift velocity is a factor of 10^3 smaller than the electron drift velocity, they slowly diminish in concentration away from the leading edge.

Several effects can lead to modifications of Eq. 9.14. Positive ions striking the cathode can liberate additional electrons. In addition, energetic photons emitted anywhere in the gap may strike the cathode and produce electrons via the photoelectric effect. These processes cause an increase in the current. An opposite effect occurs if the gas is effective in attaching electrons.

The coefficient α can be used to predict the onset of discharge in a chamber. According to the Raether condition, breakdown occurs when [1]

$$\alpha d \gtrsim 20$$

where d is the electrode separation. Notice that increasing the electrode separation for constant α increases the probability of a breakdown.

Figure 9.4 Ionization in the gap between parallel plate electrodes.

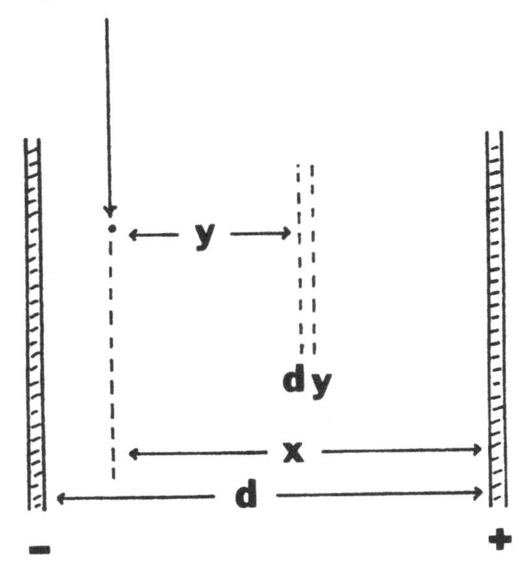

9.3 Proportional chamber gases

A large number of gases have been investigated for use in proportional chambers. Desirable properties of the gas include low working voltage, high gain, good proportionality, high rate capability, long lifetime, high specific ionization, and fast recovery [1]. It is necessary to use mixtures of gases to optimize as many desirable features as possible. The main gas component is usually chosen to be argon, since it has a large multiplication at a relatively low working voltage, has a high total specific ionization, and is economical. Pure argon chambers are limited to gains $\sim 10^3$ since the photons emitted in the deexcitation of excited argon atoms can cause secondary emission at the cathode. For this reason, a polyatomic gas is usually mixed with the argon.

Excited polyatomic molecules have many rotational and vibrational levels. Examples of some commonly used gases are Freon, CO_2, and isobutane. These gases absorb photons over a wide range of energies, and they deexcite through elastic collisions or dissociation into simpler molecules. Secondary emission is also unlikely when ionized polyatomic molecules are neutralized at the cathode. Mixtures of argon with a gas of polyatomic molecules are capable of achieving gains ($\sim 10^6$) much higher than pure argon.

Still higher gains ($\sim 10^7$) can be obtained with the addition of a small amount of an electronegative gas, such as Freon or ethyl bromide. A particular gas mixture in common use ("magic gas") consisting of argon, isobutane, and Freon-13 B1 roughly in the proportion 70/29.6/0.4 gives amplifications of 10^8. Unfortunately, the dissociation products of the isobutane can recombine to form solid or liquid residues that contaminate the chamber electrodes [2]. Positive space charge can build up on this layer for high rate operation and cause serious distortions of the electric field. Some other characteristics of high rate operation are a drop-off of signal size and efficiency, an increase in singles counting rate (noise), and a decrease of the extent of the high voltage operating plateau [1]. This aging process can be reduced by adding a fourth, nonpolymerizing gas, such as methylal, $CH_2(OCH_3)_2$, into the mixture. It is advantageous to work with lower gains if suitable low noise readout electronics is available. Problems associated with chemical deposition on the wires, broken wires, and spark breakdown are all reduced in low gain operation.

Behrends and Melissinos [6] have used a simple, single wire proportional chamber to examine properties of argon–ethane and argon–methane gas mixtures. Measurements of the electron multiplication in argon–ethane at 1 atm, due to the irradiation of the chamber with X-rays, are shown in Fig. 9.5. The multiplication is seen to increase with increasing high voltage and to decrease with increasing ethane concentration. The multiplication M for a given gas composition is found to vary exponentially with high voltage V,

$$M = e^{\gamma(V - V_0)} \tag{9.15}$$

where γ is the slope and V_0 is a threshold voltage. The exponential multiplication was found to be valid for $10^2 < M < 10^4$. The secondary ionization process is limited by space charge when the charge density of positive ions in the vicinity of the wire becomes comparable to the charge on the anode wire. We note from Fig. 9.5 that the slopes of the curves are

Figure 9.5 Electron multiplication as a function of high voltage for mixtures of argon–ethane. (S. Behrends and A. Melissinos, Nuc. Instr. Meth. 188: 521, 1981.)

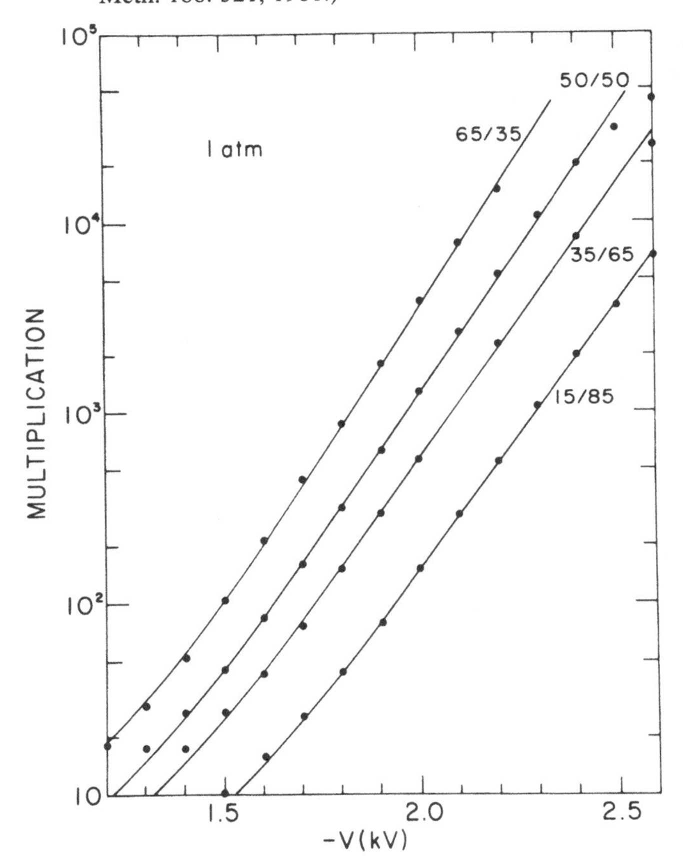

insensitive to the gas composition. Gases with high ethane concentration require a large threshold V_0 before overcoming its quenching ability.

The resolution of the chamber signal was found to be optimum for $M \sim 10^3$. The resolution deteriorates at lower M due to fluctuations in the multiplication process and the stability of the preamplifier gain at low signal levels. The deterioration of the resolution at high multiplication is thought to be due to secondary production of photons, which can photo-emit electrons from the cathode. The resolution at high multiplication improves with increasing ethane concentration.

9.4 Multiwire proportional chambers

A very important development in high energy instrumentation was the construction of multiwire proportional chambers (MWPC) by Charpak and his collaborators at CERN [3, 7]. In a MWPC a plane of anode wires is separated from two cathode planes. Each anode wire acts as an individual proportional counter, thereby allowing a big improvement in spatial resolution. These chambers can be used as trigger elements and have shorter deadtime, better resolving time, and higher efficiency than spark chambers. On the other hand, the signal is small, so that independent amplification of each anode is required.

The geometry of a typical chamber is shown in Fig. 9.6. The anode–cathode gap L is usually chosen to be 3 or 4 times the anode wire spacing s [1]. Some typical values for the dimensions are $s = 2$ mm, $L = 8$ mm, and $d = 30$ μm.

In the limit of zero wire diameter the potential in a chamber with anode wires equidistant from cathode planes is [1]

$$V(x, y) = \frac{CV_0}{4\pi\epsilon_0}\left\{\frac{2\pi L}{s} - \ln\left(4\sin^2\frac{\pi x}{s} + 4\sinh^2\frac{\pi y}{s}\right)\right\} \tag{9.16}$$

Figure 9.6 Geometry of a multiwire proportional chamber.

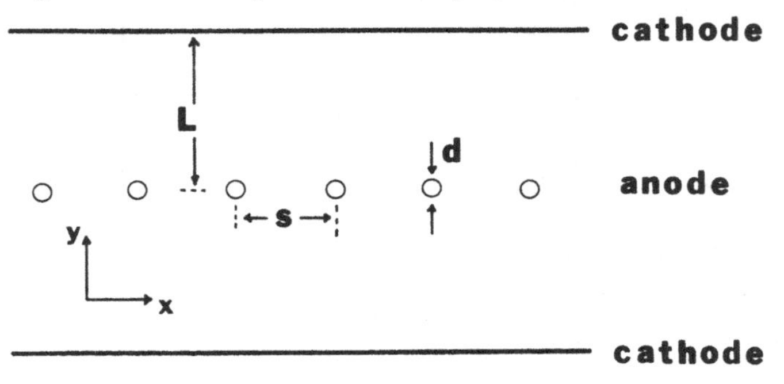

where V_0 is the potential at the anode wires and $V = 0$ at the cathodes. The equipotential lines are approximately circular near the anode wires and planar several wire spacings from the anode plane. The potential for cathode planes consisting of separated cathode wires is more complicated and may deviate significantly from Eq. 9.16 for large cathode wire spacings [8]. The capacitance per unit length is given by

$$C = \frac{2\pi\epsilon_0}{\pi L/s - \ln(\pi d/s)} \tag{9.17}$$

where d is anode wire diameter. The capacitance per unit length of the typical chamber mentioned above is 3.56 pF/m.

The electric field distribution is shown in Fig. 9.7. For most of the time following the creation of an electron–ion pair, the electron merely drifts in the field. Almost all the secondary ionization occurs within several wire diameters from the anode, where the electric field goes as $1/r$. As a result, the amount of multiplication is a strong function of the wire diameter. Thick wires have a very steep gain versus voltage dependence [1]. On the other hand, the wire diameter must not be too small or else it would be incapable of maintaining the tension necessary to resist the electrostatic forces acting on the wire.

Measurements of the avalanche distribution in a MWPC operating in the proportional regime have shown that the avalanche is localized to one

Figure 9.7 Typical electric field pattern in a MWPC (solid lines); equipotential surfaces (dotted lines).

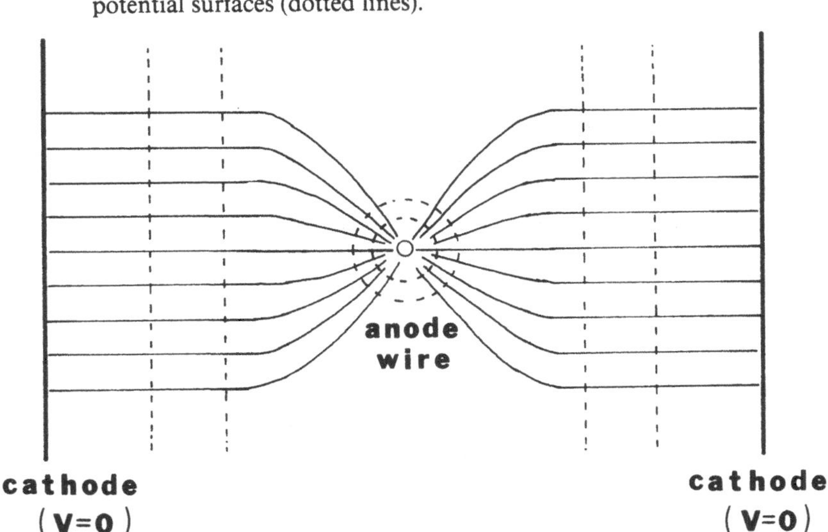

side of the anode wires [9]. However, as the collected charge is increased above 10^7 equivalent electrons, the avalanche begins to surround the wire. Under normal conditions the avalanche is also localized along the wire to within 0.5 mm [1].

The negative pulse on the anode wire induces positive pulses on its neighboring wires, although for $s = 2$ mm the induced signals are only about 20% the size of that on the active wire. There is also a complementary pulse induced on the cathode [3].

Sometimes it is desirable to operate the chamber in the saturated gain regime. Magic gas is frequently used in this way. Under these conditions the anode pulse height is almost independent of the energy lost by the incident ionizing particle. Although the chamber has no energy resolution, it is highly efficient, and the large amplitude and reduced dynamic range of the output pulses simplifies the readout electronics.

It is important that the wire spacing be very uniform throughout the detector. A small displacement of a wire or irregularities in the diameters of the wires can lead to a large change in the charge on the displaced wire and on adjacent wires. The wires tend to stagger themselves out of the plane due to electrostatic repulsion. Vinyl-coated support wires perpendicular to the anode wires have been used to prevent this. Mechanical variations can be expected to lead to variations of 30–40% in the gain [1].

The voltage V_0 applied to the wires must be such that the electrostatic force on the wire is smaller than the restoring force due to the tension T of the wire, or that [1]

$$V_0 \leq \frac{s\sqrt{4\pi\epsilon_0 T}}{LC} \tag{9.18}$$

where L is the length of the wire, s is the wire spacing, and C is the capacitance per unit length. The breaking strength available depends strongly on the wire diameter. For example, for 5-μm-diameter tungsten wires a tension up to 0.04 N can be used, while for 30-μm wires this increases to 1.45 N. Electrostatic forces also cause a net attraction of the cathode electrodes toward the anode plane. This can cause a variation of the gap separation across the chamber and is another cause of gain variation.

The high voltage power supply used with the chamber should be equipped with circuitry to shut off the voltage and discharge the chamber when a sudden surge of current is detected. Otherwise the stored energy in the circuit may cause a gap that breaks down to arc repeatedly, with the danger that a wire may break.

Detailed descriptions of all aspects of chamber construction can be found in the latest proceedings of the International Wire Chamber Conference [10]. For the sake of illustration we will describe here some details of the construction and performance of two actual chambers. Crittenden et al. [11] constructed a set of 1-mm pitch MWPCs capable of operating at high rates. A cross section of one edge of the chamber is shown in Fig. 9.8. Each chamber contained a plane of vertical anode wires with a 1-mm separation and two planes of anode wires with 2-mm spacing oriented $\pm 45°$ from the vertical. The anode wires were 15-μm-diameter gold-plated tungsten, tensioned to 25 g and glued onto a G-10 frame. The chamber was desensitized in the beam region by attaching a small piece of Mylar to the anode wires. Cathode planes were located 3.2 mm on either side of the anode plane. The cathodes were made of stainless steel wire cloth with a wire diameter of 50 μm and a spacing of 0.5 mm. The modules were held in an aluminum frame using RTV silicone rubber to form the gas seal. Alignment pins were used so that it was possible to accurately reassemble the chamber. The gas was confined in the chamber using heat shrunk Mylar windows.

The gas mixture used with the 1-mm chamber was 58% argon, 25% isobutane, 16% methyal, and 0.6% Freon-13 B1. The efficiency of the chamber for various beam intensities is shown in Fig. 9.9 as a function of the chamber high voltage. At low intensity the efficiency rapidly increases to a wide plateau near 100%. The efficiency at a fixed high voltage drops off as the intensity is increased. The efficiency can be partially restored by increasing the high voltage, although the plateau region becomes nar-

Figure 9.8 Cross section of a high rate MWPC. (R. Crittenden, S. Ems, R. Heinz, and J. Krider, Nuc. Instr. Meth. 185: 75, 1981.)

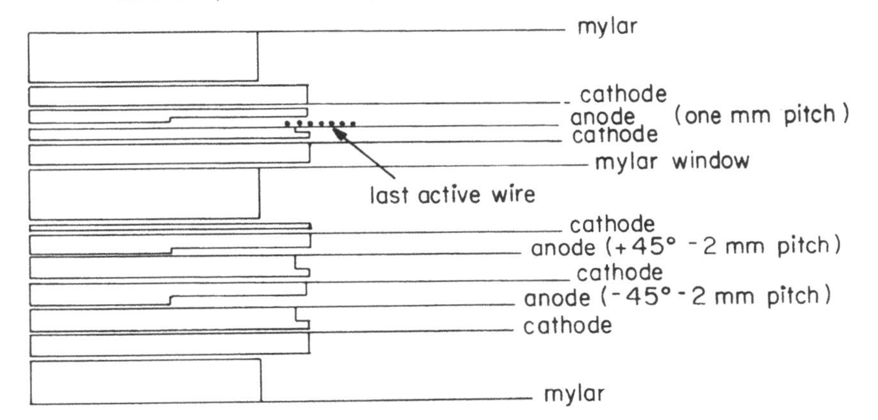

rower. The efficiency at a given high voltage is also strongly dependent on the gas composition.

A set of large (~ 8 m^2) MWPCs were built for the NA3 Spectrometer at the SPS [12]. The chambers were required to handle a large particle flux and to have good efficiency for multitrack events. The anodes were made of 20-μm-diameter gold-plated tungsten. The cathodes consisted of 25-μm-thick plastic foils coated with graphite. The cathode surface was subdivided electrically so that the beam region of the chamber could be desensitized for high rate running. The electrodes were strung over a reinforced epoxy frame, which was accurately pinned into a steel support frame. The aluminized Mylar chamber windows were glued to the steel frame. The long chamber wires were stabilized against electrostatic forces using zigzag-shaped pieces of plastic called garlands.

MWPCs have nearly 100% efficiency for the detection of single, minimum ionizing particles. Because each of the wires is essentially an independent detector, the efficiency for detecting many simultaneous tracks is also very high. Among the more important factors affecting the efficiency are the specific ionization of the gas, the pathlength in the chamber, the high voltage, the angle of incidence of the ionizing particle, and the

Figure 9.9 Particle detection efficiency as a function of high voltage and incident particle rate for the chamber of Fig. 9.8. (R. Crittenden, S. Ems, R. Heinz, and J. Krider, Nuc. Instr. Meth. 185: 75, 1981.)

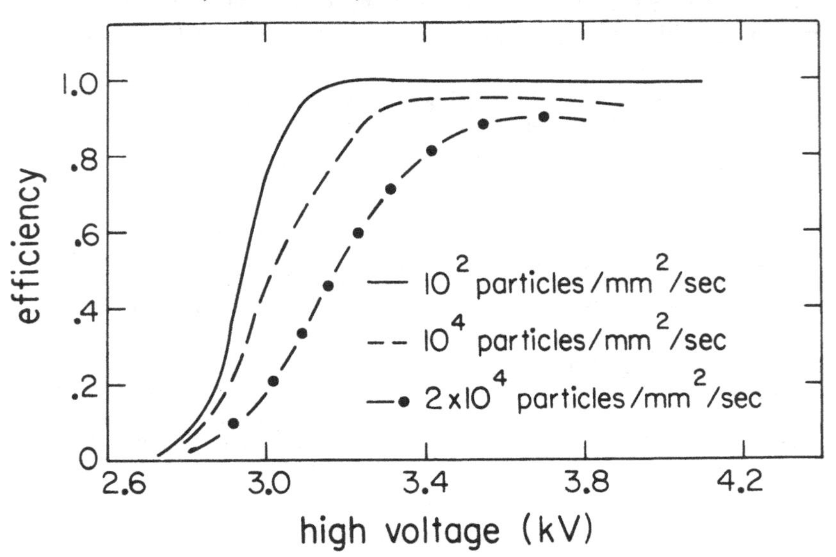

width of the timing gate for accepting pulses after the ionizing particle has passed [3].

The spatial resolution of a MWPC is on the order of half the wire spacing. Improving the spatial resolution by decreasing the anode spacing below 2 mm is difficult because, according to Eq. 9.17, this decreases the capacitance per unit length. Thus, to maintain a fixed gain, the voltage must be correspondingly increased, thereby diminishing the stability of the chamber operation. The spatial resolution can be significantly improved by segmenting and reading out the signal on the cathode. This is discussed further in the following section. With analog readout the resolution can approach the ultimate localization accuracy of the chamber determined by the physical extent and distribution of ionization clusters and the diffusion of the produced charges before they are collected.

Tracks inclined at large angles will fire a cluster of wires [3]. This can be seen in Fig. 9.10, which shows that a 40° track can fire six or seven adjacent wires. The effect can be minimized at the cost of some loss of efficiency by the addition of an electronegative gas such as ethyl bromide. This reduces the sensitive region to a small cylinder around the wire. Alternatively, the timing gate could be shortened so that only pulses from

Figure 9.10 Wire cluster size as a function of the direction of the incident particle. (G. Charpak, reproduced with permission from the Annual Review of Nuclear Science, Vol. 20, © 1970 by Annual Reviews, Inc.)

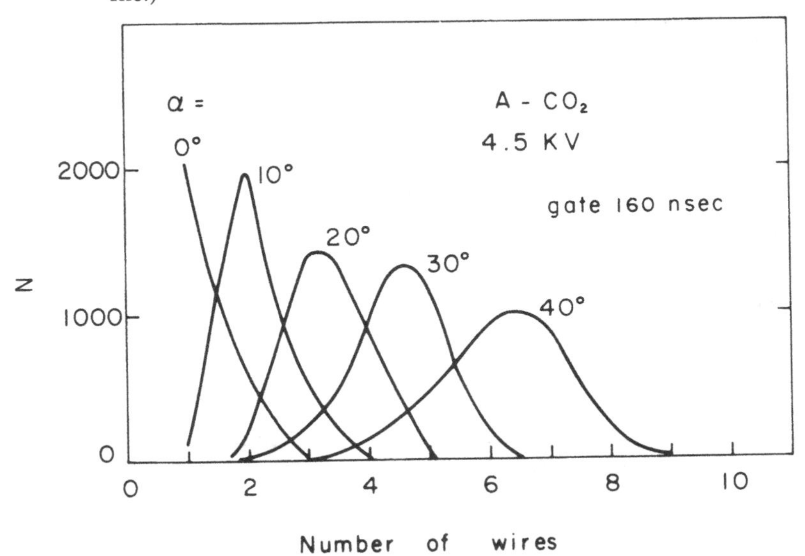

electrons liberated near the wire would be accepted or by only accepting the first pulse to arrive. Production of delta rays also leads to large clusters.

The uncertainty in the arrival time of a pulse at the logic electronics after the passage of an ionizing particle is the time resolution of the chamber. A value of around 30 ns is typical for 2-mm wire spacing. The actual arrival time of the pulse is correlated with the position at which the track crosses the chamber. A typical distribution of time delays on a single wire consists of a large peak arising from tracks that pass within several wire diameters of the wire and a long tail from electrons produced outside this region. The time resolution may be improved at the expense of some inefficiency by the addition of an electronegative gas. Another technique is to use several chambers with the wire spacings staggered, so that the wires of the second chamber are behind the interwire gap of the first chamber. Using the first pulse to arrive from any of the chambers reduces the time jitter.

An anode wire is dead for a period of several hundred nanoseconds following a pulse [2]. The actual deadtime depends on the wire spacing and amplifier electronics. During this time the electrons are collected, the positive ions drift away from the region of the anode, and the amplifier circuit recovers.

Figure 9.11 Effect of magnetic field on the counting rate of several adjacent MWPC wires. (After G. Charpak, D. Rahm, and H. Steiner, Nuc. Instr. Meth. 80: 13, 1970.)

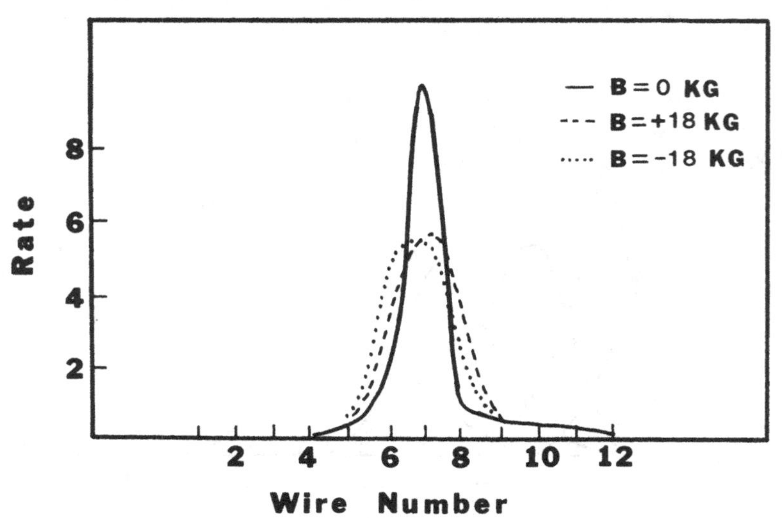

When a MWPC is operated inside a magnetic field oriented parallel to the anode wires, the electrons drift in the direction $\mathscr{E} \times B$ before they are collected at the anode wire. However, the drift is in opposite directions for ions produced on opposite sides of the central wire plane. As a result, if the readout electronics is sensitive to the start of the ionization pulse, the mean of the collected charge distribution is not very sensitive to B [7]. This can be seen in Fig. 9.11, which shows the counting rate on 12 adjacent wires as the magnetic field is varied. Note that the rate on the central wire decreases as the field is increased, but the rate on the adjacent wires increases with the result that the total counting rate remains constant. Thus, the magnetic field only gives a small displacement and a slight loss of resolution. No effect is observed when the magnetic field is parallel to the electric field.

The MWPC wire arrangement can be modified for operation in very high flux environments. In the multistep avalanche chamber [13] the incident particles first enter a preamplification region. The ionization electrons drift toward a pair of wire meshes, which act as a gate to the second half of the chamber. The potential on the gate can be biased so that all the charge is collected, and no electrons make it to the anodes in the second half of the chamber. During the time the electrons are drifting in the preamplification region, external trigger logic can decide if the particle should be detected. In that case a potential of the opposite polarity is applied to the gate. The major problem with the scheme appeared to be capacitive coupling between the gate meshes and the anode.

9.5 Readout electronics

The task of the readout electronics is to provide sufficient information so that the positions where ionizing particles traversed the chamber may be determined as accurately as possible. The time dependence of the anode signal v can be found by assuming that the signal is totally due to a group of positive ions that leave the surface of the anode wire with constant mobility at $t = 0$ [1]. Integrating Eq. 9.4, we find

$$v(t) = \frac{-Q}{2\pi\epsilon_0 L} \ln \frac{r(t)}{r_w} \tag{9.19}$$

where $r(t)$ gives the distance of the positive ions from the axis of the cylinder. Using the definition of mobility, $r(t)$ can be written

$$r(t) = \left(r_w^2 + \frac{\mu^+ C V_0}{\pi\epsilon_0} t \right)^{1/2} \tag{9.20}$$

Substituting back into Eq. 9.19, we find

$$v(t) = -\frac{Q}{4\pi\epsilon_0 L} \ln\left(1 + \frac{\mu^+ C V_0}{\pi\epsilon_0 r_w^2} t\right) \tag{9.21}$$

This important result shows that the voltage has $\ln(1 + t/t_0)$ time dependence.

The output pulse from the anode has a fast initial rise (~6% of maximum in 10 ns). When used with an RC pulse differentiation network, this allows MWPCs to be used as trigger devices. The total undifferentiated pulse length may be several hundred microseconds. The electrons liberated by the ionizing particle drift through varying distances before reaching the high field region near the anode. Thus, the output pulse is the resultant of many pulses arriving over a period of time and has an approximately linear rise at first.

Under typical proportional chamber conditions the initial deposited energy is amplified by 10^4 or 10^5. Even with this amplification individual amplifiers are necessary for each wire to obtain reasonable signal levels (~1 V). The output pulses deviate from strict proportionality because of fluctuations in the number of initial ion pairs and in the amplification.

The design of the chamber readout electronics is as varied as the design of the chambers themselves. One possible readout arrangement for each wire is shown in Fig. 9.12. The amplified signal from the anode wire is sent to a discriminator with an adjustable threshold. The threshold setting (~0.5 mV) is optimized to reject noise, yet maintain high efficiency. In general, full efficiency is obtained for minimum ionizing particles with a threshold that is about one-tenth the peak amplitude [1]. A loss of effi-

Figure 9.12 Simple readout scheme for an anode wire.

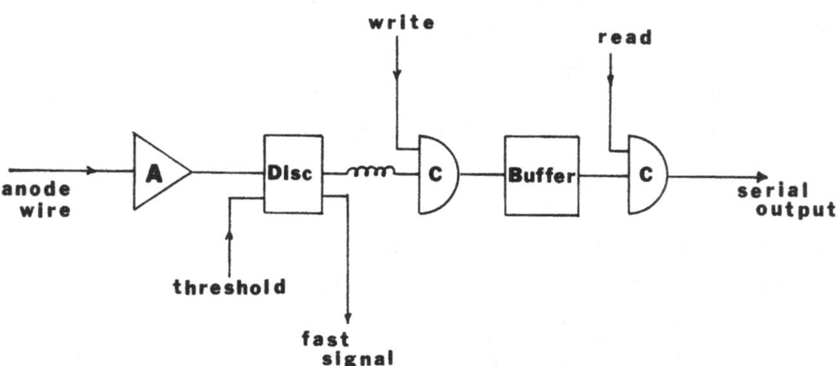

ciency from increasing the discriminator threshold can be offset to some extent by increasing the high voltage.

Following the discriminator a fast signal can be extracted for use in the trigger logic. The signals from all the wires can be combined in order to make multiplicity tests. Another signal from each of the discriminators is then delayed while the trigger decision is being made. For good events the trigger initiates a WRITE signal that allows the anode signals to pass into a buffer. At an appropriate time in the readout cycle, the data handler sends a READ pulse, which allows the signal for each wire to enter the readout data stream. A typical readout scheme would specify the address of one of the hit wires in a cluster and the width of the cluster. Readout electronics systems similar to this are available commercially [14].

A block diagram of the readout electronics for the NA3 MWPCs discussed earlier is shown in Fig. 9.13. The major features of the system include amplifier and readout cards that plug into connectors on the edge

Figure 9.13 Block diagram of the NA3 MWPC readout electronics. (A) amplifier card, (R) readout card, (T) trigger signal, (C) control bus, (D) data bus, (PS) power supply, and (DA) data acquisition system. (After R. Hammarstrom et al., Nuc. Instr. Meth. 174: 45, 1980.)

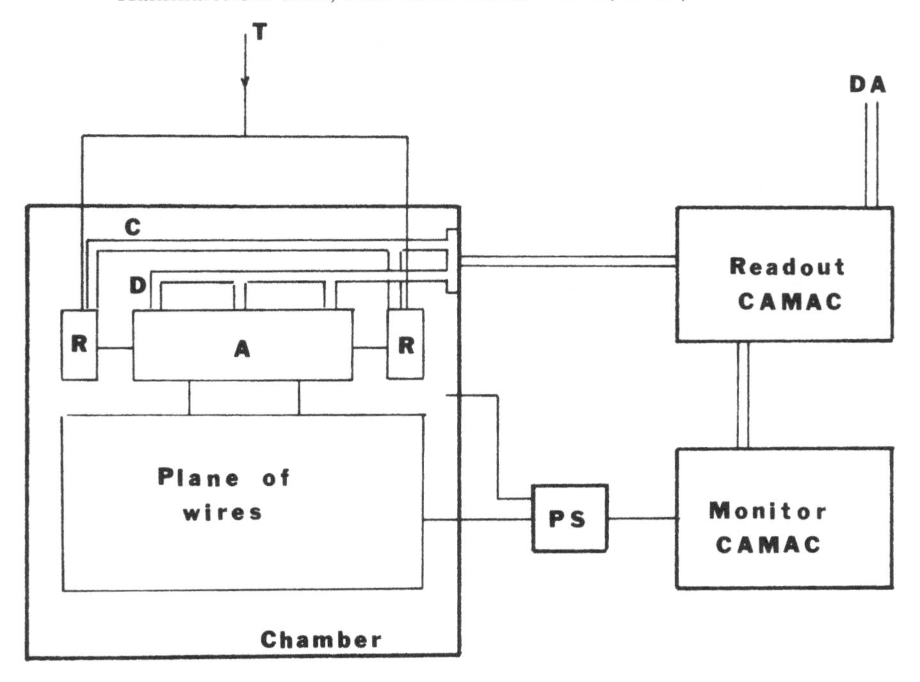

of the chamber, CAMAC modules to enter the MWPC information into the data acquisition system, a computer-controlled power supply system, and a dedicated CAMAC control and monitoring system [12]. The amplifier cards used hybrid integrated circuits that contained differential amplifiers, discriminators, one-shot delays, and an output gate capable of driving the external data bus. The discriminator threshold could be varied from 1 to 10 μA, while the time delay could be adjusted from 330 to 750 ns. During the readout cycle the amplifier card was addressed, causing it to put the contents of the 16 output gates onto the data bus. The data was converted by the CAMAC modules into the corresponding addresses of the hit wires. The readout of a complete chamber plane took about 60 μs.

Improved spatial resolution can be obtained if analog signals from the wires are measured. In this case signals from nearby wires can be averaged in an appropriate manner to interpolate the value of the actual coordinate. The most important analog techniques that have been used for this purpose are listed in Table 9.4. When using these methods, the readout electronics is necessarily more complicated than that shown in Fig. 9.12. The methods listed in Table 9.4 are also used for drift chambers.

Analog signal methods are frequently used to obtain a second coordinate of the particle trajectory without introducing additional planes of wires at different angles. The second coordinate is typically the direction along the anode wire and orthogonal to the particle's trajectory for planar chambers, or the coordinate along the axis for cylindrical chambers.

A common construction uses the amplifier–discriminator readout system mentioned previously for the anode signals and a perpendicular, segmented cathode readout to give the coordinate along the anode wire direction. The cathode strip (or cathode pad) method utilizes the induced charge on a segmented cathode [15]. Chambers with this feature can combine excellent spatial resolution with inherently good time resolu-

Table 9.4. *Analog methods for position measurements in wire chambers*

Method	Requirements
Cathode strips	segmented cathode, each with readout
Delay line	supplementary transmission lines, dual readout
Charge division of anode signal	resistive anode, dual readout
Risetime of anode signal	resistive anode, dual readout
Direct timing of anode signal	high conductivity anode, dual readout

tion. Figure 9.14 illustrates the basic principle. One cathode of the MWPC is segmented perpendicularly to the anode wires. The width of the segments is roughly equal to the anode–cathode separation distance. As the ions are drifting between the electrodes, pulses are induced on nearby cathodes. A charge Q on an anode situated between two cathode planes will induce a charge distribution [16]

$$\sigma(z) = -\frac{Q}{4L}\,\mathrm{sech}\,\frac{\pi z}{2L} \tag{9.22}$$

on the perpendicular cathode, where z is the distance along the anode wire and L is the anode–cathode separation. The charge on each cathode pad is measured using an accurate ADC.

The simplest method for determining the position of the particle is to compute the center of gravity

$$\bar{x} = \frac{\sum Q_i X_i}{\sum Q_i} \tag{9.23}$$

of the charges Q_i deposited on the strips [16, 17]. However, the deposited charge must be integrated over the finite width of the strip. This introduces nonlinear correlations and can lead to systematic shifts between the avalanche position and the calculated centroid. Additional nonlinearities may be caused by mutual capacitance between the strips, electronic cross

Figure 9.14 Two-dimensional readout using cathode strips. (A. Breskin et al., Nuc. Instr. Meth. 143: 29, 1977.)

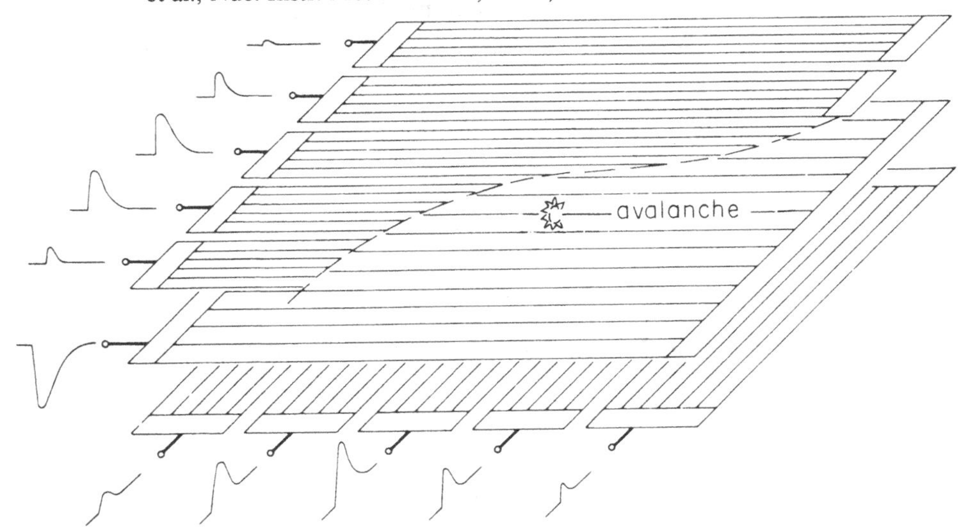

talk, and gain variations between channels. The accuracy of the calculation can be improved by subtracting a bias level from each strip before averaging [17]. Small chambers instrumented with cathode strip readout have achieved 40 μm rms spatial resolution along the anode wires. The resolution is slightly worse for the direction perpendicular to the anode wires.

A delay line is a transmission line that allows the position of a particle to be determined from the difference in arrival times of the signals at the two ends of the line [18]. The delay line may be either tapped directly to the cathode wires, or capacitatively coupled if the signal is sufficiently strong. Delay lines parallel to the anode can limit the rate capability of the chamber.

As an example we mention the drift chambers of the NA-1 Spectrometer at CERN [19]. The chambers used delay lines to provide a measurement of the second coordinate and to aid in the resolution of the left–right ambiguity. The delay lines consisted of 2-mm-diameter aluminum pipe surrounded by Teflon and copper wire. The delay lines ran parallel to the drift chamber sense wires. The signal induced on the delay line propagated with a velocity of 1.6 mm/ns to the edge of the chamber, where a coupling transformer matched the signal to an amplifier. The characteristic impedance was 1800 Ω. The FWHM in the length determination was 4.5 mm.

Analog methods that use the anode signal itself have the advantage of not complicating the chamber construction or introducing additional scattering material. The charge division method determines the position of the avalanche caused by a passing particle by measuring the amount of

Figure 9.15 Principle of the charge division method.

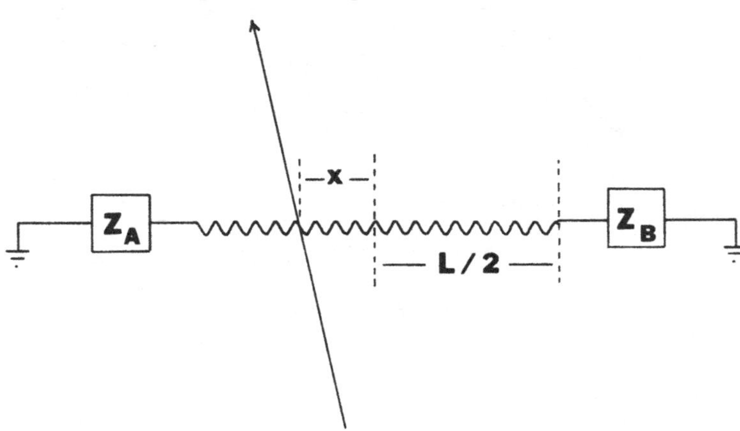

charge reaching the two ends of a resistive wire [20, 21]. Consider a wire chamber constructed with high resistance anode wires of length L as shown in Fig. 9.15. The two ends of the wire are connected to charge sensitive preamplifiers with input impedances Z_A and Z_B. Suppose the avalanche from the passage of a charged particle hits the anode at a distance $\frac{1}{2}L - x$ from the A end. The charges that are induced at the two ends of the wire are inversely proportional to the total impedances along the two paths, so long as the integration time for the signal is long compared to $\rho L C_W$, where ρ and C_W are the resistivity and distributed capacitance of the anode wire. Thus, the ratio of the collected charges is given by

$$\frac{Q_A}{Q_B} = \frac{(\frac{1}{2}L + x)\rho + Z_B}{(\frac{1}{2}L - x)\rho + Z_A} \tag{9.24}$$

It is desirable that the preamplifier input impedances be small compared to the resistance of the wire. The quantities that are actually measured are the ADC channels of the signals. Thus, the position determination may require corrections for any differences in amplifier gain, pedestal corrections, and so on. Resolutions of ~0.4 mm have been achieved using charge division [22].

The other two analog methods mentioned in Table 9.4 measure either the risetime of the pulses or the arrival time of the pulses at the two ends of the anode. The readout electronics for the risetime method is simpler than for the charge division method [23]. The method has the disadvantage that it is dependent on all parasitic capacitances and inductances in the chamber and so may not be suitable for large chambers. The direct timing method uses the anode and surrounding electrodes as a delay line [24]. The position of a particle is then determined from the difference of arrival times at the two ends. The resolution of this method is expected to be independent of the length of the wires.

The pulse height of MWPC anode signals have been used to measure the energy loss of particles. This subject will be discussed in the following chapter.

References

A description of the construction and operation of classical ionization, proportional, and Geiger counters can be found in S. Korff, *Electron and Nuclear Counters,* New York: Van Nostrand, 1955.

[1] F. Sauli, Principles of operation of multiwire proportional and drift chambers, CERN Report 77-09, 1977. References to most of the original work on proportional chambers and MWPCs can be found in this and the following reference.

[2] P. Rice-Evans, *Spark, Streamer, Proportional, and Drift Chambers,* London: Richelieu, 1974.

[3] G. Charpak, Evolution of the automatic spark chambers, Ann. Rev. Nuc. Sci. 20: 195–254, 1970.

[4] A. von Engel, *Ionized Gases,* London: Oxford, 1955.
[5] V. Davidenko, B. Dolgoshein, V. Semenov, and S. Somov, Measurements of the relativistic increase of the specific primary ionization in a streamer chamber, Nuc. Instr. Meth. 67: 325–30, 1969.
[6] S. Behrends and A. Melissinos, Properties of argon-ethane/methane mixtures for use in proportional chambers, Nuc. Instr. Meth. 188: 521–34, 1981.
[7] G. Charpak, D. Rahm, and H. Steiner, Some developments in the operation of multiwire proportional chambers, Nuc. Instr. Meth. 80: 13–34, 1970.
[8] S. Beingessner and L. Bird, An extension of the standard formulae for electric fields in MWPCs, Nuc. Instr. Meth. 172: 613–6, 1980.
[9] J. Fischer, H. Okuno, and A. Walenta, Spatial distributions of the avalanches in proportional chambers, Nuc. Instr. Meth. 151: 451–60, 1978.
[10] Proceedings of Wire Chamber Conference, Nuc. Instr. Meth. 176: 1–432, 1980; Nuc. Instr. Meth. 217: 1–381, 1983.
[11] R. Crittenden, S. Ems, R. Heinz, and J. Krider, A design for one mm pitch MWPCs operating at high rates, Nuc. Instr. Meth. 185: 75–9, 1981.
[12] R. Hammarstrom, P. Kristensen, R. Lorenzi, G. Matthiae, A. Michelini, O. Runolfson, J. Timmermans, and M. Uldry, Large multiwire proportional chambers for experiment NA3 at the CERN SPS, Nuc. Instr. Meth. 174: 45–52, 1980.
[13] A. Breskin, G. Charpak, S. Majewski, G. Melchart, G. Petersen, and F. Sauli, The multistep avalanche chamber: A new family of fast, high rate particle detectors, Nuc. Instr. Meth. 161: 19–34, 1979; F. Sauli, Possible improvements in the performance of gaseous detectors, Physica Scripta 23: 526–33, 1981.
[14] For example, the PCOS III MWPC System, LeCroy Research Systems, Spring Valley, NY.
[15] G. Charpak, G. Melchart, G. Petersen, and F. Sauli, High accuracy localization of minimum ionizing particles using the cathode induced charge centre-of-gravity readout, Nuc. Instr. Meth. 167: 455–64, 1979.
[16] I. Endo, T. Kawamoto, Y. Mizuno, T. Ohsugi, T. Taniguchi, and T. Takeshita, Systematic shifts of evaluated charge centroid for the cathode readout proportional chamber, Nuc. Instr. Meth. 188: 51–8, 1981.
[17] G. Charpak and F. Sauli, High resolution electronic particle detectors, Ann. Rev. Nuc. Part. Sci. 34: 285–349, 1984.
[18] E. Beardsworth, J. Fischer, S. Iwata, M. Levine, V. Radeka, and C. Thorn, Multiwire proportional chamber focal-plane detector, Nuc. Instr. Meth. 127: 29–39, 1975.
[19] S. Amendolia, G. Batignani, E. Bertolucci, L. Bosisio, U. Bottigli, C. Bradaschia, M. Budinich, F. Fidecaro, L. Foa, A. Giazotto, M. Giorgi, F. Liello, P. Marrocchesi, A. Menzione, M. Quaglia, L. Ristori, L. Rolandi, A. Scribano, R. Stanga, A. Stefanini, and M. Vincelli, A set of drift chambers built for the FRAMM-NA1 Spectrometer, Nuc. Instr. Meth. 176: 461–8, 1980.
[20] P. Rehak, Detection and signal processing in high energy physics, in G. Bologna (ed.), *Data Acquisition in High Energy Physics,* Amsterdam: North-Holland, 1983, pp. 25–89.
[21] V. Radeka and P. Rehak, Charge dividing mechanism on resistive electrode in position sensitive detectors, IEEE Trans. Nuc. Sci. NS-26: 225–38, 1979.
[22] J. Buskens, B. Koene, L. Linssen, P. Rewiersma, H. Schuijlenburg, R. Van Swol, J. Timmermans, M. Haguenauer, G. Roiron, and J. Velasco. Small high precision wire chambers for the measurement of $\bar{p}p$ elastic scattering at the CERN Collider, Nuc. Instr. Meth. 207: 365–78, 1983.
[23] E. DeGraaf, J. Smits, A. Paans, and M. Woldring, The rise time method for the readout of a proportional chamber: Linearity, sensitivity, and noise, Nuc. Instr. Meth. 200: 311–31, 1982.

[24] R. Boie, V. Radeka, P. Rehak, and D. Xi, Second coordinate readout in drift chambers by timing of the electromagnetic wave propagating along the anode wire, IEEE Trans. Nuc. Sci. NS-28: 471–7, 1981.

Exercises

1. Derive an expression for the radius at which ion multiplication begins in a cylindrical proportional chamber in terms of the applied voltage and the properties of the chamber and of the gas.

2. Use the data of Table 9.3 to estimate the ionization cross for electrons in argon in a 100-V/cm field at 0.5 atm.

3. Derive Eq. 9.12. Use the kinetic theory and assume that any electron whose energy exceeds the ionization potential of the gas will cause ionization.

4. For voltages $V \gg V_0$ the gain factor M can be written $M = ke^Q$, where k is a constant and Q is the charge per unit length on the wire. Find the gain variations due to variations in the anode wire diameter, anode wire spacing, and the anode–cathode separation.

5. Estimate the expected signal for a 3-GeV/c proton in a 2-cm-radius proportional chamber containing a 50% Ar/50% C_2H_6 gas mixture at 2000 V.

6. A MWPC with argon gas at atmospheric pressure has $L = 2s$ and 10-μm-diameter anode wires. If the ion multiplication is 10^4, what wire spacing would give 100-mV pulses for minimum ionizing particles?

7. A MWPC has 12-μm-diameter anode wires, 2 mm wire spacing, and 4 mm anode–cathode spacing. Find the critical voltage for wire breakage if the 1-m-long wires are tensioned to 1 N.

8. A 1-m-long wire with resistivity 60 Ω/cm has readout circuits with input impedance 100 Ω at each end. Find the position along the wire of a passing particle if the peak signal is in channel 36 on one side and in channel 20 on the other. Assume the ADC pedestals are in channel 5.

10
Drift chambers

A drift chamber is a particle tracking device that uses the drift time of ionization electrons in a gas to measure the spatial position of an ionizing particle [1–3]. Drift chambers can achieve spatial resolutions an order of magnitude smaller than MWPCs. A typical timing measurement accuracy of 2 ns and a drift velocity of 4 cm/μs corresponds to a theoretical spatial accuracy of 80 μm. Drift chambers have found wide acceptance as charged particle tracking detectors in spectrometer systems. When covering large solid angles, drift chambers provide resolution equal to MWPCs at a much lower cost.

A typical drift chamber arrangement is shown in Fig. 10.1. A region of approximately uniform field is set up between the anode and cathode wires. A charged particle traversing the chamber liberates electrons that drift toward the anodes. The passage of the particle also generates a fast pulse in the scintillation counter that can be used to define a reference time t_0. The electrons drift for a time Δt, after which they are collected at the anode, thereby providing a signal that a particle has passed. The position where the particle traversed the chamber is then given by

$$x = \int_{t_0}^{t_0+\Delta t} w(t)\, dt \tag{10.1}$$

where $w(t)$ is the drift velocity.

10.1 Properties of drift chamber gases

The passage of ionizing radiation through the gas in a drift chamber cell leaves behind a trail of electrons and positive ions. In order to use the electron arrival times to accurately locate the position in the chamber that was traversed by the incoming particle, we must have an accurate knowledge of the drift velocity of the electrons in the gas. The liberated electrons have inelastic collisions with the surrounding gas molecules and quickly establish a Maxwellian distribution of velocities. The most probable random (or thermal) electron energy at room temperature is ~ 0.04 eV. When an electric field $\mathscr{E}$ is applied, the component of the drift velocity w parallel to the field is [4, 5]

$$w_{\parallel} = \frac{2}{3}\frac{e\mathscr{E}}{m}\left\langle\frac{\lambda}{v}\right\rangle + \frac{1}{3}\frac{e\mathscr{E}}{m}\left\langle\frac{d\lambda}{dv}\right\rangle \tag{10.2}$$

where e and m are the charge and mass of the electron and $\lambda(v)$ is the mean free path for an electron with random velocity v in the drift chamber gas. The averages must be performed over the distribution of electron random velocities. The variation of λ with v is often ignored when estimating gas properties. Note that $\langle\lambda/v\rangle$ is the mean time between collisions. Some measurements of drift velocities in pure gases are given in Table 10.1.

If a magnetic field B perpendicular to $\mathscr{E}$ is also present, the drift velocity will have a component along the direction $\mathscr{E} \times \mathbf{B}$ given by

$$w_{\perp} = \left(1 + \frac{e^2B^2\lambda^2}{m^2v^2}\right)^{-1}\left(\frac{1}{3}\frac{e\mathscr{E}}{m}\omega_{\mathrm{L}}\left\langle\frac{\lambda^2}{v^2}\right\rangle + \frac{2}{3}\frac{e\mathscr{E}}{m}\omega_{\mathrm{L}}\left\langle\frac{\lambda}{v}\frac{d\lambda}{dv}\right\rangle\right) \tag{10.3}$$

Figure 10.1 Principle of operation of a drift chamber.

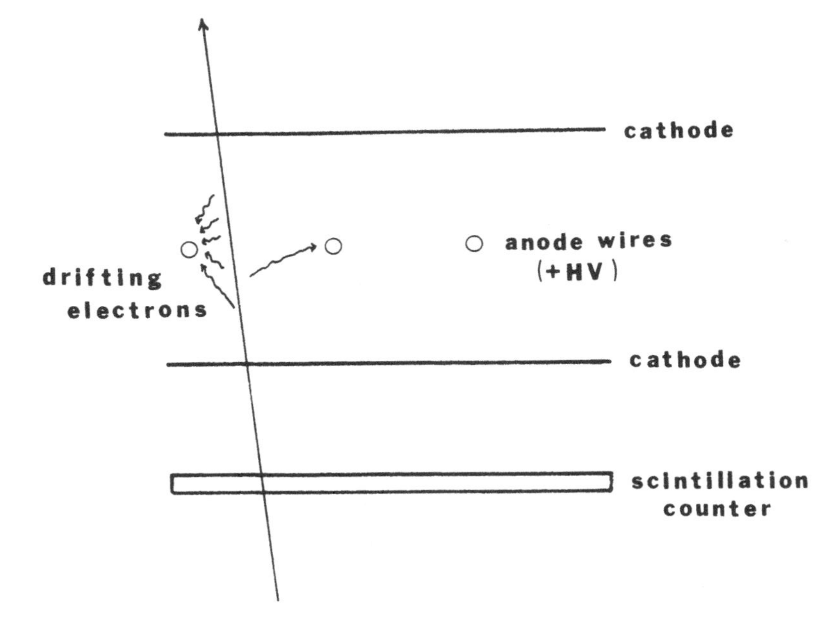

where

$$\omega_L = eB/m \qquad (10.4)$$

is the Larmor precession frequency. The net drift angle with respect to $\mathscr{E}$ in this case would be

$$\tan \alpha = w_\perp / w_\parallel \qquad (10.5)$$

Figure 10.2 shows measurements of the drift angle as a function of $\mathscr{E}$ and B. Besides electric and magnetic fields, other factors that can influence the drift velocity include the temperature and pressure of the gas and the addition of "impurity" gases. Note that the drift velocity is a function of $\mathscr{E}/P$, where P is the gas pressure.

A second important property of a drift chamber gas is the diffusion coefficient for electrons traveling through the gas. The amount of diffusion that occurs limits the spatial resolution obtainable with the chamber. According to classical kinetic theory, the electron cloud will diffuse, as it drifts along with a diffusion coefficient D, given by

$$D = \tfrac{1}{3} \langle \lambda v \rangle \qquad (10.6)$$

Table 10.1. *Drift velocities of pure gases, STP (cm/μs)*

	$\mathscr{E}$ (kV/cm)		
Gas	1	2	3
CH_4	10.4	8.5	7.1
C_2H_6	5.2	5.5	5.5
C_3H_8	4.0	4.7	4.9
Iso-C_4H_{10}	3.3	5.1	
C_2H_4	4.8	5.1	4.8
He	1.0	1.6	2.3
Ar	0.4	0.7	1.1
Ar (liquid)	0.2	0.3	0.3
Xe	0.2	0.4	0.6
H_2	1.1	1.8	2.0
N_2	0.9	1.4	2.0
O_2	2.4	3.1	3.8
CO_2	0.7	1.5	2.6
$CH_2(OCH_3)_2$	0.7	1.5	2.1

Source: A. Peisert and F. Sauli, CERN Report 84-08, 1984; B. Jean-Marie, V. Lepeltier, and D. L'Hote, Nuc. Instr. Meth. 159: 213, 1979; F. Sauli, CERN Report 77-09, 1977.

After drifting a distance x, each component of the electron's position becomes dispersed by an amount

$$\sigma_x = \sqrt{\frac{2Dx}{w}} \tag{10.7}$$

Note, however, that the diffusion coefficient need not be isotropic.

Simple estimates for the magnitudes of the transport coefficients have been given by Palladino and Sadoulet [4] using conservation of energy arguments. The electron cloud tends to be accelerated by the electric field. On the other hand, collisions between the electrons and the gas atoms occur with a mean time between collisions of v/λ. After a time $\sim 10^{-11}$ sec the electron cloud reaches equilibrium with the gas and travels with an approximately constant drift velocity. If $\Gamma(E)$ is the mean fractional energy loss per collision, then the energy gained per unit time from the electric field must equal the energy lost per unit time in atomic collisions, or

$$e\mathscr{E}w_{\parallel} = \left\langle \Gamma E \frac{v}{\lambda} \right\rangle$$

If we assume λ is independent of v and that the energy distribution is narrow, then

$$e\mathscr{E}w_{\parallel} = \tfrac{1}{2}mv^2\Gamma\frac{v}{\lambda}$$

Figure 10.2 Drift angle as a function of electric and magnetic fields. (U. Becker, M. Capell, D. Osborne, and C. Ye, Nuc. Instr. Meth. 205: 137, 1983.)

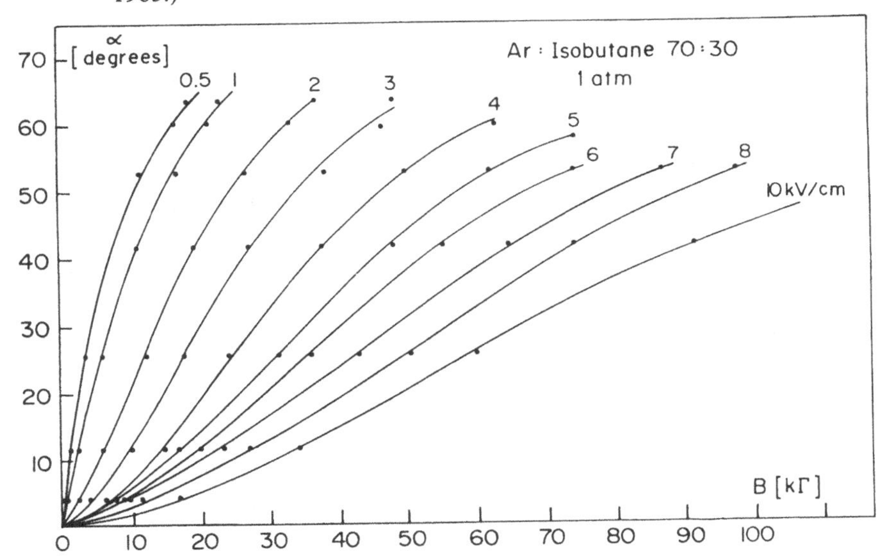

Substituting $w_{\parallel}$ from Eq. 10.2, we obtain the following expression for v:

$$v = \left[\frac{2e\mathscr{E}\lambda}{(3\Gamma)^{1/2}m}\right]^{1/2} \tag{10.8}$$

It is convenient to use Eq. 10.8 to express the various transport coefficients in terms of λ and Γ. Thus, we find that the components of the drift velocity are

$$w_{\parallel} = \left(\frac{2}{3}\frac{e\mathscr{E}\lambda}{m}\left(\frac{\Gamma}{3}\right)^{1/2}\right)^{1/2} \tag{10.9}$$

$$w_{\perp} = \frac{eB\lambda}{6m}(3\Gamma)^{1/2} \tag{10.10}$$

The mean energy of the electrons is

$$\begin{aligned}\langle E\rangle &= \tfrac{1}{2}mv^2\\ &= e\mathscr{E}\lambda(3\Gamma)^{-1/2}\end{aligned} \tag{10.11}$$

while the diffusion coefficient is given by

$$D = \left[\frac{2e\mathscr{E}\lambda^3}{9(3\Gamma)^{1/2}m}\right]^{1/2} \tag{10.12}$$

Note that the combination of terms

$$\begin{aligned}E_k &= \frac{eD\mathscr{E}}{w_{\parallel}}\\ &= \langle E\rangle\end{aligned} \tag{10.13}$$

gives a directly measurable indication of the characteristic energy of the electrons in the gas.

Examination of Eqs. 10.9–10.13 shows that they all vary monotonically with the mean free path for electrons in the gas. The dependence is strongest for the diffusion coefficient, which is proportional to $\lambda^{3/2}$. The components of the drift velocity also increase for increasing fractional energy loss per collision, while the mean energy and diffusion coefficient decrease.

Accurate calculations of the properties of mixtures of gases can be made using transport theory [4–6]. The theory uses density and energy conservation to obtain a set of equations for the distribution function $F(E)$ for a free electron in an external electric or magnetic field. Elastic and inelastic cross sections for electrons in the relevant gases are used as input. Once the equations have been solved for $F(E)$, the transport coefficients, such as w and D, can be obtained from appropriate integrals over the energy.

The optimum gas mixture for a wire chamber is strongly dependent on the intended application [6]. Table 10.2 indicates gas mixtures that give various desired properties. Gases with high specific energy loss are used for efficient detection of charged particles and photons. A gas with low specific energy loss would be used if it was important to minimize multiple scattering. High drift velocity gases are used in high rate environments, while a lower drift velocity improves the spatial resolution obtainable with electronics of a given speed. Small diffusion is desirable for good spatial resolution. Chambers are typically operated in a region where the drift velocity is independent of $\mathscr{E}$ (saturation region). The choice of the actual drift velocity that is used is influenced by the available drift path, particle intensity, and speed of the recording electronics.

Argon is widely used as a drift chamber gas. The argon may be mixed with gases consisting of heavy organic molecules, such as carbon dioxide or isobutane. These gases, known as quenchers, have many degrees of freedom and can efficiently absorb energy from the gas [7]. As a result, the effective temperature of the electron is reduced, the drift velocity is increased, and the diffusion is decreased. Sometimes a small amount of a third gas, such as dimethoxymethane (DMM), is also added. A typical composition is 68% Ar/30% C_4H_{10}/2% DMM.

Figure 10.3 shows measured drift velocities for argon–propane mixtures as a function of the applied electric field [8]. The curves have been displaced by 10 μm/ns from each other for clarity. The dotted line associated with each curve corresponds to a velocity of 50 μm/ns. Note that certain gas proportions result in drift velocities which are independent of $\mathscr{E}$. This phenomenon is quite important in achieving linear space–time correlations, since it minimizes the change in drift velocity as the electrons approach the sense wire. The drift chamber gas should not attach electrons, or else the detection efficiency will depend on the drift distance.

Table 10.2. *Properties of gas mixtures*

Desired property	Gas components
High specific mass	Xe
Low specific mass	H_2, He
High drift velocity	CH_4, CF_4
Low drift velocity	CO_2, dimethylether, He-C_2H_6
Small diffusion	hydrocarbons, CO_2, dimethylether, NH_3
Small magnetic deflection angle	CO_2, CO_2-isoC_4H_{10}
Electron capture	O_2, H_2O, CF_3Br, Freons

Source: A. Peisert and F. Sauli, CERN Report 84-08, 1984.

Figure 10.3 Drift velocities of argon – propane gas mixture as a function of electric field. (After B. Jean-Marie, V. Lepeltier, and D. L'Hote, Nuc. Instr. Meth. 159: 213, 1979.)

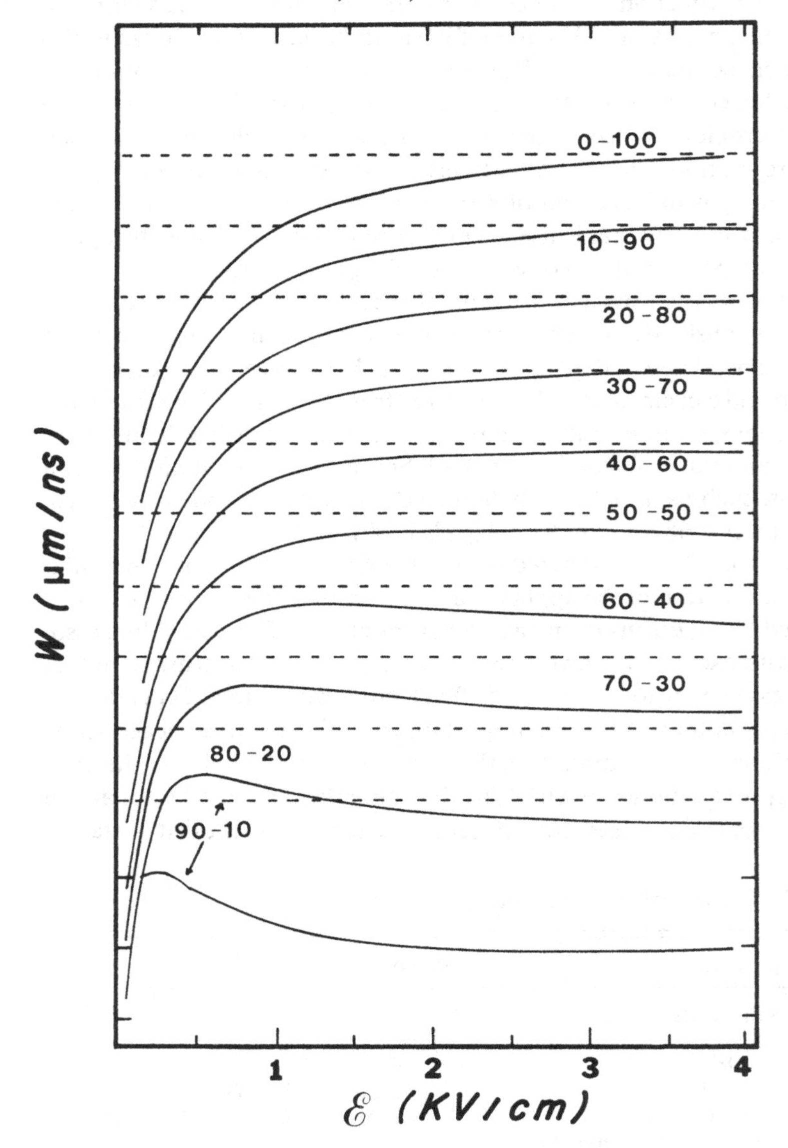

10.2 Construction

Three major types of drift chambers are in common use: (1) planar chambers, (2) cylindrical chambers, and (3) proportional drift tubes. Cylindrical chambers are used to surround an interaction region, particularly at colliding beam machines. Proportional drift tubes are useful in large volume applications, such as for calorimeters. The choice of the size of the drift cell is influenced by the maximum expected event rate and track multiplicity. The maximum rate that can be tolerated per wire appears to be ~ 10^4 counts/mm-s [9].

Now let us consider several examples of actual chambers. The wire arrangement in the BNL MPS planar chambers with small 3.2 mm drift space is shown in Fig. 10.4. The drift space is bounded by planes of cathode wires held at a large negative potential [10]. The drifting electrons are collected on the grounded anode wires. Gold-plated tungsten wire 25 μm diameter is used for the anode, while 75-μm diameter stainless steel is used for the field and cathode wires. The electric field uniformity is improved by using field wires between the anodes. These wires are held at a slightly higher potential than the cathode.

The MPS chamber wires were strung on strong fiberglass–polyester frames and stacked together to form modules. The anode spacing of the x' measuring planes are displayed by half a drift cell width in order to help resolve the ambiguity of whether a track has passed to the right or left side of a given anode wire. Each module also contains y measuring planes and u and v ($\pm 30°$ to y) measuring planes. This enables each module to measure a vector on the particle's trajectory, which simplifies the pattern recognition. The chambers use a 79% Ar/15% C_4H_{10}/6% dimethoxymethane gas mixture. The maximum drift time for these chambers is 70 ns, and the space–time relation shown in Fig. 10.5 is quite linear. One nice feature of small drift spaces is that the $\mathscr{E} \times \mathbf{B}$ displacement is small.

It is usually desirable for the field to be uniform in the drift space, so that x varies linearly with t. In chambers with large drift spaces located in a

Figure 10.4 Wire arrangement for the MPS planar drift chambers. (A) anode wire, (F) field shaping wire, and (C) cathode wires. The anode–cathode spacing is 6.4 mm and the anode–field wire spacing is 3.2 mm.

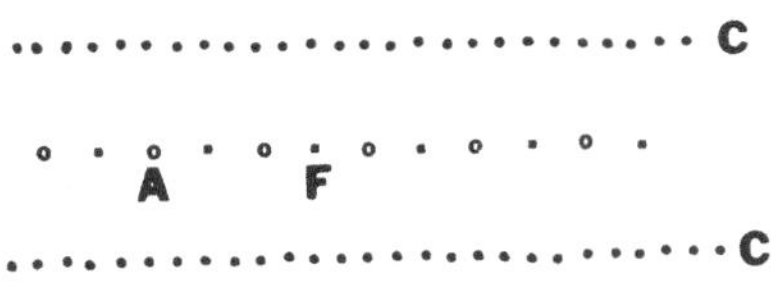

magnetic field, the potential on each cathode wire may be gradually varied by applying the high voltage through a divider chain [1].

Cylindrical drift chambers have found wide acceptance at intersecting storage rings. Wagner [11] separates the chambers that have been built so far into "classical" and "pictorial" categories. The classical chambers are defined to have 15–20 planes of wires, approximately half of which are oriented parallel to the magnetic field axis. These planes are used for pattern recognition. The other wire planes are rotated by a small angle with respect to the field axis and are used to resolve ambiguities and to determine the z coordinates of the reconstructed tracks. In a pictorial drift chamber, on the other hand, 40–200 points are recorded for each track. The high measurement density is essential for good separation of closely spaced tracks and for finding secondary vertices and decays. The term *pictorial* arises since there are sufficient hits in the chamber to recognize tracks without fitting. Such chambers typically put a much larger burden on the readout electronics. The TPC, discussed in Section 4, is an example of this type of chamber. Table 10.3 lists characteristics of some large cylindrical drift chamber systems.

Figure 10.5 Drift time versus track position in the MPS planar drift chambers in a 10-kG field. (After E. Platner, Brookhaven National Laboratory report BNL 30898.)

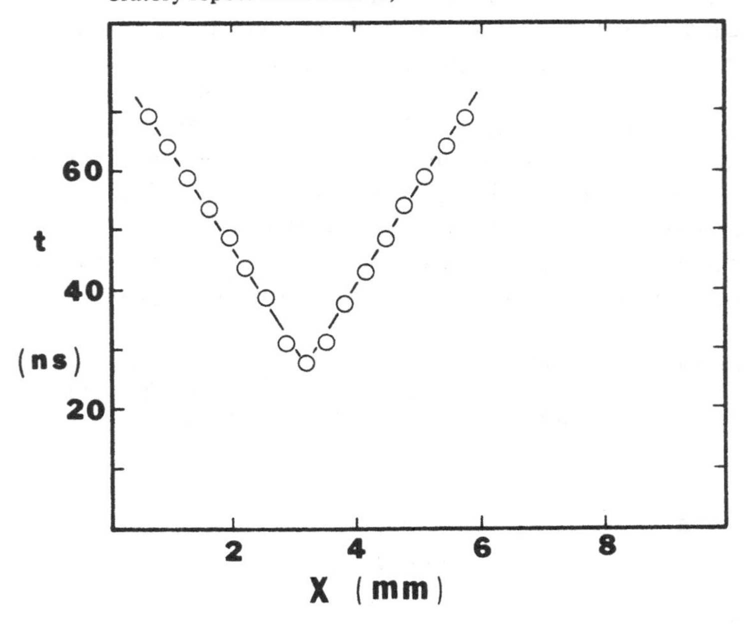

Table 10.3. *Representative cylindrical drift chamber systems*

Detector	Number of sense wires	Maximum wire length (cm)	Sense wire diameter (μm)	Field wire diameter (μm)	Gas mixture	σ (μm)	Comments[a]
UA-1	6110	220	35Ni-Cr	100Cu-Be	40 Ar/60 ethane	250	CD, dE/dx
AFS	3444	140	30SS-Ni	100Cu-Be	50 Ar/50 ethane	220	CD, dE/dx
HRS	2448	253	37W	125Be-Cu	89 Ar/10 CO_2/1 CH_4	250	
JADE	1536	240	20W-Re	100Be-Cu	89 Ar/8 CH_4/3 isobutane	160	CD, dE/dx, HP
MARK II	3204	264	38Cu-Be	152Cu-Be	50 Ar/50 ethane	200	
TASSO	2340	352	30W	120Mo	90 Ar/10 CH_4	220	
CELLO	~3000	220	20W-Re	50,100Cu-Be	90 Ar/10 CH_4	210	

[a] Abbreviations: CD, charge division; HP, high pressure (4 atm).

Source: M. Barranco Luque et al., Nuc. Instr. Meth. 176: 175, 1980; O. Botner et al., Nuc. Instr. Meth. 196: 315, 1982; D. Rubin, J. Chapman, D. Nitz, A. Seidl, and R. Thun, Nuc. Instr. Meth. 203: 119, 1982; W. Davies-White et al., Nuc. Instr. Meth. 160: 227, 1979; H. Boerner et al., Nuc. Instr. Meth. 176: 151, 1980; W. deBoer et al., Nuc. Instr. Meth. 176: 167, 1980; W. Farr et al., Nuc. Instr. Meth. 156: 283, 1978.

The central detector of the TASSO spectrometer at PETRA ues a cylindrical drift chamber 2.56 m in diameter and 3.52 m in length inside a solenoidal magnetic field [12]. The chamber contains 2340 drift cells. The sense wires are 30-μm diameter gold-plated tungsten, while the cathode uses 120-μm gold-plated molybdenum. The sense wires were tensioned to 0.8 N and the cathode wires to 3 N. All the wires were stretched between two aluminum end plates, which were spaced apart by tubes on the inside and outside radii of the drift region. The chamber used a 90% Ar/10% CH_4 gas mixture.

Figure 10.6 Electric field lines in the TASSO drift cell (H. Boerner et al., Nuc. Instr. Meth. 176: 151, 1980.)

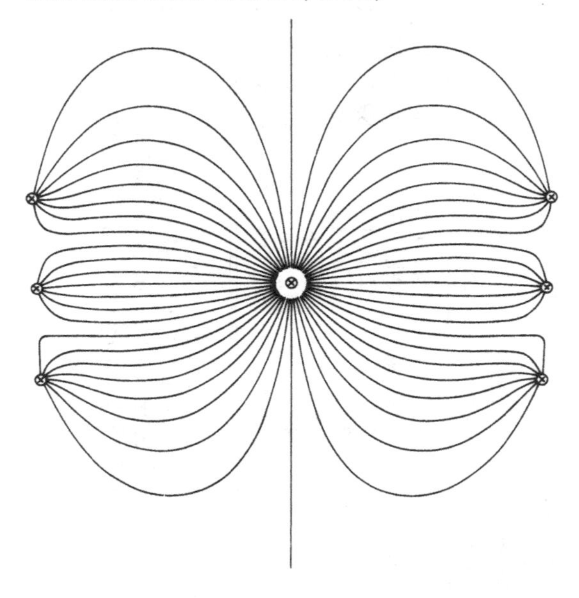

The field shape in the 16-mm drift space is shown in Fig. 10.6. The high voltage distribution system contains both active and passive circuits to prevent any wire from drawing an excessive current. The high voltage efficiency curve had a plateau above 2400 V. The efficiency below the plateau is adversely affected by a magnetic field. For drift distances larger than about 6 mm the space – time relationship is strongly affected by the angle of the particle trajectory. This is largely caused by the deflection of the electron trajectories in the magnetic field. A typical event in the chamber is shown in Fig. 10.7.

Proportional drift tubes offer a number of advantages over wire chambers for certain applications, particularly those involving large solid angle coverage. They are easier to construct, cheaper, modular, and have a self-supporting structure [13]. A broken wire only eliminates a single cell instead of a whole plane. They have the disadvantages of poorer resolution, a nonuniform electric field, and an enhanced sensitivity to impurities in the gas.

Figure 10.7 A typical event in the TASSO cylindrical drift chamber (H. Boerner et al., Nuc. Instr. Meth. 176: 151, 1980.)

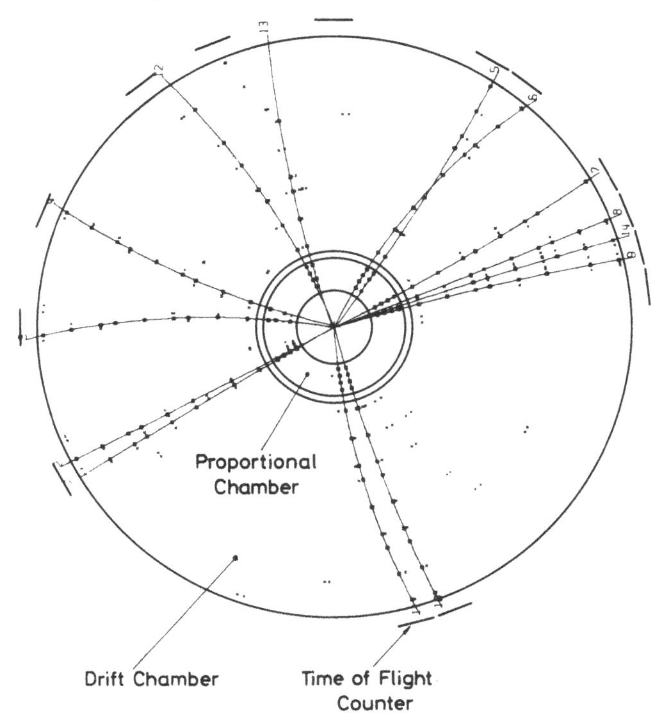

The CHARM neutrino detector at CERN [14] contains a system of 12,600 proportional tubes. Figure 10.8 shows the details of the mechanical construction. The 50-μm stainless steel sense wires are mounted in a 29 mm $\times$ 29 mm $\times$ 4 m extruded aluminum tube. The wires are stretched to a tension of 1 N. The tubes contain a 95% Ar/5% C_3H_8 gas mixture. The gas amplification depends strongly on the high voltage, gas density, and gas composition. A gas control system was used to stabilize the pressure and thus reduce gain variations.

The European Muon Collaboration at CERN [13] have constructed a system containing 1100 PDTs. The tubes are circular with a 5-cm inner diameter and are up to 2 m in length. The tubes are constructed from 1-mm-thick bakelight paper. The cathode consists of a conductive paint on the inside of the tube. The anode is a 20-μm gold-plated tungsten wire. The tubes contained an 80% Ar/20% ethane gas mixture and had a spatial resolution of 0.18 mm.

Figure 10.8 Construction details of the CHARM proportional drift tubes. (1) sense wire, (2) aluminum tube, (3) wire support, (4) gas inlet/outlet, (5) insulator, (6) support and manifold, and (7) cover plate. (A. Diddens et al., Nuc. Instr. Meth. 176: 189, 1980.)

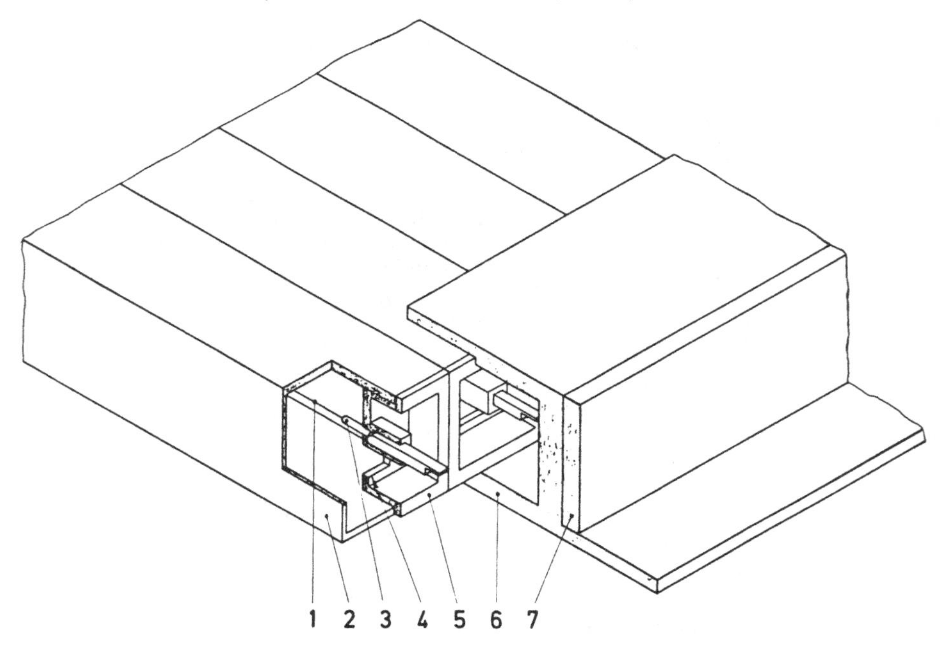

10.3 Readout electronics

The minimum requirement for a drift chamber electronics is to specify which wire has detected a pulse and to measure the elapsed time between some reference signal t_0 and the time the pulse is collected by the sense wire. Figure 10.9 gives the fundamental components of a system to measure the drift time for a single wire. The reference time is usually provided by a scintillation counter upstream of the drift chambers. This signal can be used as a start signal to a TDC. The anode wire from the drift chamber gives the stop pulse.

A multiwire readout scheme contains a number of independent digitizing channels [1]. A start signal enables the first channel scaler to count oscillator pulses from a clock. This is stopped by a pulse from any of the anodes. The wire number of the active wire is then stored in a latch. The data acquisition electronics for detectors with large arrays of wires are generally designed to record data only from the active wires. Commercial preamplifier–discriminator integrated circuits and special purpose TDC

Figure 10.9 Simple readout system for drift chamber signals. (HV) high voltage, (A) preamplifier, (D) discriminator, (TDC) time to digital convertor, and (DA) data acquisition.

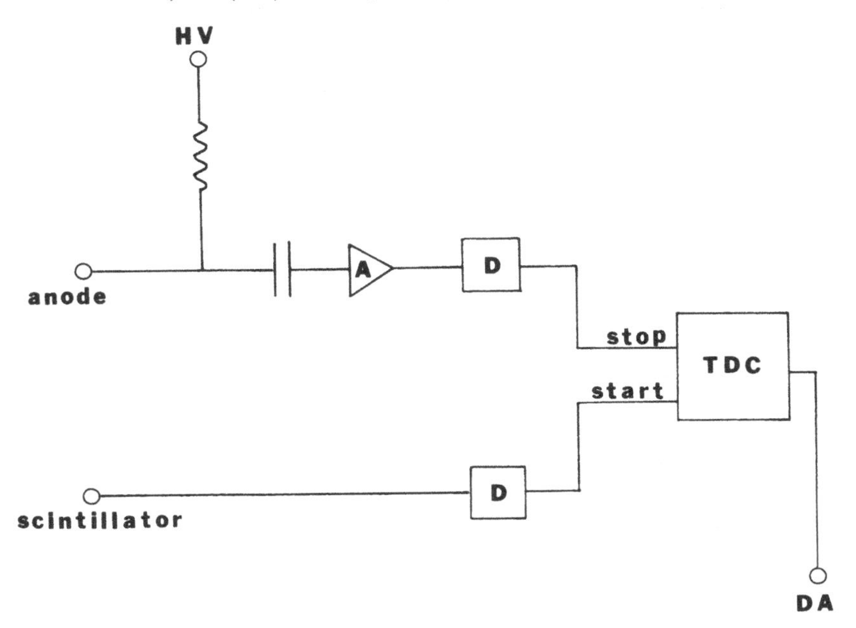

and controllers are also available to make drift chamber signals available to CAMAC data systems.

A block diagram of the readout electronics for the CHARM PDTs [14] is shown in Fig. 10.10. The input signal from each sense wire is divided between the charge amplifier and the timing discriminator. Since the drift velocity in the argon – propane gas mixture is slow, the readout electronics can be built using TTL integrated circuits. The electronics could function as an 8-bit ADC or TDC. The maximum deadtime was less than 15 μs.

Pictorial drift chambers measure the drift time and signal charge using fast electronics capable of responding to more than one hit [11]. This can be done using an analog storage device, such as a CCD, which samples the input signal at a rate determined by an external clock. Another technique is to use flash ADCs, which sample the input signal with short (~32-ns) integration times.

Fast electronics are required to take full advantage of the intrinsic time resolution of drift chambers. A "time stretcher" circuit can be used to expand the time interval and then read it out with slower electronic circuits [1]. The readout system employed with the MPS planar chambers uses amplifiers with a 4-ns risetime [10]. The signal is shaped and fed into a discriminator – comparator. Here the reference voltage can be varied to change the signal threshold. A 256-bit shift register acts as a digital delay and time digitizer. The discriminator output is continuously fed into the

Figure 10.10 Readout system for CHARM proportional drift tubes. (A. Diddens et al., Nuc. Instr. Meth. 176: 189, 1980.)

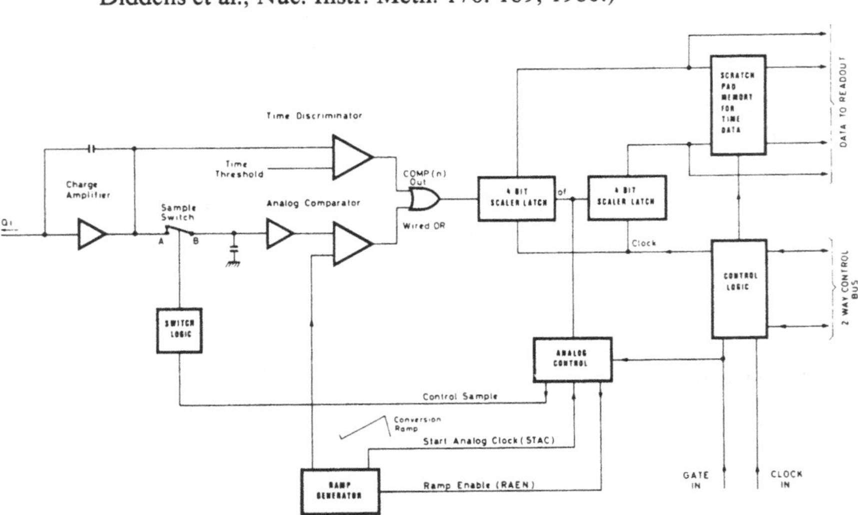

shift register. An appropriately delayed trigger signal stops the register after about 1 μs. Active channels contain YES bits somewhere is the last 32 bits of the register. The position of the furthest YES bit provides a measure of the drift time to within 4 ns.

The analog signal methods listed in Table 9.4 have also been employed in drift chambers to measure a second coordinate of a particle's trajectory through the chamber. The European Muon Collaboration [13] used delay lines with their proportional drift tubes, shown in Fig. 10.8. The delay line assembly was mounted adjacent to the last plane of PDTs in each assembly in order to measure the position of the interaction along the wire. The positive signal on the resistive cathode induced a signal on the delay line electrodes. The delay line followed a folded zigzag path to the electronics at each end. The position resolution using the time information from both ends was 6.3 mm.

An auxiliary system of electronics may be necessary in large chamber arrays to monitor the performance of the chambers. For example, Daum et al. [15] have used a minicomputer to monitor the performance of their chambers at the CERN SPS. The minicomputer can be used between beam spills to check for electronic faults, inefficient modules, timing errors, and missing wires.

10.4 Performance

There are a number of important performance characteristics for drift chambers, including efficiency, double track resolution, and operation in a magnetic field. However, the most important property is usually the spatial resolution.

A large number of effects influence the ultimate spatial resolution that can be achieved [16]. First, there are limitations due to physical processes. We have seen that a charged particle passing through the gas in a drift cell leaves behind a trail of ions. There are about 30 electron–ion pairs created per centimeter in argon at STP. However, the actual number and size of these ion clusters and their distribution along the path of the particle is subject to statistical fluctuations. In addition, as the electrons begin to drift toward the anode under the influence of the electric field, the cloud will spread out due to diffusion, multiple scattering, and secondary ionization.

The spatial uncertainty arising from diffusion is [1]

$$\sigma^2 = 2\int_0^x \frac{D}{w}\,dx \tag{10.14}$$

and thus one wants a gas with a small value for the ratio D/w. The diffusion coefficient is not isotropic. Fortunately, the coefficient $D_{\parallel}$ for diffusion along the direction of the electric field is much smaller than the coefficient $D_{\perp}$ for transverse diffusion. It is $D_{\parallel}$ that affects the arrival time of the electron cloud at the sense wire. Equation 10.7 can be written in terms of the characteristic energy as

$$\sigma_x^2 = 2E_k x/e\mathscr{E} \tag{10.15}$$

The addition of a multiatomic molecule like isobutane to pure argon increases the fractional energy loss per collision Γ because energy can then be absorbed into vibrational and rotational modes. According to Eqs. 10.11 and 10.13, this reduces E_k and thus the dispersion σ_x. The effect that this dispersion has on the actual spatial resolution depends on the detection method [7].

Another physical process is fluctuations in ion multiplication near the anode wire. These processes affect the arrival time of electrons at the anode.

A second set of limitations are due to inaccuracies in the chamber construction. The major contribution arises from errors in the wire positions. These may be random errors due to the tolerances in the support structure, or systematic errors such as gravitational sagging or electrostatic deflections.

Table 10.4. *High resolution drift chambers*

Chamber	Principle
Scintillating drift chamber	detects secondary gas scintillation near anode with PMT; fast response; no space charge distortion
Time projection chamber; jet chamber; imaging chamber	track segments continuously sampled and recorded electronically; large volume
Microjet chamber	thin anode wires
Precision drift imager	only accepts small segments of ionization trail; good two-track separation
Time expansion chamber	low velocity drift region; flash ADC readout
Parallel plate avalanche chamber	uniform electric field in gain region
Multistep avalanche chamber	amplifying regions separated by wire grids; single electron imaging

Source: G. Charpak and F. Sauli, Ann. Rev. Nuc. Part. Sci. 34: 285, 1984.

A third set of limitations are due to the electronics. There will be uncertainties in the time of arrival of the scintillator reference pulse and anode signal due to jitter and slewing in the discriminator thresholds. High frequency noise may be present on the chamber signal. In addition, the drift time cannot be measured more accurately than the size of the time bins in the timing circuitry. The electronics should be designed so that these errors do not limit the chamber's performance.

Finally there are limitations in converting the time interval into a spatial distance. There may be uncertainties in the drift velocity due to variations in the electric field, the magnetic field, and the pressure, temperature, and composition of the chamber gas. Table 10.4 lists a number of drift chambers specially constructed to obtain high resolution (10–50 μm). These devices frequently use high pressure gas, which improves the spatial localization by increasing the primary ionization and decreasing diffusion.

The resolution of one in a series of chambers can be determined by fitting a track through all the chambers but one and then measuring the deviation of the measured position in that chamber from the fitted value. Figure 10.11 shows a set of measurements of spatial accuracy obtained this way as a function of the distance of the particle from the sense wire.

Figure 10.11 Contributions to drift chamber spatial resolution. (F. Sauli, CERN report 77-09.)

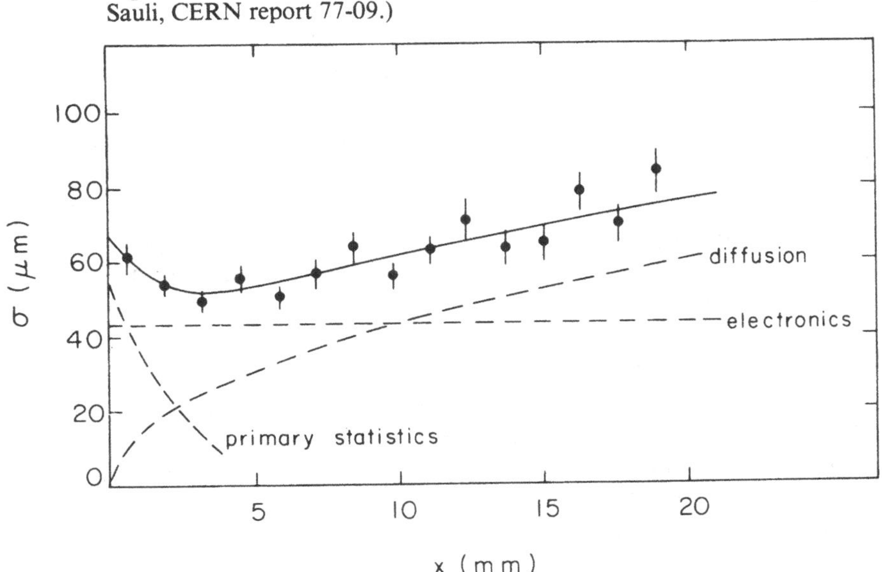

Also shown are the contributions to the resolution arising from primary ion statistics, electronics, and diffusion. Fluctuations in the primary ionization are chiefly important near the sense wire. The electronics contribution is independent of distance, while the time spread from diffusion increases with distance.

The space–time relation $x(t)$ can be determined by measuring the time spectrum of a uniformly distributed beam since [2]

$$\frac{dN}{dt} = \frac{dN}{ds}\frac{ds}{dt} = kw(t) \tag{10.16}$$

where k is constant. Thus, the time distribution gives a measure of $w(t)$, and by Eq. 10.1 the integral of the distribution gives the space–time relation.

Wire chambers suffer a degradation in performance after prolonged exposure to radiation. Turala and Vermeulen [17] observed a significant drop in drift chamber efficiency when the accumulated number of charges collected on the anode wires reached 10^{17} cm^{-1}. The loss of efficiency was nonuniform and nonsymmetrical. Whitish-gray deposits were found on the irradiated anode wires. The deposits are believed to contain silicon and hydrocarbon fragments. They could be removed by cleaning the wires with acetone.

10.5 Particle identification

A number of wire chambers have been built that combine the functions of measuring particle coordinates with identifying the type of particle that made the track [18]. The particle identification is achieved by making measurements of dE/dx in the drift chamber gas. In the region of the relativistic rise the expected energy loss differs slightly depending on the mass of the particle. The design goal for any dE/dx detector is to maximize the separation D in the relativistic rise for the particles of interest, at the same time obtaining the best possible resolution σ in the measurement of dE/dx. The usual figure of merit, or resolving power, is D/σ.

Measurements of dE/dx in the gas of a wire chamber are very difficult because of the small difference expected between different mass particles, the saturation of the effect at large $\beta\gamma$ or gas density, and the large fluctuations in energy loss for a given gas sample. The energy loss in a thin absorber follows a Landau-like distribution, which as we have seen is characterized by a long high energy loss tail. The width of this distribution is much broader than the size of the relativistic rise difference. The size of the fluctuations can be reduced by increasing the gas sample thickness up to a point, after which the width of the distribution is constant at about 30% FWHM. The sample thickness may be increased by either increasing the gas thickness or by operating the chamber at higher pressure. However, while higher pressure improves the single measurement resolution, it decreases the size of the relativistic rise effect.

The usual method is to sample the ionization loss many times. This can be done by providing many planes of transverse drift regions or by using a longitudinal drift geometry, where the electric field is approximately parallel to the particle's trajectory. When many energy loss samples are taken, the mean energy loss can be calculated using the truncated mean technique. The largest 40% or so of the energy loss measurements are discarded before calculating the mean. This method gives better resolution than averaging all the data.

A number of systematic effects must also be understood before the energy loss resolution can be reduced below the 10% level [18]. The response can depend on track position since (1) different wires may be involved, (2) there may be a variation in gas gain along the length of the wire, (3) the amount of collected charge depends on the initial distance of the incident particle from a given anode wire, (4) electrons are more likely to be attached by impurities such as O_2 in the gas the further the incident particle was from the anode wire, and (5) saturation of gas gain for tracks with a small angle with respect to the anode wires. The response of the chamber depends on the incident particle conditions since, first, the incident rate affects the space charge from the buildup of positive ions and reduces the gas gain and, second, there can be an overlap in the signals of two closely spaced tracks. Other systematic effects include cross talk between neighboring cells and baseline shifts.

Table 10.5 shows some gas properties that are important for designing a dE/dx detector [19]. The resolution in measuring dE/dx in a gas is given by the percentage FWHM. For a single measurement this is $\sim 30-60\%$. The resolution from taking the truncated mean of 64 independent measurements is much smaller, typically $5-11\%$. Note that the resolution improves with increasing pressure and that the pure hydrocarbons have better resolution than the noble gas mixtures. The next column, labeled relativistic rise, gives the ratio of the truncated mean energy losses for electrons and protons. The largest effect is observed in the xenon gas mixture. Unlike the resolution, the relativistic rise effect decreases at

Table 10.5. *Gas mixtures for dE/dx measurements*

Gas mixture	P (atm)	$\mathscr{E}/P$ (kV/cm-atm)	FWHM (%) Single sample	Truncated mean	Relativistic rise	D/σ $e\pi$	πp
90% Ne + 10% C_2H_6	1.0	0.31	57	9.6	1.392	2.51	6.66
	2.0	0.32	47	8.1	1.335	2.37	6.80
95% Ar + 5% CH_4	1.0	0.18	59	11.1	1.435	3.02	5.29
96.4% Xe + 3.6% C_3H_8	0.9	0.40	62	9.8	1.513	4.77	6.39
CH_4	1.0	0.31	45	7.4	1.210	1.44	4.98
C_2H_4	1.0	0.34	41	6.6	1.209	1.75	5.57
C_2H_6	1.0	0.36	38	6.3	1.181	1.57	4.98
	2.0	0.37	31	5.7	1.139	1.30	4.35
	3.0	0.31	29	5.7	1.120	1.14	3.86
C_3H_8	1.0	0.36	34	5.3	1.163	1.65	5.49

Source: I. Lehraus, R. Matthewson, and W. Tejessy, Nuc. Instr. Meth. 200: 199, 1982.

increasing pressure. The last two columns give the resolving powers for e$-\pi$ and $\pi-$p separation. The noble gas mixtures clearly give the best performance.

In the longitudinal drift geometry technique the particles pass through the chamber parallel to the drift field. Special readout electronics are then used to sample the charge collected over short time intervals. Maximum resolution should be obtained using a gas with small drift velocity and small diffusion coefficient, so that the arriving ionization clusters are well separated. For example, Imanishi et al. [20] have used this technique to make dE/dx measurements in a drift chamber with a 51-mm drift space filled with 90% Ar/10% CH_4. The pulse heights were measured using flash ADCs with a 40-ns sampling time. The time-sliced pulse height spectrum showed a broad plateau over a region of 1.4 μs, which corresponds to the gas sampling thickness. The long time-constant tail in the current pulse was filtered out to prevent pileup. The truncated mean method was used to calculate the most probable energy loss. The resolving power was independent of the fraction of samples retained over the range 50–80%.

Some dE/dx detectors rely on counting the number of ionization clusters instead of measuring the total amount of liberated charge [18]. This method reduces the large fluctuations in single measurements, since the number of clusters that are created follows a Poisson distribution instead of a Landau-like one. The mean number of clusters is given by

$$\bar{n} = \frac{\alpha t}{\beta^2} \frac{a}{E_{\text{ion}}} \left(\ln \frac{2mc^2}{E_{\text{ion}}} + 2 \ln \beta\gamma - \beta^2 + b \right) \tag{10.17}$$

where E_{ion} is the ionization energy, $a \sim 0.3$ and $b \sim 3$ are gas dependent shell corrections, and

$$\alpha t = 0.153(Z/A\rho t)$$

where αt is in MeV when ρt is in g/cm^2. The mean number of clusters varies from 5 to 45 for different gases at atmospheric pressure. A second advantage of the method is that the information is inherently digital, so that systematic errors should be smaller. One problem with cluster counting is that the smallest cluster, involving a single electron, must be detected with high efficiency since such clusters are the most common. Another disadvantage is that the relativistic rise in the number of clusters reaches its plateau value faster than the rise in energy loss.

A time projection chamber (TPC) is a drift chamber that provides spatial information in three dimensions, as well as measurements of dE/dx for particle identification [21–23]. The essential feature of a TPC is that the electric field in the chamber is parallel to the magnetic field from a spectrometer magnet. This arrangement reduces the diffusion of electrons in the gas because the magnetic field forces the drifting electron to wind around the electric field lines. As a consequence, the electron can drift over a long distance and still provide good spatial resolution. The detector has a large solid angle acceptance and can simultaneously measure a large number of tracks.

The first such chamber was designed for use around an intersection point at the PEP colliding beam facility at SLAC [23]. An 80% Ar/20% CH_4 gas mixture is used to provide high mobility and low electron attachment. After the passage of an ionizing particle the liberated electrons cross a large (~1-m) drift space in the direction of the sense wire plane, as shown in Fig. 10.12. It was necessary to carefully shape the electrostatic drift field in order to eliminate track distortions. Grounded grid wires focus the electrons toward the sense wires and collect stray positive ions. Field wires located between the sense wires reduce cross talk among neighboring sense wires due to the motion of positive ions.

The trajectory of the particle can be reconstructed using the distribution of sense wire hits, the drift times to the sense wires, and the charge distribution on cathode pads beneath some of the sense wires. These 7.5-mm-wide strips run parallel to the sense wires and are subdivided into a number of individual parts, each with its own readout. The spatial

Figure 10.12 Principle of TPC readout. (Courtesy of University of California, Lawrence Berkeley Laboratory.)

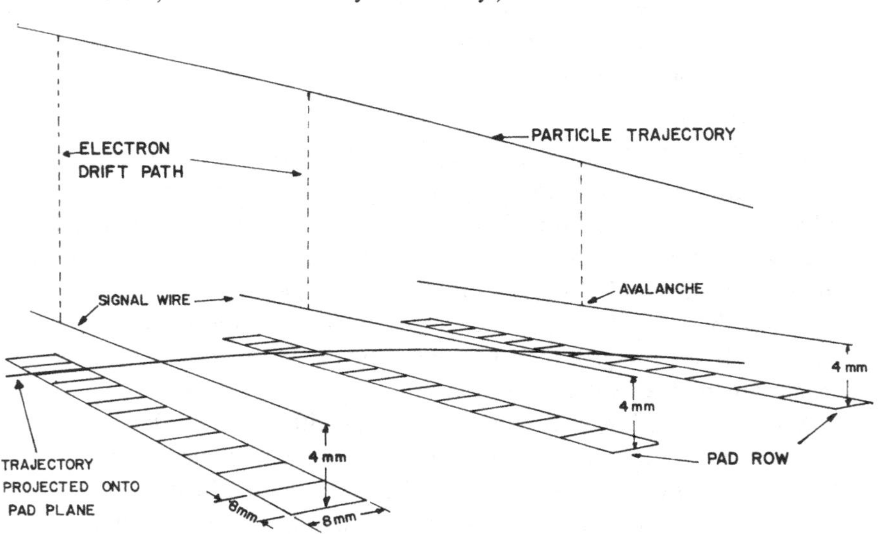

resolution of the chamber was 160 μm in the radial direction and 350 μm along the axis.

The particle identification is achieved by measurements of energy loss. The magnitude of the signal on each sense wire and cathode pad is sampled in ≤100-ns time intervals using a linear charge coupled device. The chamber is operated at a pressure of 8.5 atm to provide sufficient path length in the gas. Good resolution (~3% rms) on the determination of the energy loss can be obtained by taking the truncated mean of many independent measurements.

References

A good introduction to drift chambers can be found in references 1 and 2. The latest developments for all aspects of drift chamber technology can be found in the proceedings of Wire Chamber Conferences given in reference 3.

[1] P. Rice-Evans, *Spark, Streamer, Proportional, and Drift Chambers,* London: Richelieu, 1974, Chap. 10.
[2] F. Sauli, Principles of operation of multiwire proportional and drift chambers, CERN Report 77-09, 1977.
[3] Nuc. Instr. Meth. 176: 1–432, 1980; Nuc. Instr. Meth. 217: 1–381, 1983.
[4] V. Palladino and B. Sadoulet, Application of classical theory of electrons in gases to drift proportional chambers, Nuc. Instr. Meth. 128: 323–35, 1975.
[5] J. Townsend, *Electrons in Gases,* London: Hutchinson, 1947.
[6] A. Peisert and F. Sauli, Drift and diffusion of electrons in gases: A compilation (with an introduction to the use of computing programs), CERN Report 84-08, 1984. This reference contains measurements of drift velocities and diffusion coefficients for a large variety of gas mixtures.
[7] B. Sadoulet, Fundamental processes in drift chambers, Physica Scripta 23: 434–45, 1981.
[8] B. Jean-Marie, V. Lepeltier, and D. L'Hote, Systematic measurement of electron drift velocity and study of some properties of four gas mixtures: A-CH_4, A-C_2H_4, A-C_2H_6, A-C_3H_8, Nuc. Instr. Meth. 159: 213–9, 1979.
[9] G. Charpak, Multiwire and drift proportional chambers, Phys. Today, Oct. 1978, pp. 23–30.
[10] S. Eiseman, A. Etkin, K. Foley, R. Longacre, W. Love, T. Morris, S. Ozaki, E. Platner, V. Polychronakos, A. Saulys, C. Wheeler, S. Lindenbaum, M. Kramer, and Y. Teramoto, The MPS II drift chamber system, IEEE Trans. Nuc. Sci. NS-30: 149–52, 1983.
[11] A. Wagner, Central detectors, Physica Scripta 23: 446–58, 1981.
[12] H. Boerner, H. Fischer, H. Hartmann, B. Lohr, M. Wollstadt, D. Cassel, U. Kotz, H. Kowalski, B. Wiik, R. Fohrmann, and P. Schmuser, The large cylindrical drift chamber of TASSO, Nuc. Instr. Meth. 176: 151–7, 1980.
[13] C. Broll, A. Charveys, Y. DeClais, J. Favier, M. Lebeau, M. Moynot, and G. Perrot, Large drift tube arrays with external delay line readout, Nuc. Instr. Meth. 206: 385–95, 1983.
[14] A. Diddens, M. Jonker, J. Panman, F. Udo, J. Allaby, U. Amaldi, G. Barbiellini, A. Baroncelli, V. Blobel, G. Cocconi, W. Flegel, W. Kozanecki, E. Longo, K. Mess, M. Metcalf, J. Meyer, R. Orr, W. Schmidt-Parzefall, F. Schneider, A. Wetherell, K. Winter, F. Busser, P. Gall, H. Grote, P. Heine, B. Kroger, F. Niebergall, K. Ranitzsch, P. Stahelin, V. Gemanov, E. Grigoriev, V. Kaftanov, V. Khovansky, A.

Rosanov, R. Biancastelli, B. Borgia, C. Bosio, A. Capone, F. Ferroni, P. Monacelli, F. de Notaristefani, P. Pistilli, C. Santoni, and V. Valente, Performance of a large system of proportional drift tubes for a fine grain calorimeter, Nuc. Instr. Meth. 176: 189–93, 1980.

[15] C. Daum, L. Hertzberger, W. Hoogland, R. Jongerius, J. Knapik, W. Spierenburg, and L. Wiggers, Some features of a system of drift chambers for experiment NA11 at the SPS, Nuc. Instr. Meth. 176: 119–27, 1980.

[16] F. Sauli, Limiting accuracies in MWPC and drift chambers, Nuc. Instr. Meth. 156: 147–57, 1978.

[17] M. Turala and J. Vermeulen, Ageing effects in drift chambers. Nuc. Instr. Meth. 205: 141–4, 1983.

[18] A. Walenta, Performance and development of dE/dx counters, Physica Scripta 23: 354–70, 1981.

[19] I. Lehraus, R. Matthewson, and W. Tejessy, dE/dx measurements in Ne, Ar, Kr, Xe, and pure hydrocarbons, Nuc. Instr. Meth. 200: 199–210, 1982.

[20] A. Imanishi, T. Ishii, T. Ohshima, H. Okuno, K. Shiino, F. Naito, and T. Matsuda, Particle identification by means of fine samplng dE/dx measurements, Nuc. Instr. Meth. 207: 357–64, 1983.

[21] J. Macdonald (ed.), *The Time Projection Chamber,* American Institute of Physics Conf. Proc. No. 108, New York: AIP, 1984.

[22] R. Madaras and P. Oddone, Time projection chambers, Phys. Today, Aug. 1984, pp. 36–47.

[23] D. Fancher, H. Hilke, S. Loken, P. Martin, J. Marx, D. Nygren, P. Robrish, G. Shapiro, M. Urban, W. Wenzel, W. Gorn, and J. Layter, Performance of a time projection chamber, Nuc. Instr. Meth. 161: 383–90, 1979; J. Marx and D. Nygren, The time projection chamber, Phys. Today, Oct. 1978, pp. 46–53.

Exercises

1. Estimate the mean free path of electrons in an atmospheric drift chamber with a 2000-V/cm electric field. Use a typical drift velocity and assume that the random velocity of the electrons is $\sim 10^7$ cm/sec. What is the corresponding coefficient of diffusion?

2. Estimate the mean number of ion clusters left by a 50-GeV/c proton in a 2-mm argon drift chamber cell at a pressure of 2 atm. What is the FWHM of the cluster distribution?

3. What is the mean ionization energy loss for exercise 2? What is the FWHM of the energy loss distribution assuming a Landau distribution?

4. A drift chamber uses methane gas in a 2-kV/cm electric field. If the cell size is 1 cm, what should the least count of the readout electronics be, so that it is smaller than the time spread arising from diffusion?

5. Prove Eq. 4 in Appendix F.

11

Sampling calorimeters

A device that measures the total energy deposited by a particle or group of particles is known as a calorimeter, in analogy with the laboratory instrument that measures the amount of deposited heat. We have already encountered several devices, such as sodium iodide scintillation counters and total absorption Cerenkov counters, that can be used as calorimeters for photon detection. We will consider properties of these "continuous" calorimeters again in Chapter 14. In this chapter we consider a class of calorimeters that periodically sample the development of a shower initiated by an incident particle. There are two major types of sampling calorimeters, depending on whether the incident particle initiates an electromagnetic or hadronic shower. Each type of calorimeter is optimized to maximize the rejection of the other type of shower.

Calorimeters have found wide use in particle physics experiments. Neutral particles can only be detected by using this method. Sampling calorimeters of very large size have been used as neutrino detectors. We have seen that at high energy, particle multiplicities grow with increasing energy, and the angular distribution of groups of the produced secondaries are highly collimated (jet effect). Under these conditions calorimeters can provide a useful trigger for interesting events based on the total energy deposited in a localized area. Calorimeters can easily be modularized and made to cover large solid angles. In addition, we shall see that the size of a calorimeter needed to measure the energy of a particle scales like $\ln(E)$, whereas the size of a magnetic deflection device would scale like $E^{1/2}$ [1].

11.1 Electromagnetic showers

Consider an electron or positron with several GeV energy traversing a slab of some material. As far as shower development is concerned, the electron and the positron behave almost identically, and unless stated otherwise, the discussion of electron behavior in what follows also refers to positrons. We saw in Chapter 2 that for energies above 100 MeV electrons lose energy almost entirely through bremsstrahlung. The emitted photons typically carry off a large fraction of the electron's initial energy. For photons with energy greater than 100 MeV the major interaction is pair production, which gives another energetic electron or positron. In this manner a single initial electron or photon can develop into an electromagnetic shower, consisting of many electrons and photons. The shower continues until the particles' energy falls below 100 MeV, at which point dissipative processes, such as ionization and excitation, become more important.

The mathematical description of the development of a completely general electromagnetic shower is extremely complex and cannot be solved in closed form [2]. The theory has been worked out with several simplifications. First, one only considers the average behavior of the shower, or fluctuations from the average behavior. Second, since at high energy the angles of emission of the electrons and photons are small, the shower will develop primarily in the forward direction. It is then customary to treat separately the longitudinal and transverse developments.

The unit of distance traversed is typically measured in radiation lengths, given approximately by [1]

$$X_{\rm rad} = 180A/Z^2 \qquad {\rm g/cm^2} \tag{11.1}$$

Values of the radiation lengths of many materials were given in Table 2.1. Shower calculations usually make a number of approximations. (1) At high energy the probabilities for bremsstrahlung and pair production are assumed to be independent of Z when distances are measured in radiation lengths. (2) The theoretical cross sections are based on the Born approximation and are most reliable for low Z materials. Deviations in large Z materials are proportional to Z^2. (3) The difference in cross sections for high energy electrons and positrons is neglected. (4) The asymptotic formulas for radiation and pair production are assumed valid. (5) The Compton effect and collisional processes are neglected at high energy.

A particularly subtle point is how to handle the transition to the region where the collisional energy loss is important. Sometimes a sharp cutoff η is assumed such that whenever the energy of a shower particle falls below η, it is considered to lose the remainder of its energy to collisions. This can be taken to be the critical energy E_c from Eq. 2.54, which we recall is the collision energy loss per radiation length of an electron with energy E_c. One can also consider a constant collision loss.

A simple model that gives the correct qualitative description of the longitudinal development of a shower has been given by Heitler [3]. The model makes the following assumptions. (1) Each electron with $E > E_c$ travels 1 radiation length and then gives up half its energy to a bremsstrahlung photon. (2) Each photon with $E > E_c$ travels 1 radiation length and then undergoes pair production with each created particle receiving half of the energy of the photon. (3) Electrons with $E < E_c$ cease to radiate and lose their remaining energy to collisions. (4) Neglect ionization losses for $E > E_c$.

Suppose we begin with an electron of energy $E_0 \gg E_c$, as shown in Fig. 11.1. After the first radiation length there will be one electron and one photon, each with energy $E_0/2$. In the second radiation length the electron emits a second photon, while the first photon undergoes pair production into an electron–positron pair. Thus, after 2 radiation lengths there will be two electrons, one positron, and one photon in the shower, each with energy $E_0/4$.

This simple model predicts a number of features of electromagnetic showers. (1) In the development of the shower the total number of parti-

Figure 11.1 Simple model for the development of an electromagnetic shower. Solid lines (with +) indicate electrons (positrons) and wavy lines indicate photons. The numbers at the bottom show the distance measured in radiation lengths.

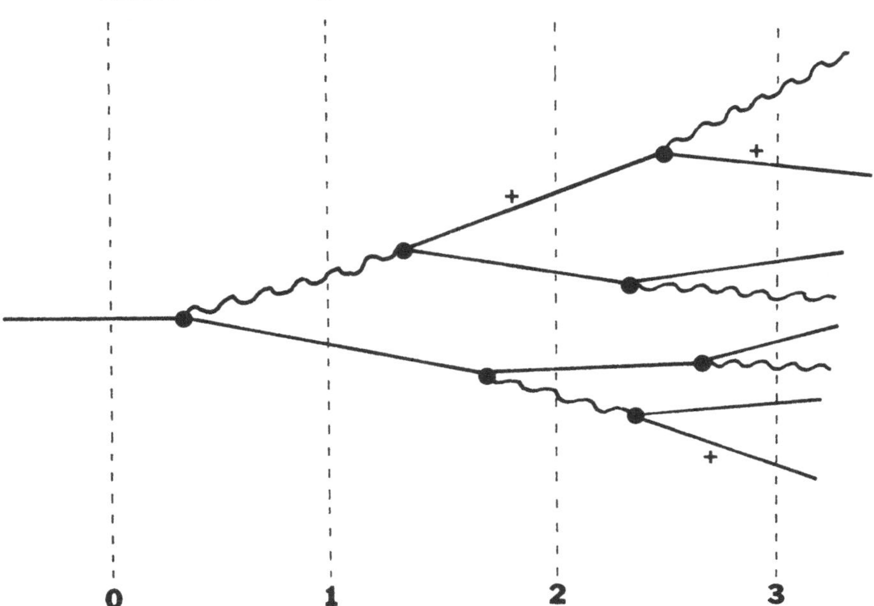

cles present after t radiation lengths is

$$N(t) = 2^t = e^{t \ln 2} \tag{11.2}$$

Thus, the number of shower particles increases exponentially with t. (2) There will be approximately equal numbers of electrons, positrons, and photons in the shower. (3) The average energy of a shower particle at the depth t is

$$E(t) = E_0/2^t \tag{11.3}$$

(4) The depth at which the shower energy equals some value E' occurs when $E(t) = E'$, and

$$t(E') = \frac{\ln(E_0/E')}{\ln 2} \tag{11.4}$$

(5) The shower has the maximum number of particles when $E(t) = E_c$. This occurs at the depth

$$t_{max} = \frac{\ln(E_0/E_c)}{\ln 2} \tag{11.5}$$

At this point the shower abruptly stops. Note that the maximum shower depth increases logarithmically with primary energy. (6) The number of particles at the maximum is

$$N_{max} = e^{t_{max} \ln 2} = E_0/E_c \tag{11.6}$$

Thus, the maximum number of shower particles is directly proportional to the incident energy. (7) The number of particles in the shower that have energy greater than some value E' is ($E' \ll E_0$)

$$\begin{aligned} N(E > E') &= \int_0^{t(E')} N(t)\, dt \\ &\simeq \frac{1}{\ln 2} \frac{E_0}{E'} \end{aligned} \tag{11.7}$$

Thus, the energy spectrum dN/dE' falls like $1/E'^2$. (8) The sum of all the track lengths of all charged particles in the shower is

$$\begin{aligned} L &= \frac{2}{3} \int_0^{t_{max}} N(t)\, dt \\ &\simeq \frac{E_0}{E_c} \end{aligned} \tag{11.8}$$

We see that the total charged track length is directly proportional to the incident energy.

The discontinuous behavior at t_{max} is the result of the overly simplified assumptions. Quantitatively accurate treatments of shower development are generally performed using Monte Carlo techniques. These calculations can take into account the energy dependence of the cross section, the lateral spread of the shower due to multiple scattering, statistical fluctuations, and other complications.

Shower data for the number of particles with energy greater than E' as a function of depth in the shower is shown in Fig. 11.2, along with the results of more accurate calculations. We see that the discontinuity at t_{max} has been broadened into a long tail. The data [4] indicate that the following empirical formulas are accurate from 2 to 300 GeV. The maximum number of electrons occur at the depth

$$t_{max} = 3.9 + \ln E_0 \tag{11.9}$$

where t_{max} is measured in radiation lengths and E_0 is in GeV. The number

Figure 11.2 Shower profiles in lead. The number of electrons should be multiplied by a normalization factor of 0.79. (D. Müller, Phys. Rev. D 5: 2677, 1972.)

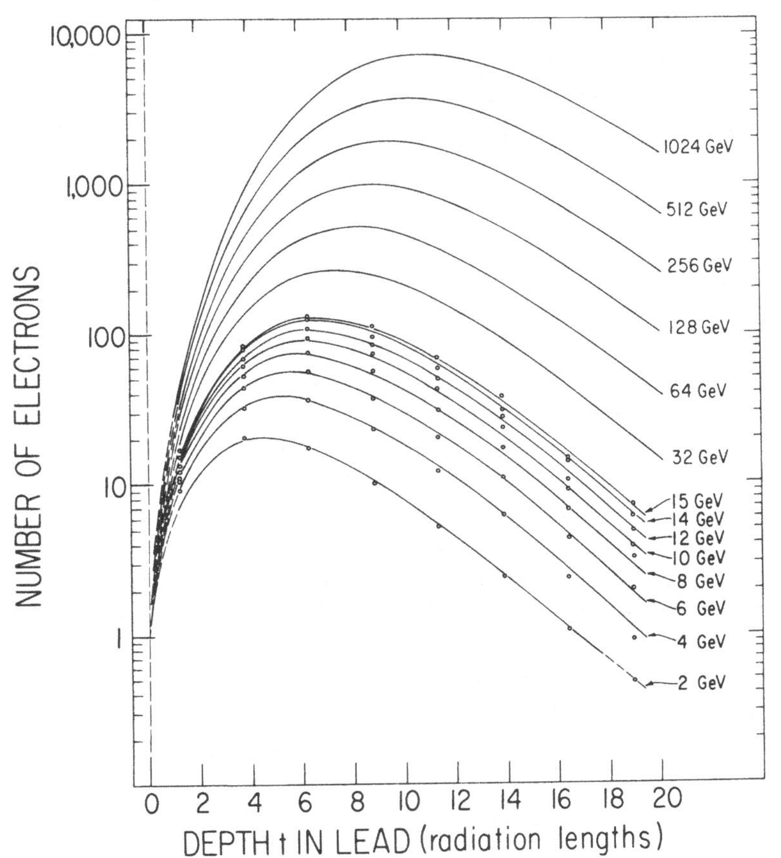

of electrons at the maximum is given by

$$N_{\text{max}} = 8.46E_0^{0.935} \tag{11.10}$$

Let $N(E_0, t)$ be the number of shower electrons at depth t resulting from a particle of energy E_0. The total path length of electrons in the shower is

$$L(E_0) = \int_0^\infty N(E_0, t)\, dt$$
$$= 60.2E_0 \tag{11.11}$$

The energy dependence of the last three equations are in qualitative agreement with the simple model (Eqs. 11.5, 11.6, and 11.8).

The longitudinal development of electromagnetic showers in different materials is found to scale if distances are measured in radiation lengths [1]. Figure 11.3 shows showers created by 6-GeV/c electrons in aluminum, copper, and lead. The scaling is seen to hold rather accurately over the first 15 radiation lengths. Also shown is the shower radius at each depth that contains 90% of the shower particles. Up to the shower maximum the shower is contained in a cylinder with radius < 1 radiation length. Beyond that point the electrons are increasingly affected by multiple scattering, and the lateral size scales in the "Moliere radius," which is given approximately as [1]

$$\rho_m \simeq 7A/Z \qquad \text{g/cm}^2 \tag{11.12}$$

The propagation of photons in the shower causes deviations from Moliere radius scaling [5]. However, roughly 95% of the shower is contained laterally in a cylinder with radius $2\rho_m$.

The number of shower tracks found inside a ring of radius r at the depth of the shower maximum is found to increase as a power of the variable rE_0, where E_0 is the energy of the electron initiating the shower [6]. The result is independent of E_0 over the range 50 – 300 GeV. This implies that a cylinder containing a fixed number of shower particles is found closer and closer to the shower axis as the incident energy is increased.

Since the processes that lead to a shower are statistical in nature, there will be fluctuations in the number of shower particles at any depth. The behavior of the particles in the first few radiation lengths is particularly important because of the amplification resulting from the later stages of the shower. For this reason, Poisson statistics do not accurately describe the fluctuations in the number of shower particles. Rossi [2] has given an informative illustration of this. First consider the probability that an incident photon will convert. On the average, photons convert after traversing a distance $t_{av} = 1.3$ radiation lengths. We know that photon conversion is governed by Poisson statistics. Therefore, the probability that the photon has not converted after traveling a distance $3t_{av}$ is $e^{-3} \sim 0.05$, independent of the photon energy. Thus, there is a 5% probability of finding no electrons at a depth of 3.9 radiation lengths into the shower. However, we see from Fig. 11.2 that for a 10-GeV shower at a depth of 3.9 radiation lengths, there are on the average about 50 electrons in the shower. If the fluctuations in the number of electrons were also governed by Poisson statistics, the probability of finding no electrons would be $e^{-50} \ll 0.05$. This shows that the fluctuations observed in showers greatly exceed what is expected from Poisson statistics.

Measurements [6] of the fluctuations in the number of shower particles

Figure 11.3 Longitudinal development of electromagnetic showers in different materials. Right scale shows radii for 90% shower containment. (C. Fabjan and T. Ludlam, adapted with permission from the Annual Review of Nuclear and Particle Science, Vol. 32, © 1982 by Annual Reviews, Inc.)

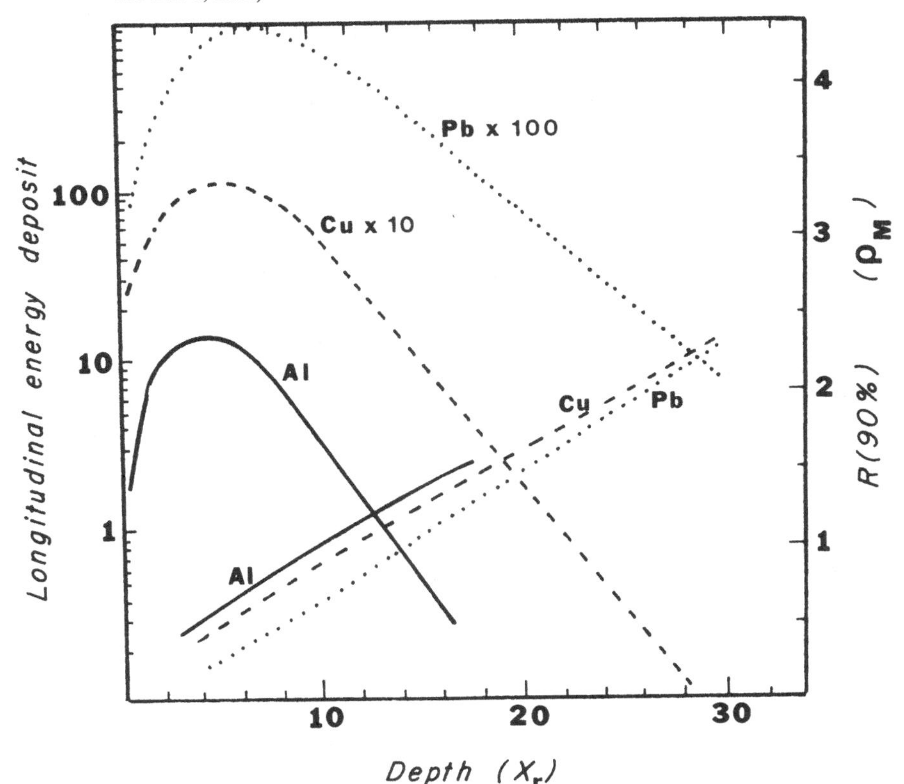

versus the longitudinal depth in the shower show that the spread in the distributions falls to a minimum at t_{max} and then increases again as the number of particles in the shower begins to decrease. For Poisson statistics $\sigma_N/N = N^{-1/2}$. The measured fluctuations are much larger. The relative fluctuations in electron number decreases as the energy is increased.

According to Eq. 11.7, we can obtain a crude measurement of E_0 by counting the number of particles in the shower whose energy exceeds some value. A more accurate measurement comes from measuring the total ionization produced, which should be proportional to the total charged track length. The fluctuations σ_L/L of this quantity are smaller than that of the number distribution and are found to decrease inversely with depth into the shower [6].

11.2 Electromagnetic shower detectors

The shower phenomenon is utilized in electromagnetic shower detectors to detect high energy electrons and photons. Once they are properly calibrated, these devices also measure the energy of the shower and hence that of the incident particle. Detectors may consist of either a homogeneous absorber or a sandwich structure that periodically samples the energy loss. Homogeneous detectors include NaI and bismuth germanate (BGO) scintillators, scintillating glass, lead-glass blocks, thallium-doped heavy liquid counters, and liquid argon [1].

The most common types of electromagnetic sampling calorimeters are

1. metal–scintillator sandwich,
2. metal–liquid argon ionization chamber, and
3. metal–gaseous PWCs.

Each device consists of alternate layers of a metal radiator to enhance photon conversions and some active substance to sample the energy loss. The metal radiator is usually made of lead. Some typical detector arrangements are shown in Fig. 11.4. The calorimeter cells are typically arranged in either a strip or a tower arrangement [7]. In a strip structure the profile of the shower is sampled at various depths. Usually two or three profile orientations are measured. In a tower structure the calorimeter is divided into many narrow, deep units that point to the interaction region.

The calorimeter readout may be either digital or proportional. In a calorimeter with digital readout the active region is finely divided into channels, each of which can provide a yes or no signal. The energy deposition is proportional to the number of yes channels. A detector using flash tubes is an example of a digital calorimeter. In a proportional calorimeter, on the other hand, the analog signals from the active regions are summed to produce a total signal that is proportional to the deposited energy. The lead-scintillator sandwich is an example of a proportional calorimeter.

In the scintillator devices charged particles in the shower produce light in the scintillator planes. Both solid and liquid scintillator have been used. The signal produced is usually large and very fast, making it useful in a high rate environment. Problems arise when the detector must be finely segmented or operated inside a magnetic field. The PMTs must be located in a weak field, and this may require a complicated light pipe system. Some large calorimeters have used scintillators made from polymethyl methylacrylate to reduce costs.

One solution to the readout problem illustrated in Fig. 11.4 involves passing the light from a number of scintillator layers into a bar of a second scintillator doped with a wavelength shifter such as BBQ. The emission spectrum of the POPOP in the scintillator layers is well matched to the absorption spectrum of the BBQ. The absorbed light is then reemitted isotropically at a longer wavelength. Part of the emitted light is internally

Figure 11.4 Typical readout techniques for calorimeters: (a) lead-scintillator sandwich, (b) lead-scintillator sandwich with wavelength shifter bars, (c) liquid argon ionization chamber, and (d) lead-MWPC sandwich. (C. Fabjan and T. Ludlam, adapted with permission from the Annual Review of Nuclear and Particle Science, Vol. 32, © 1982 by Annual Reviews, Inc.)

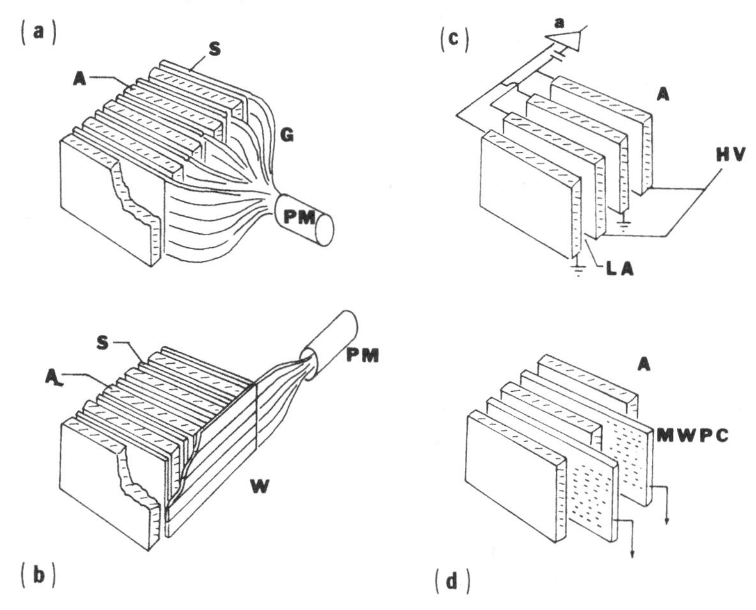

reflected to the PMT. This scheme allows a simpler construction with a minimum amount of dead space.

The signal from the PMT is usually fed into an ADC, which gives an output proportional to the area under the PMT signal. The time stability of the PMT gain can be checked by monitoring with a built-in LED flasher. The energy scale can be calibrated by measuring the response of the device to a muon or other minimum ionizing particle.

A liquid argon calorimeter consists of a series of metal plates immersed in liquid argon. A typical plate separation is 3 mm. The plates are maintained at a positive high voltage so that electrons produced in the liquid argon are collected at the plates. Since there is no charge multiplication, the collected charge is quite small (~0.6 pC/GeV), and the detector requires a preamplifier and associated electronics for each channel. It is important that the detector be free of any electronegative gases. The chamber must be operated at liquid argon temperatures (80 K) and thus requires a cryogenic system. The detector is relatively slow. The TASSO liquid argon calorimeter takes ~900 ns to give a pretrigger signal [8]. On the other hand, it is stable, is not adversely affected by the presence of a magnetic field, and is easily segmented. The detector has uniform sensitivity, and it is possible to make a highly accurate charge calibration.

The gaseous PWC detectors are cheaper than the other types of detectors discussed previously and can operate satisfactorily in a magnetic field. With 2-dimensional readout they can also give the correlation between the projections of the shower. This is especially important for resolving ambiguities in multishower events. Disadvantages of the PWC detectors include worse energy resolution, slow response, and lower density than the other calorimeters.

Calorimeters are widely used in particle physics experiments, particularly at the large spectrometers [1, 8]. Characteristics of some representative electromagnetic calorimeters are listed in Table 11.1.

Improved resolution has been observed with MWPCs operating in the Geiger, limited streamer, and saturated avalanche modes. Atac et al. [9] studied the response of a PWC calorimeter operating in the saturated avalanche regime to positrons with energy up to 17.5 GeV. The calorimeter consisted of 34 units, each containing a 2.8-mm-thick lead plate and a proportional chamber plane with anode wires in 9.5 $\times$ 9.5-mm cells. The counters used a 49.3% Ar/49.3% ethane/1.4% ethanol gas mixture. The rate of growth of the proportional chamber gain decreases continuously in the regime between proportional and streamer operation. With the high voltage set for operations in the region between these regimes, the gain is sufficiently high so that the calorimeter may be operated without amplifiers between the anode wires and the ADCs in the readout electronics. In addition, there is a greater concentration of primary ionization, and the tail of the Landau distribution is greatly suppressed. As a result, the calorimeter has good linearity, good energy resolution ($\sigma/E \sim 16\%/\sqrt{E}$), and no measurable systematic effects.

The shower that actually develops in the detector depends on the incident particle type, the incident particle energy, the incident particle angle of incidence, and the spatial distribution and nature of the radiator and active layers. The response of a given detector to two identical incident particles will differ because of statistical fluctuations in the shower development. However, a properly designed detector should be capable of returning a signal that is proportional to the energy of the incident particle. The energy spectra of electrons of various energies in a liquid argon calorimeter is shown in Fig. 11.5. We see that the observed energy grows with incident electron energy [10]. The widths of the distributions do not increase linearly, so the energy resolution improves with E. We expect the response of the detector to be linear if all the energy is absorbed in the device.

Table 11.1. *Examples of electromagnetic calorimeters*

Detector	Absorber thickness (mm)	Sensitive material thickness[a] (mm)	Total depth (radlens)	σ/E ($\%/\sqrt{E}$)
ARGUS	1 Pb	5 Sc	12.5	8
CELLO	1.2 Pb	3.6 LA	20	13
MARK II	2 Pb	3 LA	15	13
MARK III	2.8 Pb	12.7 PWC	12	18
Tagged γ	1.6 Pb	12.7 LSc	19	12
TASSO	2 Pb	5 LA	14	11
UA-1	3 Pb	2 Sc	26	15
AFS	1.6 U	2 × 2.5 Sc	19.5	11
UA-2	3.5 Pb	4 Sc	17	14

[a] Abbreviations: Sc, plastic scintillator; LSc, liquid scintillator; LA, liquid argon.
Source: A. Drescher et al., Nuc. Instr. Meth. 205: 125, 1983 (ARGUS); V. Bharadwaj et al., Nuc. Instr. Meth. 155: 411, 1978 (tagged γ); S. Wu, Phys. Rep. 107: 59, 1984; W. Toki et al., Nuc. Instr. Meth. 219: 479, 1984 (Mark III); R. Carosi et al., Nuc. Instr. Meth. 219: 311, 1984 (AFS); A. Beer et al., Nuc. Instr. Meth. 224: 360, 1984 (UA-2).

The development of the shower with depth in a liquid argon detector is shown in Fig. 11.6. We see that a lower energy electron deposits most of its energy in the early layers [11]. The 4-GeV/c electrons deposit most of their energy in the middle of the detector. The lateral spread of the shower is primarily caused by multiple scattering and is independent of the incident energy over this energy interval.

The final signal from any detector is the number of electrons that are registered in the electronic circuits. The resolution of the device for detecting an incident particle of energy E_0 is determined by fluctuations in the number N of these electrons. These fluctuations can arise from

1. the actual energy deposited in the active layers of the detector (sampling fluctuations),
2. leakage of energy out of the calorimeter,
3. noise in the active layers,
4. photocathode statistics or gain variations,
5. electronic noise, and
6. more than one event within the time resolution (pileup).

If these fluctuations follow a Poisson distribution, the standard deviation is $\sigma = \sqrt{N}$, and the resolution is

$$\sigma(N)/N = 1/\sqrt{N} \tag{11.13}$$

Figure 11.5 Observed electron energy distributions in a liquid argon calorimeter. The number by each curve gives the incident electron energy. (After J. Cobb et al., Nuc. Instr. Meth. 158: 93, 1979.)

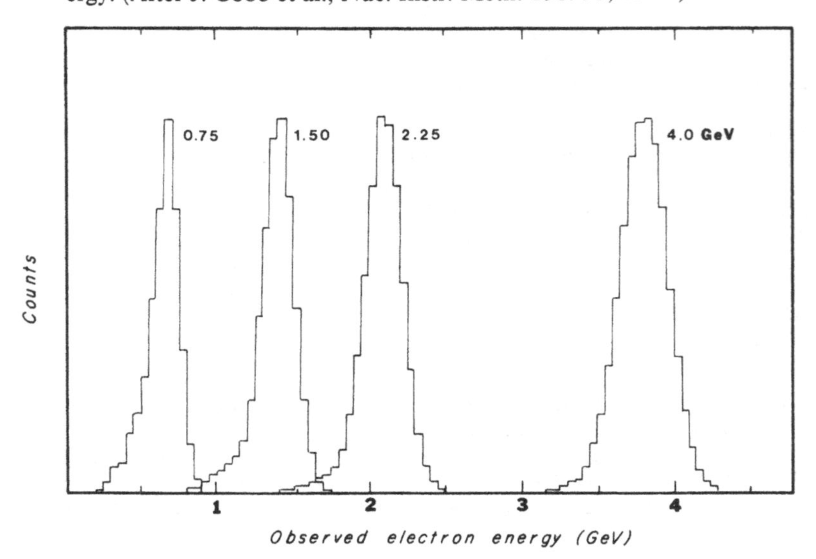

The major contribution to the energy resolution of electromagnetic calorimeters is usually sampling fluctuations.

It is instructive to expand the variance of the energy distribution into a power series in E,

$$\sigma^2(E) = \sigma_0^2 + \sigma_1^2 E + \sigma_2^2 E^2 + \cdots$$

Dividing by E^2, we obtain an expansion of the energy resolution [12].

$$\left(\frac{\sigma(E)}{E}\right)^2 = \frac{\sigma_0^2}{E^2} + \frac{\sigma_1^2}{E} + \sigma_2^2 + \cdots \tag{11.14}$$

Figure 11.6 Shower profiles at five sampling depths in a liquid argon calorimeter. The left-hand curves are for a low energy electron, while the right-hand curves are for a higher energy electron. The numbers associated with each set of curves is the range of sampling depths in radiation lengths. (After D. Hitlin et al., Nuc. Instr. Meth. 137: 225, 1976.)

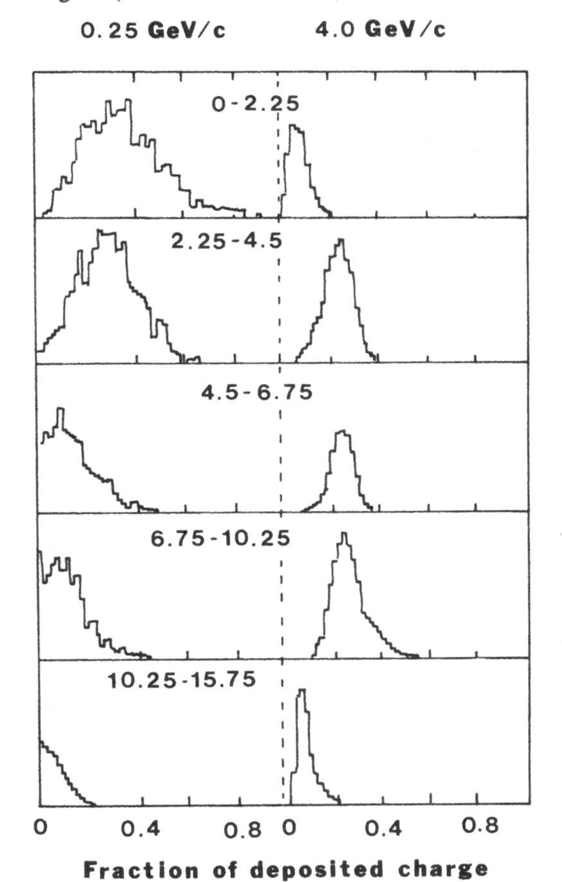

The constant σ_0 represents contributions to the resolution that are only important at low energy, primarily the ADC pedestal widths. The σ_1 term is usually dominant. Any process governed by Poisson statistics will contribute to this term. These include fluctuations in energy loss (sampling) and fluctuations in the number of photoelectrons released at the face of a PMT. Contributions to σ_2 affect the resolution curve as a whole. Hence they include calibration errors.

The signals from individual scintillator layers in a lead-scintillator detector fall off roughly like $1/\sqrt{N}$, but with a coefficient larger than the 1.0 expected from Poisson statistics [4]. The signal from the sum of all the layers more closely approximates Poisson statistics for large signals. If the detector response is linear with incident energy, the resolution should improve like $1/\sqrt{E}$.

The resolution also depends on the frequency of sampling in the detector. Fluctuations in measurements of the number of particles in the shower at any depth should go like

$$\sigma/E = 1/\sqrt{N_e} = \sqrt{\Delta E/E}$$

where N_e is the average number of sampled electrons, and ΔE is the average energy loss of electrons per sampling layer. Then

$$\Delta E \simeq E_c \, \Delta x/X_{rad} = E_c t_s$$

where t_s is the sampling thickness measured in radiation lengths. It follows that

$$\sigma/E = k\sqrt{t_s/E} \tag{11.15}$$

where k is a constant. This is roughly confirmed by the data [13]. If t_s is reduced below 0.1, the width falls off faster than $t_s^{1/2}$ due to the increased probability of detecting low energy particles in the shower [11].

The leakage of energy out the sides or end of the detector adversely affects its resolution. Leakage can be effectively studied by adding modules to the shower detector and watching the improvement in the energy resolution. In general, leakage out the end of the calorimeter is more harmful than leakage out the sides [5].

The primary reason for the worse energy resolution in PWC calorimeters is that they are subject to additional fluctuations, which are more important in gases than in liquids or solids [5]. These include fluctuations from the asymmetric Landau-like energy deposition and path length fluctuations due to low energy electrons.

Another important feature of a detector is hadron rejection. It may be necessary to detect electrons in a beam with a large pion background. In this case it is necessary to exploit the different characteristics of hadronic and electromagnetic showers. The principal difference is the amount of deposited energy. Figure 11.7 shows the total energy deposited in a lead–scintillator detector by 4-GeV/c electrons and pions [14]. The electron distribution is peaked at the beam energy, showing that all its energy was contained in the calorimeter. The low energy tail results from pion contamination in the beam. The pion distribution shows a peak correspond-

Figure 11.7 Deposited energy spectrum of electrons and pions in a lead–scintillator calorimeter. Note the different energy scales for the two curves. (After G. Abshire et al., Nuc. Instr. Meth. 164: 67, 1979.)

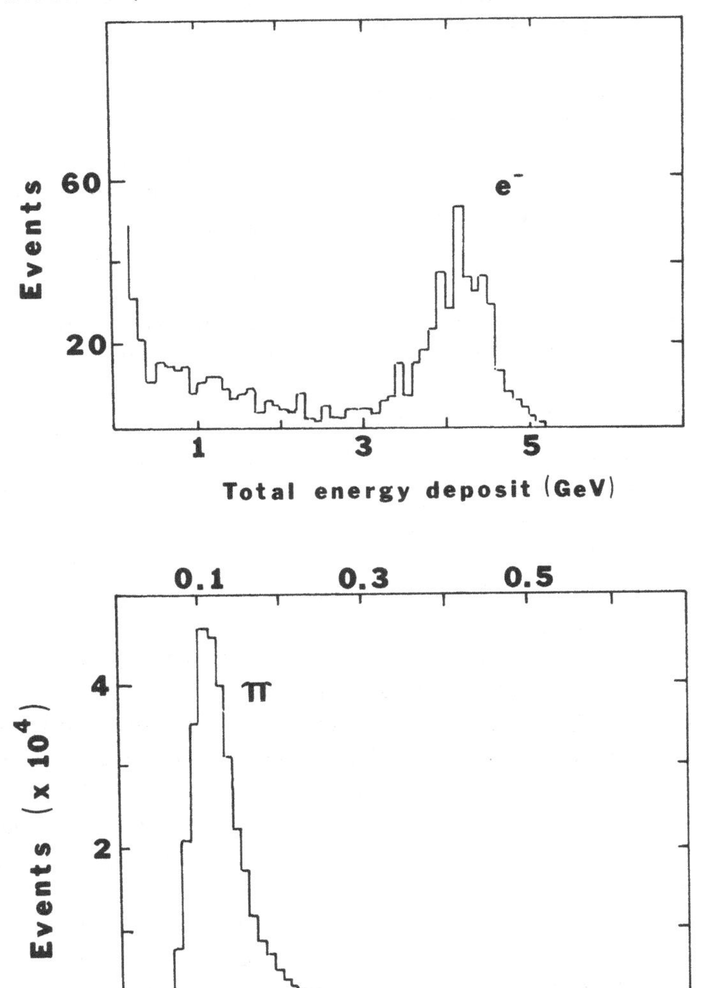

ing to minimum ionizing particles that passed straight through the detector. The tail of the pion distribution extends all the way to the full beam energy, corresponding to events where the beam pion interacted and produced a π^0, which then underwent an electromagnetic decay.

Further small improvements in the hadron rejection may be made from cuts on the fraction of the total energy deposited in a given longitudinal or lateral segment of the detector. We will discuss this further in the next section. It should be noted that improvement of the hadron rejection may require a loss of energy resolution or electromagnetic detection efficiency. Thus, the various parameters have to be optimized for the particular experiment under consideration.

11.3 Hadronic shower detectors

Sampling detectors for showers initiated by hadrons are similar in construction to the electromagnetic sampling detectors. Basically the problem of hadron rejection in the electromagnetic detector becomes one of hadron optimization in the hadronic detector. In this section we will further emphasize the differences in the development of the two types of showers.

The physical processes that cause the propagation of a hadron shower are considerably different from the processes in electromagnetic showers. About half the incident hadron energy is passed on to additional fast secondaries [1]. The remainder is consumed in multiparticle production of slow pions and in other processes. A typical secondary hadron is produced with a transverse momentum of ~ 350 MeV/c, so that hadronic showers tend to be more spread out laterally than electromagnetic ones.

Table 11.2 shows the results of a calculation by Gabriel and Schmidt of the most important processes involved in the loss of energy of a 10-GeV proton in iron [15]. We see that most of the energy is dissipated by the ionization losses of scattered secondary protons. The second largest loss arises from the production of π°'s from nuclear interactions. These decay immediately into two photons and thus give rise to an electromagnetic shower within the hadronic one. The third largest loss is due to the binding energy used to break up nuclei and from the production of neutrinos.

Sophisticated Monte Carlo codes that simulate high energy hadronic showers are very useful in calorimeter design. The codes take into account relevant processes such as energy loss, particle decay, scattering, Fermi motion, neutron absorption, noise, and so on. The programs can be checked against measured low energy shower characteristics and then used to predict the expected shower behavior at high energy. The Monte

Carlo program of Gabriel et al. [16] predicts that the fraction of energy going into electromagnetic processes increases with increasing energy. As a result, fluctuations due to binding energy losses decrease, and the resolution should improve faster than extrapolations of low energy measurements. An overall fit to experimental and Monte Carlo data between 1 and 250 GeV gives

$$\frac{\sigma(E)}{\mathrm{E}} = \frac{33.4\%}{\sqrt{E}} + 5.07\%$$

where E is in GeV.

The longitudinal development of hadronic showers scales with the nuclear absorption (or interaction) length

$$\lambda = A/N_{\mathrm{A}}\sigma_{\mathrm{abs}} \tag{11.16}$$

where σ_{abs} is the absorption cross section discussed in Chapter 3. Figure 11.8 shows hadron-induced showers in four different materials. The lateral shower development does not scale with λ [1].

We saw in Section 2 that the electromagnetic and hadronic showers could be distinguished by the amount of energy deposition. Another important characteristic of a hadronic shower is that it takes longer to develop than an electromagnetic one [17]. This can be seen by comparing the number of particles present versus depth for pion- and electron-initiated showers. Long tails are present in the pion distributions. As a consequence, hadronic shower detectors must be deeper than electromagnetic

Table 11.2. *Average fractional energy deposition for a 10-GeV proton in an iron/liquid argon calorimeter*

Process	Percent of total
Secondary proton ionization	31.6
Electromagnetic cascade	21.0
Nuclear binding energy plus neutrino energy	20.6
Secondary $\pi^{\pm}$ ionization	8.2
Neutrons with $E > 10$ MeV	4.9
Neutrons with $E < 10$ MeV	3.9
Residual nuclear excitation energy	3.7
$Z > 1$ ionization	2.4
Primary proton ionization	2.3
Other	1.4

Source: T. Gabriel and W. Schmidt, Oak Ridge National Laboratory report, ORNL/TM-5105, 1975.

detectors of the same type to completely contain the shower. The mean depth of the hadronic shower energy deposition may be parameterized as [18].

$$\lambda_{\max} = 0.90 + 0.36 \ln E \tag{11.17}$$

where $\lambda_{\max}$ is in interaction lengths and the energy is in GeV.

Next let us consider the integral energy deposition. Define L to be the thickness of absorber beyond which only 10% of the energy remains. Figure 11.9 shows the total deposited energy as a function of the depth in the shower. We see that up to 40 GeV essentially all the shower energy is contained in the first five interaction lengths. The data can be parameterized as [18]

$$L = -1.26 + 1.74 \ln E \tag{11.18}$$

where L is in interaction length and E is in GeV.

For finely segmented calorimeters the width of the shower can be defined in terms of the number of detector elements in which the deposited energy exceeds some minimum cutoff value. The shower grows wider as it develops due to large angle nuclear processes and multiple scattering. If

Figure 11.8 Longitudinal development of hadronic showers in different materials. Right scale shows radii for 90% shower containment. Distances are measured in absorption lengths. (C. Fabjan and T. Ludlam, adapted with permission from the Annual Review of Nuclear and Particle Science, Vol. 32, © 1982 by Annual Reviews, Inc.)

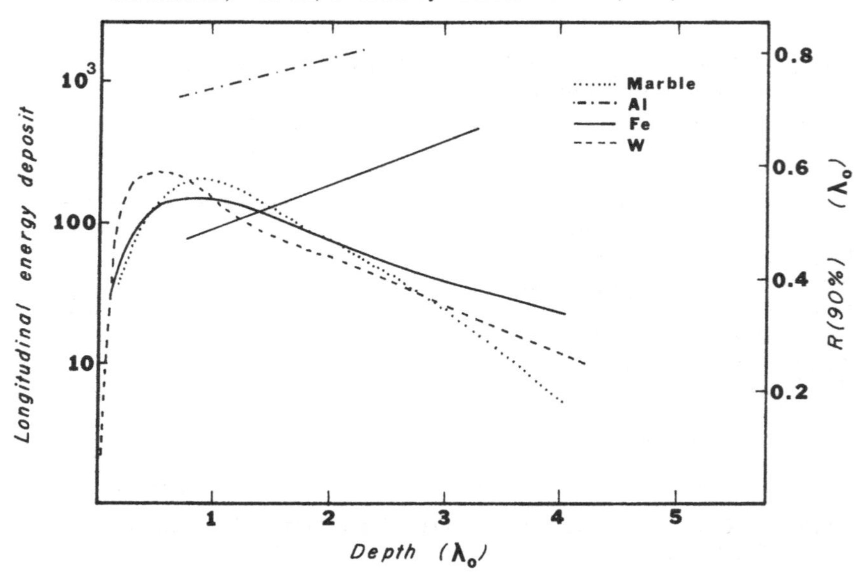

only nuclear processes were involved, we would expect the shower width to steadily increase until the energy of the shower particles falls below the cutoff, at which point the shower would peter out. Actually the width of the shower grows at first and then contracts. This can be seen in Figure 11.10, where the shower profiles in an iron–liquid argon calorimeter are plotted versus the depth in the shower. The narrower profiles deep in the shower are believed to be due to late developing electromagnetic processes, which tend to be produced at small angles. The shower envelope can be defined to be the width that contains 99% of the shower energy. The width of the envelope increases slowly with energy [18]

$$W(E) = -17.3 + 14.3 \ln E \tag{11.19}$$

where the width is in centimeters and the energy is in GeV.

Figure 11.9 Integral energy deposition versus total sampling depth for hadronic showers. (After A. Sessoms et al., Nuc. Instr. Meth. 161: 371, 1979.)

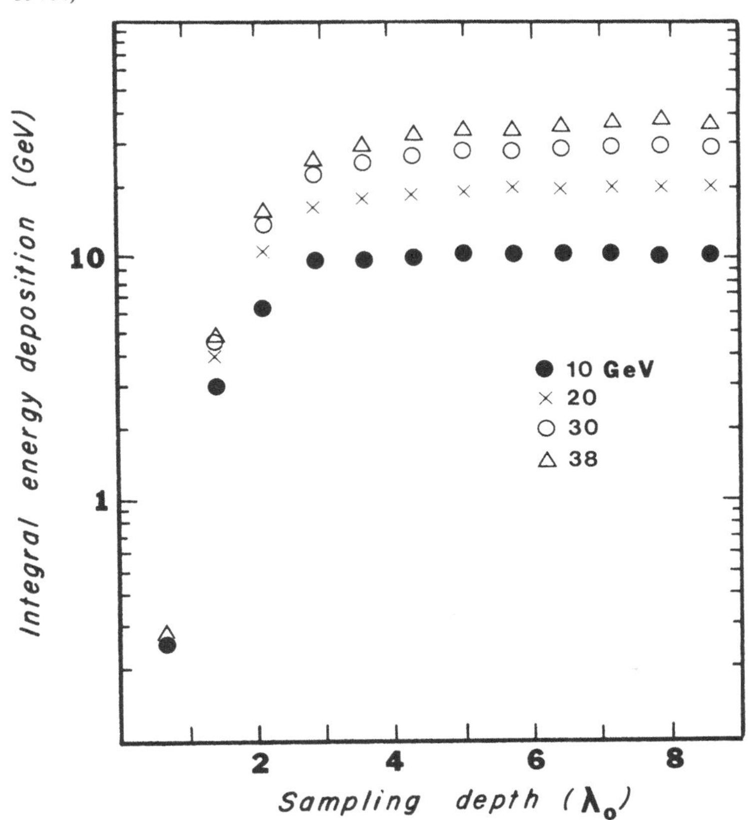

Some examples of large hadronic calorimeters are listed in Table 11.3. The metal radiators most often used for hadron calorimeters are iron and lead. In general, one wants a high density since the shower depth is roughly inversely proportional to ρ. The energy resolution improves with the use of uranium plates [15]. The reason for this seems to be related to the conversion of incident energy into binding energy in nuclear interactions. This energy is ordinarily not detected. However when using uranium absorbers, some fraction of the shower energy results in the production of low energy neutrons. These neutrons can induce fission in ^{238}U, compensating for some of the lost energy through the release of fission

Figure 11.10 Transverse shower profiles at various depths in a hadronic shower. The showers originated from the interaction of a 20-GeV hadron in layer 2. (After A. Sessoms et al., Nuc. Instr. Meth. 161: 371, 1979.)

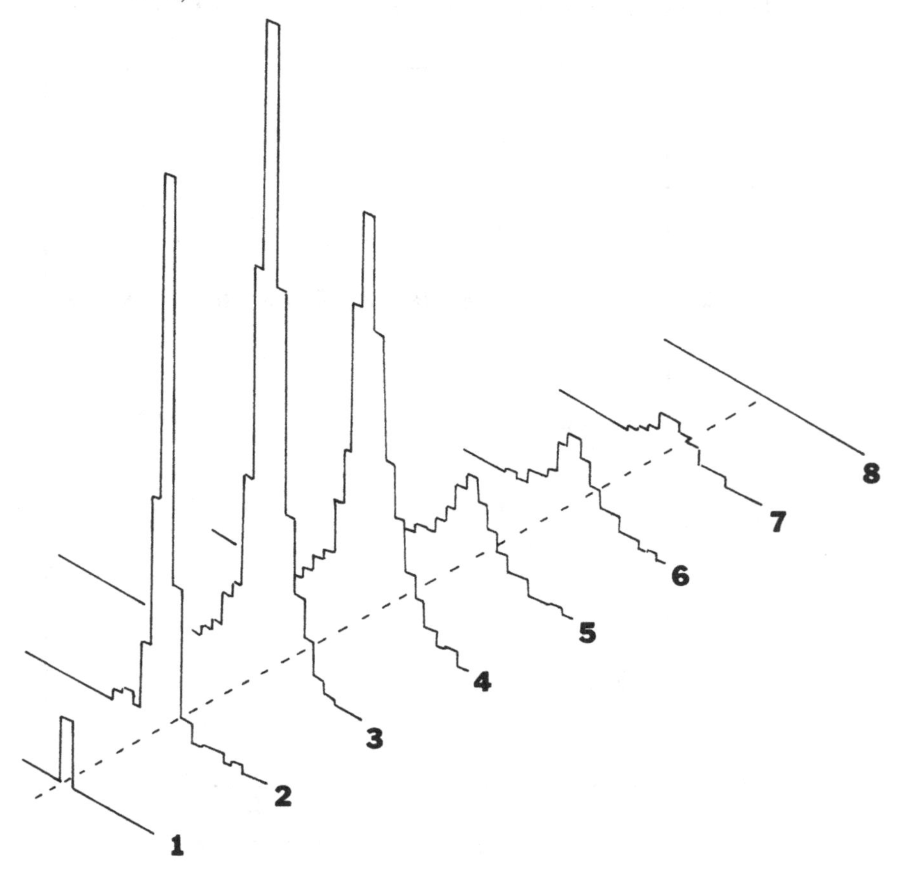

energy. This produces photons and electrons, which can radiate and be detected, and fast neutrons, which can scatter elastically, so that the recoil proton can be sampled.

Important aspects of calorimeter design include the solid angle coverage, operation in a magnetic field, rate characteristics, position and angular resolution, particle identification capability, resolving time, and energy resolution. The position and angular resolutions depend on the degree of segmentation of the active regions. Fine segmentation is necessary to reduce the ambiguity in multiparticle events or to separate nearby showers. This is particularly important for π^0 identification.

Some methods for particle identification using calorimeters are given in Table 11.4. Electron or photons can be identified depending on whether a charged or neutral particle initiates an electromagnetic shower. A neutral pion can be identified from the invariant mass of its decay photons. If momentum analysis is available, protons, neutrinos, and muons can also be identified in certain kinematic regions by comparing the visible energy in the calorimeter with the measured momentum.

In general, the energy resolution of hadron calorimeters is worse than that of electromagnetic calorimeters. Hadronic calorimeters are subject to all the fluctuations discussed in Section 2 in connection with electromagnetic calorimeters and to additional fluctuations due to nuclear interactions. The energy used for the production of neutrinos and high energy

Table 11.3. *Examples of hadronic calorimeters*

Detector	Absorber thickness (mm)	Sensitive thickness[a] (mm)	σ/E (%/$\sqrt{E}$)
AFS	6 U + 5 Cu	2.5 Sc	35
CDHS	25 Fe	5 Sc	58
CHARM	80 marble	PDT, 30 Sc	45
FNAL ν	16 Fe/sand	flashtubes	80
MAC	27 Fe	PWC	75
Tagged γ	25 Fe	Sc	70
UA-1	50 Fe	10 Sc	80
UA-2	15 Fe	5 Sc	60

[a] Sc, plastic scintillator.

Source: H. Gordon et al., Nuc. Instr. Meth. 196: 303, 1982 (AFS); H. Abramowicz et al., Nuc. Instr. Meth. 180: 429, 1981 (CDHS); M. Jonker et al., Nuc. Instr. Meth. 200: 183, 1982 (CHARM); D. Bogert et al., IEEE Trans. Nuc. Sci. NS-29: 363, 1982 (FNAL ν); A. Beer et al., Nuc. Instr. Meth. 224: 360, 1984 (UA-2); A. Astbury, Physica Scripta 23: 397, 1981.

muons and binding energy losses represents energy that "escapes" from the calorimeter. Fluctuations in the importance of these processes represent a major contribution to the energy resolution of hadron calorimeters. Other fluctuations considered by Fabjan et al. [15] in their study of hadronic energy resolution include saturation of the amplifier response to highly ionizing particles, the influence of dead regions on the sampling error, and detection of slow neutrons from a previous event. We show in Figure 11.11 the contributions to the standard deviation of the collected charge signal in liquid argon calorimeters. For electrons the width is almost entirely due to sampling fluctuations. On the other hand, the total width for hadrons incident upon iron plates is much larger than the width due to sampling variations. The total width with uranium plates comes closer to the sampling limit. In each case the hadronic resolution is worse than the electromagnetic one.

11.4 Neutral particle detectors

All neutral particle detectors require that the neutral particle have an interaction that results in the liberation of a charged particle somewhere within its sensitive volume. The detector then responds to the charged particle in the normal manner.

Slow neutrons can be captured by a nucleus, which then emits an alpha particle or a proton or undergoes fission. An important reaction of this type that occurs with boron nuclei is $^{10}B(n, \alpha)^7Li$, which has a cross section of 3830 barns for thermal neutrons. Boron may be used to line a

Table 11.4. *Particle identification using calorimeters*

Particle	Technique	Background
$e^{\pm}$	charged particle initiating electromagnetic shower	$\pi^{\pm}N \rightarrow \pi^0 x$
γ	neutral particle initiating electromagnetic shower	γ's from meson decays
π^0 and other vector mesons	invariant mass of $\gamma\gamma$ or decay products	
ν	compare visible energy with missing momentum	
μ	compare visible energy with momentum; range	noninteracting π
n or K_L^0	neutral particle initiating hadronic shower	

Source: C. Fabjan and T. Ludlam, Ann. Rev. Part. Sci. 32: 335, 1982.

chamber, or a boron-containing gas such as BF_3 may be used in a proportional counter. Coatings of ^{235}U or ^{209}Bi can fission when bombarded with slow neutrons. The fission fragments often produce a large ionization in a short range.

Fast neutrons can often be slowed down using a moderator. These are hydrogen rich compounds, such as paraffin or H_2O. Neutron capture cross sections are largest at small velocities. High energy neutrons are detected primarily by the ionization caused by the recoil proton in np elastic scattering or by a charged secondary in an inelastic reaction. They often interact in scintillator since it is hydrogen rich. However, some

Figure 11.11 Standard deviation of the collected charge distribution for different types of calorimeters as a function of the available energy. The solid curve shows the total width, the dashed curve gives the contribution from sampling fluctuations, and h refers to an incident hadron. (After C. Fabjan et al., Nuc. Instr. Meth. 141: 61, 1977.)

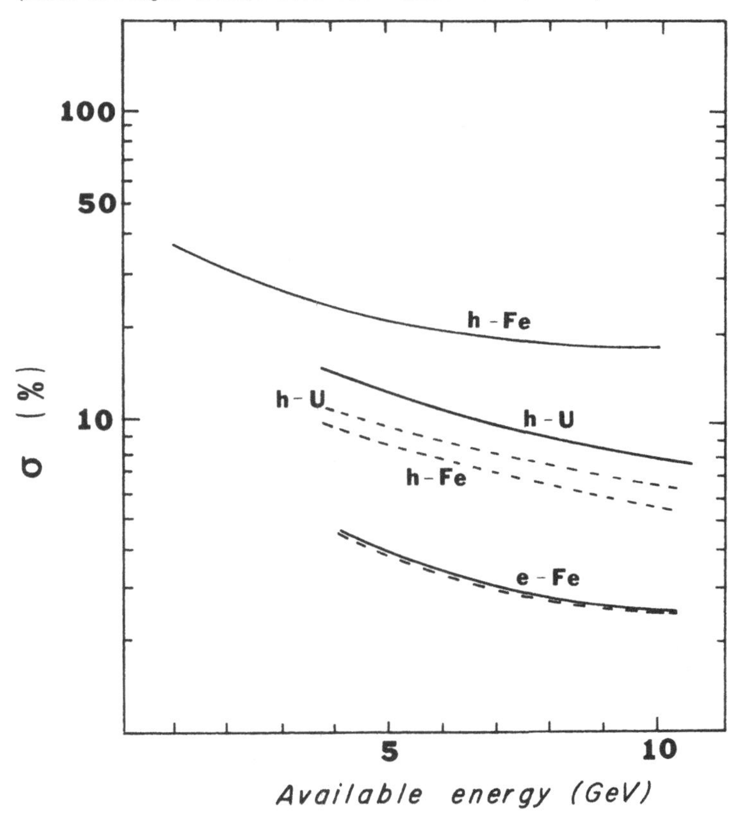

method, such as pulse height analysis or time of flight, may be necessary to discriminate against the detection of photons.

Steel – scintillator sampling calorimeters have been designed with good efficiency for detecting neutrons. As an example, consider the detector of Marshak and Schmuser [19], which consisted of six $\frac{1}{2}$-in.-thick layers of scintillator interspersed among sets of steel plates. The center to center separation between the scintillators was 4 in., allowing the amount of steel to be adjusted from 0 to 3 in. The measured detection efficiency, shown in Fig. 11.12, rose from 12% with no steel present to a plateau at 62% with 12 or more total inches of steel. The measured values agreed quite well with Monte Carlo calculations, based primarily on inelastic scattering. The efficiency was relatively flat over the center ±5 in. of the scintillator and for incident neutron momenta between 2 and 5 GeV/c.

Experiments with neutrinos require very large targets because of the extremely small interaction cross section for neutrinos in matter. This

Figure 11.12 Neutron detection efficiency in a steel – scintillator neutron detector. (After M. Marshak and P. Schmuser, Nuc. Instr. Meth. 88: 77, 1970.)

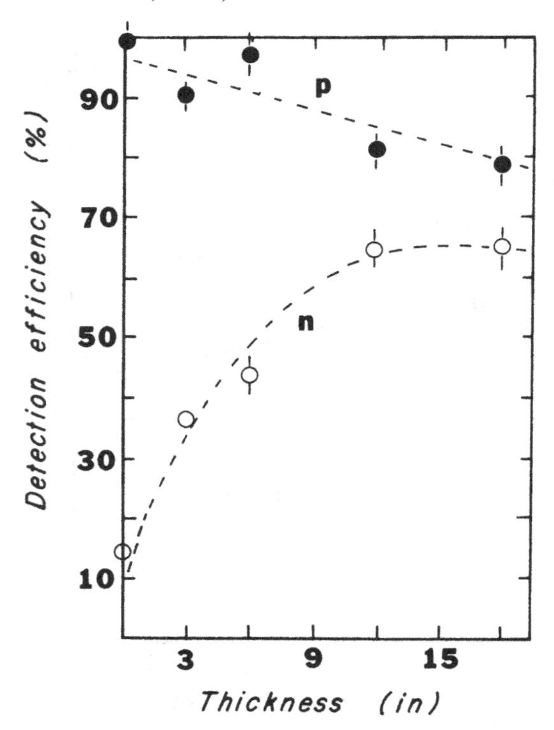

large sensitive target volume is customarily provided by a bubble chamber or a sampling calorimeter [1, 20]. The detector must be capable of providing a large mass of material to maximize the probability that the neutrino will interact and frequent track measurements so that the neutrino interaction vertex and primary interaction products may be measured. Uniform sensitivity and the ability to resolve multiparticle events require a certain minimum degree of segmentation.

Neutrino calorimeters have been constructed using scintillators, proportional drift tubes, drift chambers, and flash tubes as the active medium. The calorimeter may be partially surrounded by scintillation counters to veto events initiated by charged beam particles or cosmic rays. However, sometimes it is convenient to use the cosmic ray muons as a calibration monitor.

References

[1] C. Fabjan and T. Ludlam, Calorimetry in high energy physics, Ann. Rev. Nucl. Part. Sci., 32: 335–89, 1982.
[2] B. Rossi, *High Energy Particles,* New York: Prentice-Hall, 1952, Chap. 5.
[3] W. Heitler, *The Quantum Theory of Radiation,* 3rd ed., Oxford: Clarendon Press, 1953.
[4] D. Muller, Electron showers of high primary energy in lead, Phys. Rev. D 5: 2677–83, 1972; comparisons of the predicted shower profiles with measurements up to 300 GeV can be found in T. Prince, The energy spectrum of cosmic ray electrons between 9 and 300 GeV, Astro. J. 227: 676–93, 1979.
[5] U. Amaldi, Fluctuations in calorimetry measurements, Physica Scripta 23: 409–24, 1981.
[6] N. Hotta, H. Munakata, M. Sakata, Y. Yamamoto, S. Dake, H. Ito, M. Miyanishi, K. Kasahara, T. Yuda, K. Mizutani, and I. Ohta, Three dimensional development of cascade showers induced by 50, 100, and 300 GeV electrons, Phys. Rev. D 22: 1–12, 1980.
[7] W. Schmidt-Parzefall, Calorimeter readout methods, Physic. Scripta 23: 425–33, 1981.
[8] A. Astbury, Calorimeter facilities, Physica Scripta 23: 397–408, 1981.
[9] M. Atac, S. Kim, M. Mishina, W. Chinowsky, R. Ely, M. Gold, J. Kadyk, P. Rowson, K. Shinsky, Y. Wang, R. Morse, M. Procario, and T. Schaad, Saturated avalance calorimeter, Nuc. Instr. Meth. 205: 113–24, 1983.
[10] J. Cobb, S. Iwata, D. Rahm, P. Rehak, I. Stumer, C. Fabjan, M. Harris, J. Lindsay, I. Mannelli, K. Nakamura, A. Nappi, W. Struczinski, W. Willis, C. Kourkoumelis, and A. Lankford, A large liquid argon shower detector for an ISR experiment, Nuc. Instr. Meth. 158: 93–110, 1979.
[11] D. Hitlin, J. Martin, C. Morehouse, G. Abrams, D. Briggs, W. Carithers, S. Cooper, R. Devoe, C. Friedberg, D. Marsh, S. Shannon, E. Vella, and J. Whitaker, Test of a lead/liquid argon electromagnetic shower detector, Nuc. Instr. Meth. 137: 225–34, 1976.
[12] H. Gordon, R. Palmer, and S. Smith, New ideas in calorimetry, Physica Scripta 23: 564–8, 1981.
[13] S. Stone, J. Poucher, R. Ehrlich, R. Poling, and E. Thorndike, Characteristics of electromagnetic shower sampling counters, Nuc. Instr. Meth. 151: 387–94, 1978.

[14] G. Abshire, M. Adams, C. Brown, E. Crandall, J. Goldberger, P. Grannis, J. Stekas, G. Donaldson, H. Gordon, G. Morris, and L. Cormell, Measurement of electron and pion cascades in a lead/acrylic scintillator shower detector, Nuc. Instr. Meth. 164: 67–77, 1979.
[15] C. Fabjan, W. Struczinski, W. Willis, C. Kourkoumelis, A. Lankford, and P. Rehak, Iron/liquid argon and uranium/liquid argon calorimeters for hadron energy measurement, Nuc. Instr. Meth., 141: 61–80, 1977.
[16] T. Gabriel, B. Bishop, M. Goodman, A. Sessoms, B. Eisenstein, R. Kephart, and S. Wright, A Monte Carlo simulation of an actual segmented calorimeter: a study of calorimeter performance at high energies, Nuc. Instr. Meth. 195: 461–7, 1982.
[17] D. Bollini, P. Frabetti, G. Heiman, G. Laurenti, L. Monari, and F. Navarria, Study of the structure of hadronic and electromagnetic showers with a segmented iron-scintillator calorimeter, Nuc. Instr. Meth. 171: 237–44, 1980.
[18] A. Sessoms, M. Goodman, L. Holcomb, E. Sadowski, A. Strominger, B. Eisenstein, L. Holloway, W. Wroblicka, S. Wright, and R. Kephart, The segmented calorimeter: A study of hadron shower structure, Nuc. Instr. Meth. 161: 371–82, 1979.
[19] M. Marshak and P. Schmuser, A steel/scintillator counter to detect neutrons, Nuc. Instr. Meth. 88: 77–82, 1970.
[20] B. Barish, Experimental aspects of high energy neutrino physics, Phys. Rep. 39: 279–360, 1978.

Exercises

1. A 50-GeV electron traverses a stack of iron. (a) Using the Heitler shower model, estimate the number of positrons produced after 15 cm. (b) What is the average positron energy? (c) At what depth in the stack will the number of particles in the shower reach a maximum? (d) What is the maximum number of particles present in the shower?

2. (a) Estimate the total depth of a practical 15-GeV electromagnetic calorimeter if we allow an additional $2t_{\max}$ beyond the shower maximum to minimize the probability of escaping particles. (b) How does this compare with the total depth of a 15-GeV hadron calorimeter?

3. (a) Show that the minimum opening angle of the two photons in π^0 decay is $2m_\pi/E_\pi$, where E_π is the π^0 energy. (b) Assume that the two-photon showers can be resolved if they are separated by two Moliere radii. Estimate the maximum π^0 energy that can be resolved using a practical lead–scintillator calorimeter 2 m from the interaction point.

12
Specialized detectors

The detectors discussed to this point, scintillation counters, Cerenkov counters, proportional chambers, and drift chambers, can be found in some combination in most particle physics experiments. There are in addition other types of detectors that are used for specialized applications. These include such old standards as emulsions and bubble chambers, as well as the more recently developed transition radiation detectors and semiconductor detectors. This chapter contains a short discussion on some of the more common types of specialized detectors.

12.1 Bubble chambers

The bubble chamber was one of the most important detectors for particle physics experiments during the 1960s and early 1970s. Much of our knowledge of particle spectroscopy, strong interactions, and neutrino physics was learned with this device.

The detector consists of a cryogenic fluid, an expansion system, and usually a magnetic field. The expansion system, which typically involves a piston and bellows arrangement, is used to suddenly reduce the pressure on the chamber liquid. The liquid is initially maintained at a pressure above the equilibrium vapor pressure curve. Following a sufficiently rapid expansion, it is left in a "superheated" condition, and bubbles will form around any nucleation centers in the liquid. A charged particle will produce delta rays in the liquid, and the delta rays in turn can produce large numbers of ion pairs in a small volume. The recombination of these ion pairs produce a heat spike, which is thought to be responsible for bubble formation [1, 2].

When used in a particle physics experiment, the chamber expansion is synchronized with the arrival of a beam of particles from the accelerator.

Any charged particle that is in the beam or that is created from interactions in the chamber liquid will leave a trail of ions in the liquid. Bubbles that form around the ions typically grow to a diameter of ~ 10 μm in 0.2–50 ms. At this point flash lights illuminate the chamber, and two or more cameras simultaneously take a picture of the tracks of bubbles. A magnetic field surrounding the chamber allows the particle momenta to be measured from the curvature of the tracks. After the pictures are taken, the expansion system recompresses the liquid and collapses the bubbles. The bubble density gives a measure of the ionization energy loss and can be used to determine the mass of particles with $\beta \leqslant 0.8$.

The chamber fluid can be liquid hydrogen, liquid deuterium, or a heavy liquid mixture. Liquid hydrogen provides free protons. Tracks measured in liquid hydrogen have relatively small errors due to multiple scattering. Liquid deuterium is used to extract information about interactions on quasi-free neutrons. Heavy liquids are useful for studying interactions involving photons, electrons, or neutrinos. The greater density provides more interactions per unit volume, while the higher Z gives shorter γ ray conversion lengths. The virtues of liquid hydrogen and the heavy liquid can be combined by filling a track sensitive target with liquid hydrogen and locating it inside the volume of a heavy liquid chamber.

Bubble chambers offer a number of advantages. They have 4π acceptance, excellent tracking efficiency, good spatial resolution, mass identification for slow particles, and efficient observation of secondary decay vertices. On the other hand, several disadvantages led to a decline in the use of the method. The standard chamber cannot be triggered. Thus, many pictures are taken that do not contain useful information. Analysis of the film is a lengthy process that may take several years. Bubble chambers cannot efficiently use a high beam rate. Moderate sized chambers cannot make an accurate measurement of high momentum tracks, and they cannot be used at colliding beam machines.

Some of these difficulties may be overcome with the use of hybrid bubble chamber systems [3]. These systems involve using the bubble chamber in conjunction with counters or wire chambers. Modern neutrino experiments often use a large volume bubble chamber together with an external muon identifier. The rapid cycling bubble chamber at SLAC can cycle 20–100 times faster than standard chambers, making it suited to the fast repetition rate of the SLAC accelerator. The hybrid system includes wire chambers for accurate measurement of high momentum particles and a Cerenkov counter to aid in particle identification. The system may be operated in a triggered mode and is very effective in studying processes with a fast forward particle.

Benichou et al. [4] have constructed a small special purpose bubble chamber (LEBC) to study the production of rare, short-lived particles. The chamber was machined from Lexan plastic and had a volume of only 1.1 liters. Because the cross section for the production of the particles was small, the chamber was designed to cycle with a 33-Hz repetition rate. High resolution was necessary to clearly separate the particle production and decay vertices.

Figure 12.1 shows the pressure–temperature characteristics of low temperature hydrogen. The equilibrium vapor pressure is shown together with the foam limit. Pressure–temperature points below the foam limit curve are subject to free bubble formation. The vertical lines indicate the

Figure 12.1 The pressure–temperature characteristics of low temperature hydrogen. (J. Benichou et al., Nuc. Instr. Meth. 190: 487, 1981.)

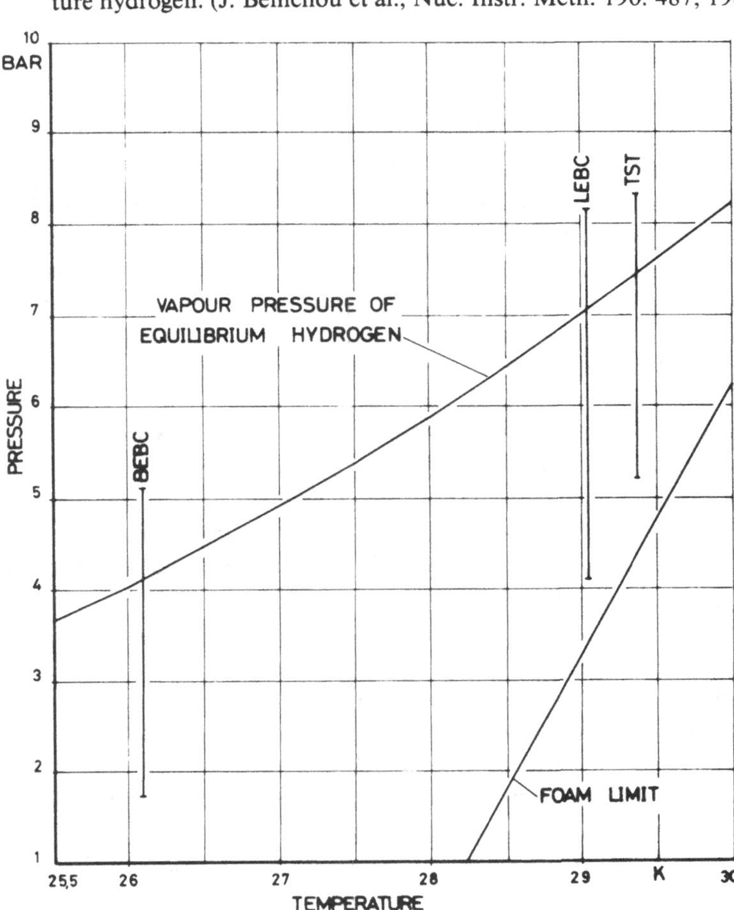

pressure range used in three bubble chambers. Normal chambers such as the Big European Bubble Chamber (BEBC) operate at lower temperatures than LEBC. The higher temperature allows LEBC to use a higher minimum pressure, resulting in slower bubble growth rate and better resolution [4].

Another possible improvement in bubble chamber technique involves the use of holography to replace the conventional photographic systems [5, 6]. This permits improvement of the optical resolution without losing depth of field and an increase in acceptable particle flux. In holography the film is illuminated with coherent light from a laser. Great care must be taken to prevent any thermal gradients or turbulence in the chamber liquid. The holographs must also be illuminated by laser on the scanning and measuring machines. Dykes et al. [5] used a pulsed ruby laser to make holograms with a spatial resolution of $\sim 8\ \mu$m and with particle fluxes up to 150 particles per picture.

12.2 Emulsions

The detection of radiation by the exposure of photographic plates dates back to the discoveries of Röntgen and Bequerel. Emulsions are still used in experiments requiring the best possible spatial resolution.

The emulsion consists of three basic components: a silver halide, a gelatin-plasticizer, and water [7]. A layer of the emulsion, typically 600 μm thick, covers a plate, a number of which may be stacked together into a sandwich. Sometimes the emulsion is "loaded" with metal wires, foil, or powdered layers to give a well-defined target.

The halide grains have a diameter of $\sim 0.2\ \mu$m. Ionization energy from a charged particle traversing the emulsion may cause the grain to develop. The production of delta rays with energy between 150 and 5000 eV plays an important role in developing the grains, since they are capable of providing a large energy deposit in a region similar in size to the grain volume. A minimum ionizing particle produces ~ 270 developed grains per millimeter of track length.

For stopping particles a measurement of the particle's range can be used to determine its energy. A measurement of the energy can also be derived from the density of delta rays. A measurement of the exposed grain density or the distribution of gaps between exposed grains can be used to measure the specific ionization or dE/dx of the track. At low energy this is proportional to $1/v^2$. Grain counts can be calibrated by measuring the distinctive electron tracks in nearby regions of the emulsion.

Grain density provides a means of particle identification for $\beta \leqslant 0.9$ but saturates for higher velocities. For relativistic particles measurements of the projected angles between successive chords on the track and the multiple scattering theory provide a determination of $p\beta$.

Events in the emulsion must be located in a tedious search using a microscope. Recent experiments [8] have used the emulsion in conjunction with a particle spectrometer. Long-lived tracks can be measured in the spectrometer and used to predict the position of the interaction vertex. This greatly reduces the volume of emulsion that must be examined.

12.3 Spark chambers

Spark chambers were used routinely during the period 1960–1975 for particle tracking measurements. They have since been almost totally replaced by MWPCs and drift chambers. One possible advantage for using spark chambers today is a smaller cost for the readout system for a large area detector.

The basic elements of a spark chamber are shown in Fig. 12.2. The chamber consists of metallic plates or wire grids separated by insulating

Figure 12.2 Principles of spark chamber operation. (S) Scintillation counter, (Co) coincidence circuit, (A) amplifier, and (sg) spark gap.

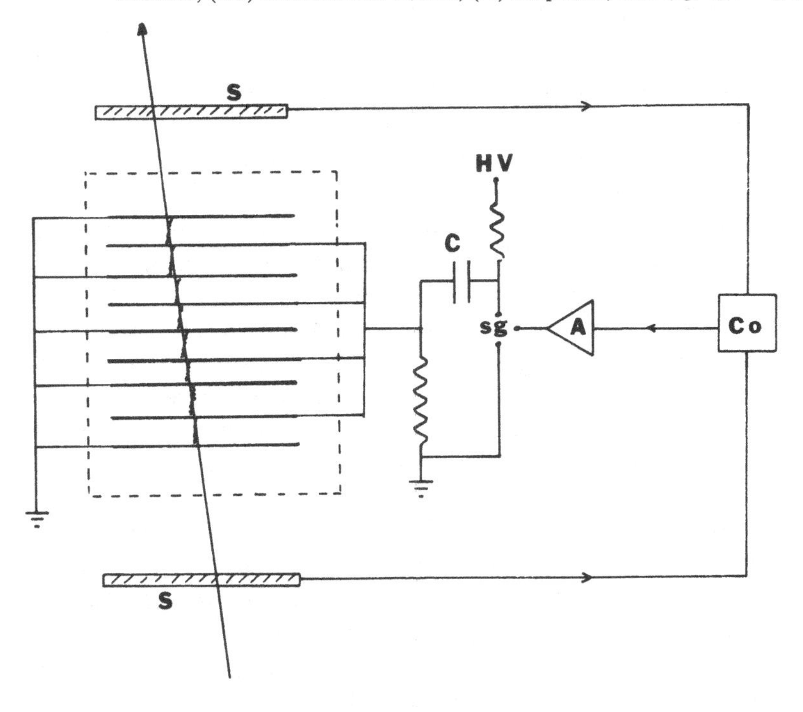

gaps and filled with a gas. A charged particle traversing the detector leaves a trail of ions in the gas. Usually a pair of scintillation counters or other fast responding devices are used in conjunction with the spark chamber. A coincidence signal from the two scintillators indicates the passage of a charged particle. The coincidence signal is amplified and applied to a spark gap or a thyratron. This produces a fast rising high voltage pulse. One plate of the capacitor C discharges quickly to ground. The charge on the opposite plate raises alternate plates of the detector to high voltage. The large potential difference causes a spark to form along the path of the ions in the gap. The locations of the sparks indicate the path of the particle. In a narrow gap chamber the sparks form at right angles to the electrodes, while in wide gap chambers the sparks tend to follow the actual path of the particle. The ions are cleared from the gap by maintaining a low voltage dc clearing field across the gap.

Spark formation requires a buildup of at least 10^8 electrons [9]. This leads to the Raether criterion: for a given chamber the applied field must be such that

$$d > 20/\alpha \tag{12.1}$$

where d is the gap separation and α is the Townsend coefficient. The formation time for the spark then is

$$\tau = 20/(\alpha w^-) \tag{12.2}$$

where w^- is the electron drift velocity. The high voltage pulse must remain on the electrodes for at least this period of time.

Important characteristics of spark chamber gases include α, the drift velocity, electron affinity, photoabsorption, specific ionization, diffusion coefficient, emission spectrum, and cost. The usual spark chamber gas is a mixture of neon and helium. Noble gases have zero electron affinity. Thus, electrons do not attach to the gas atoms, and substantial electron multiplications can occur. Small amounts of additional gases are sometimes added. Electronegative gases (e.g., Freon) help clear the chamber of residual ions. Some organic molecules (e.g., alcohol) are efficient absorbers of photons. The addition of such a gas prevents a photon from liberating an electron some distance from the site of ionization and initiating spurious sparks.

The operation of the spark chamber pulsing system is of the utmost importance. The pulse should be applied as rapidly as possible to minimize diffusion and unrelated sparks and to ensure optimum spark formation. There must be sufficient energy to allow many sparks to form simultaneously. The system must have a short recovery time so that the chamber can be used in moderately high rate environments.

The high voltage switch is a thyratron tube or spark gap. Thyratrons are capable of producing 1000-A pulses in 15–30 ns. This high voltage is applied to the chamber as symmetrically as possible using coaxial cable or insulated stripline. The chamber itself behaves like a waveguide, leading to reflections at the edges unless the electrodes are properly terminated.

Referring to Fig. 12.2, the voltage applied to the spark gap is

$$V_{\text{gap}} = CV/(C + C_{\text{sc}}) \tag{12.3}$$

where C_{sc} is the capacitance of the chamber itself. A large 1-m^2 chamber may have a capacitance $C_{\text{sc}} \sim 1000$ pF. A proper treatment of the chamber's electrical properties should include the inductances of the leads, which are important when considering very fast pulses [9].

In the absence of a magnetic field the ions are displaced across the spark chamber gap because of the clearing field. The displacement is $w\,\Delta t_1$, where w is the drift velocity and Δt_1 is the time from the creation of the ion until the high voltage (HV) pulse is applied. During the initial risetime of the HV pulse the ions undergo an additional drift of $w\,\Delta t_2$. The clearing field and high voltage pulse are ordinarily adjusted with opposite polarity so that these displacements tend to cancel. The effect leads to an angle dependent correction to the measured track position

$$\Delta x = b + m \tan\theta \tag{12.4}$$

In addition, it is sometimes desirable to alternate the direction of the clearing field in alternate gaps.

Spark chambers can routinely achieve 95% efficiency for the detection of a single charged particle. The efficiency is a function of the static clearing field and the time delay before pulsing the chamber. For a given clearing field, if one waits too long before applying the HV pulse, all the ions will be cleared from the gap and no spark will form. Multispark efficiency is enhanced with wide gaps and fast HV risetimes [9].

The spatial resolution of a wire plane spark chamber is roughly $\pm s/2$, where s is the wire spacing. This can be improved to $\pm s/4$ by adjusting the pulse so that several wires participate in the discharge [9]. Typical spatial resolutions are $\pm 300\ \mu$m. The resolution is influenced by a number of factors. Since the ions are typically produced with several electron volts of kinetic energy, they can diffuse before the application of the HV pulse. The spark channel itself may have a width of ~ 1 mm [9]. For narrow gap chambers the spark discharge is perpendicular to the electrodes and forms near the cathode, even for inclined tracks. The spatial accuracy is improved at higher gas pressures.

The time resolution, or "memory time," of the chamber must be long enough so that ions are still in the gap when the logic system determines if the event is of interest, yet short enough so that the gaps are not full of ions from unwanted events. This time is determined by the value of the clearing field and the HV pulse delay and by the addition of electronegative gases. A typical memory time is 1 μs, enabling the chamber to be used in a beam flux of 10^6 particles/sec. Following a HV pulse, the chamber requires a recovery (dead) time of ~ 1 ms to clear the gap of debris. The electrons clear from the gap first, leaving the slower positive ions and meta-stable atoms.

The occurrence of a spark leads to a number of optical, electrical, and acoustic phenomena that can be used to determine the position of the spark. A list of some readout methods and the physical principles they rely on are given in Table 12.1. We will describe the magnetostrictive readout method. For a discussion of other methods the reader is referred elsewhere [9, 10].

Table 12.1. *Spark chamber detection systems*

Method	Principle
1. Optical	picture of spark recorded on film
2. Vidicon	optical image transformed into electrical signal by TV camera
3. Sonic	propagation time in gas used to localize spark
4. Ferrite core	current flowing in wire electrode sets ferrite memory core
5. Magnetostriction	magnetic field of spark induces acoustic wave in magnetostrictive wire
6. Delay line	propagation time of signals induced by sparks in electrode or in external delay line separating wires permits localization of spark
7. Current division	amount of charge carried in spark split between two channels; difference in charge measured on transformer filter circuit gives position
8. Sparkostriction	auxiliary spark produced against a wire at small distance from electrode; time of propagation in this nonmagnetic wire gives position of spark
9. Capacitive	current in each wire charges capacitor that can subsequently be read out
10. Spark projection	low frequency potentials of wire chamber transported by external wires to second plane; auxiliary spark produced and viewed by vidicon tube
11. Magnetic tape	sparks produce signal on magnetic tape used as electrode

Source: G. Charpak, Ann. Rev. Nuc. Sci. 20: 195, 1970.

Magnetostriction is a property of certain ferromagnetic materials whereby a magnetic field induces a stress in the material. Magnetostrictive wires are placed across the leads of the spark chamber electrodes [11]. The current from the spark creates a magnetic field around the wire. The field changes the magnetization in the magnetostrictive wire and creates an acoustic pulse in the wire. The pulse has both longitudinal and torsional modes. The longitudinal mode is dispersive, but its detection is simpler. Acoustic pulses travel in both directions with velocity $\sqrt{E/\rho}$, where E is the elastic modulus and ρ is the density. For iron–cobalt wire, $v \sim 5$ mm/μs. At each end of the magnetostrictive wire is a pickup coil. Here the acoustic wire induces a current by the inverse, or Villari, effect. The arrival times of the pulses at the two detectors can be used to determine the position of the spark. Multiple sparks can be detected by counting oscillator pulses in sets of scalers, which have been simultaneously started at the arrival of the first fiducial, and stopped one by one as signals from the sparks are received from the magnetostrictive pickup coils.

Magnetostrictive readout is difficult to use in a magnetic field because the field affects the shape of the magnetostrictive pulse. This necessitates bringing the spark chamber wires outside the region of the field before being read out. For operation inside a field it is often simpler to use some other method, such as capacitive readout.

12.4 Streamer chambers

The streamer chamber is in some respects an electronic analog of the bubble chamber. Charged particles passing through the chamber gas leave the usual trail of ionization. The chamber volume is subjected to an electric field of 10–50 kV/cm for a short period of 5–20 ns. Electron avalanches develop into expanding, conducting plasmas known as streamers when their space charge field is comparable to the applied field. The streamers move at high velocity toward both electrodes. At the end of the high voltage pulse the streamers remain suspended in the chamber [9]. The ions recombine and emit light, which reveals their location and thus the particle trajectories. Some virtues of the streamer chamber are that many tracks may be followed over large distances, it has isotropic response, and unlike the bubble chamber, it can be triggered. One disadvantage is the lower density.

The basic elements of a streamer chamber are similar to the spark chamber. A coincidence signal from scintillation counters indicate the passage of a particle. A large high voltage pulse is created with a Marx generator. This consists of a set of capacitors that are charged up slowly in parallel and discharged rapidly in series. The high voltage pulse must be very short to stop the spark development at the streamer stage. Thus, a shaping network is necessary before the pulse is applied to the electrodes. Sharp pulses 20 ns in duration are typical.

The appearance of the streamer depends on the relative orientation of the camera, the electrodes, and the particle trajectory. A typical streamer chamber gas is 90% neon and 10% helium. The memory time is controlled by the addition of electronegative gases such as SF_6.

The chamber is practically 100% efficient for single tracks, since many ions are liberated along the particle trajectory. Multitrack efficiencies are also very high. The chamber can operate in a high rate environment. Some particle identification is possible below 1 GeV/c by using the streamer density.

The streamers are created with a width of about 1 mm. Particle trajectories can be located with an accuracy of about 300 μm. The resolution improves as the gas pressure is increased because the avalanche reaches the same total ionization in a smaller distance. Two effects that lead to systematic errors in locating the track are the drift of the ionization electrons during the rise of the high voltage pulse and the fact that streamers develop symmetrically about the head of the avalanche and not about the site of the original ionization. A high pressure streamer chamber constructed at Yale University [12, 13] is capable of measuring track positions with a spatial resolution of 40 μm.

12.5 Transition radiation detectors

Devices utilizing transition radiation have proven useful for the detection of electrons and show promise for providing particle identification at high energies. An important feature of such radiation is that its intensity can be made to increase linearly with the Lorentz factor γ.

A charged particle emits transition radiation when it crosses the interface between media with different dielectric or magnetic properties. When the particle is in the region of low dielectric constant, polarization effects in the surrounding medium are small, and the electric field associated with the moving charge has a large spatial extent. However, when the particle crosses the interface to a region of higher dielectric constant, polarization effects are larger, thereby reducing the extent of the electric field in the medium. The sudden redistribution of charges in the medium associated with the changing electric field of the particle gives rise to the transition radiation.

At high energy transition radiation is primarily emitted as X-rays. The total energy emitted upon crossing a single surface is [14]

$$W = \tfrac{2}{3}\,\alpha\omega_p\gamma \tag{12.5}$$

where α is the fine structure constant and ω_p is the plasma frequency of the medium,

$$\omega_p^2 = \frac{4\pi\alpha Z N_A \rho}{A m_e} \tag{12.6}$$

The mean number of X-rays emitted per surface is

$$\langle N \rangle \sim \tfrac{1}{2}\,\alpha \tag{12.7}$$

The X-rays are emitted strongly forward into a cone at an angle θ with respect to the particle direction given by

$$\theta \sim 1/\gamma \tag{12.8}$$

Practical detectors are made up of stacks of thin foils, thereby providing many interfaces. As soon as more than one surface is present, interference effects can arise from the superposition of the radiation. Foils must have a minimum thickness, known as the formation zone, so that the electromagnetic field of the particle can fully develop in the entered medium. The photon yield drops off sharply for foils thinner than the formation zone. The interference causes oscillations in the intensity as a function of ω, the photon frequency, and gives rise to γ thresholds where the intensity increases sharply.

A number of practical transition radiation detectors have been constructed. Cobb et al. [15] have built a carefully optimized transition detector for electron–hadron separation at the ISR. The device consisted of two groups of ~ 700 foil radiators followed by a MWPC detector. The foil spacing was chosen to provide constructive interference. The radiators had to have a small X-ray absorption so that the photons could reach the detector. This and other requirements lead to the choice of thin 53-μm lithium metal foils as the radiator. The threshold parameter γ_t for the device is proportional to the foil thickness d,

$$\gamma_t = \frac{d\omega_{Li}}{2c} \tag{12.9}$$

where ω_{Li} is the plasma frequency in lithium (14.2 eV/$\hbar$). One problem with a detector of this type is that the charged incident particle also passes through the MWPC and deposits ionization energy in the chamber gas. A xenon–CO_2 mixture was found to efficiently detect the transition X-rays and to give the best ratio of X-ray absorption to charged particle ionization. The detector was able to count electrons with good efficiency and with a substantial rejection for hadrons.

If a large distance is available between the radiator and detector, it is possible to improve the particle discrimination by using information about the width of the pulse. Deutschmann et al. [16] analyzed the current signals from a drift chamber using 16 analog to digital converters triggered by 16 successive 30-ns gates. Electrons with $p = 15$ GeV/c showed a considerably wider distribution than 15-GeV/c pions due to the presence of transition radiation in the electron sample.

It is also possible to improve the detector's sensitivity by counting the number of charge clusters along the particle's track in a drift chamber. Ludlam et al. [17] and Fabjan et al. [18] measured the deposited charge in 20-ns intervals of drift time. They defined a cluster to be the amount of charge deposited in any 100-ns interval corresponding to an energy of 2 keV. The 100-ns time interval was equivalent to a spatial extent of 850 μm. The number of clusters is expected to follow a Poisson distribution. Figure 12.3 shows the experimental data for the number of clusters versus γ. It can be seen that the presence of lithium radiators leads to a greater number of clusters due to interactions of the transition radiation photons in the drift chamber gas.

12.6 Short summary of some other detectors

12.6.1 Semiconductor detectors

Semiconductor counters behave like solid ionization chambers [19, 20]. Incident charged particles deposit ionization energy and dislodge electrons, which in turn produce secondary ionization. This causes a separation of the electrons and holes in the semiconductor, which then separate due to the presence of an electric field. They collect at the electrodes, giving a signal proportional to the deposited energy. The detector has better resolution than gaseous ion chambers since the average energy required to produce an ion pair in silicon is 3.6 eV, for instance, while it is typically ~30 eV in gases. Semiconductor detectors are useful when a thin, high resolution detector is required near an interaction region. Silicon surface barrier detectors have high efficiency and high rate capability and can operate inside a colliding beam vacuum chamber. The principle disadvantages are small size and radiation damage.

Amendolia et al. [21] used a stack of silicon surface barrier detectors as an active target. They identified the layer where the interaction occurred from the large pulse height resulting from the recoil nucleus. The pulse height from the subsequent layers gave information on the particle multiplicity as a function of the distance from the interaction point.

A slice of silicon can also be etched to give narrow strips of active detector [20, 22]. This provides a small area device that behaves like a MWPC but has better spatial resolution, especially for multitrack events. Such a detector is shown in Fig. 12.4. A relativistic, singly charged particle produces about 25,000 electron–hole pairs while crossing this detector [22]. All the charge is collected within 10 ns. A telescope consisting of six

Figure 12.3 Number of ionization clusters in a drift chamber as a function of the relativistic γ factor. The upper curve shows the data with lithium radiator foils. (After T. Ludlam et al., Nuc. Instr. Meth. 180: 413, 1981.)

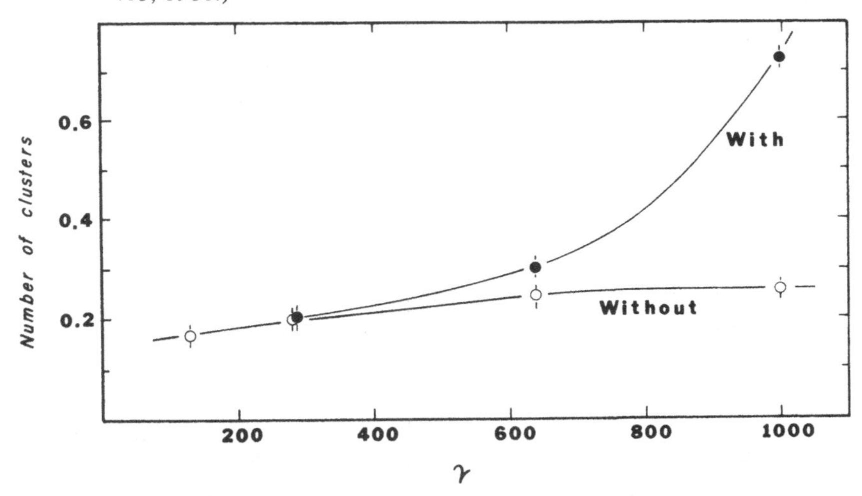

Figure 12.4 Cross section of a silicon microstrip detector. (B. Hyams et al., Nuc. Instr. Meth. 205: 99, 1983.)

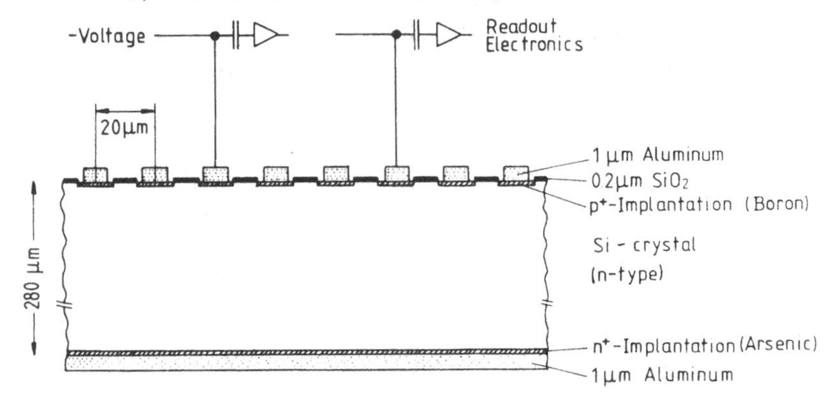

of these silicon microstrip detectors has achieved 5 μm spatial resolution for minimum ionizing particles.

12.6.2 Flashtube hodoscope chambers

The flashtube chamber is similar in operation to a spark chamber [23]. The hodoscope is made up of many tubes containing neon gas. An external trigger sensitizes each tube by initiating the application of a high voltage pulse. If any ions are in a tube due to the passage of a charged particle, a streamer will form, and the tube emits a pulse of light. This is either detected directly or causes a transducer to generate an electrical pulse.

Large arrays of flashtubes have been used in calorimeters and for studying cosmic ray showers. The chief advantage of the device is the low cost to construct a large volume array. The spatial resolution is on the order of the tube diameter ($\sim$ 2 mm). The tube can be sensitive for $\sim$ 1 μs and has $\sim$ 10 ms recovery time, depending on the gas composition and clearing field used.

12.6.3 Stopping K^+ detector

Detectors have been constructed that can efficiently indicate the presence of a K^+ [24, 25]. Astbury et al. [24] have built a detector consisting of 17 alternating layers of brass and scintillator. The range–energy relations show that for particles with $p \leqslant 600$ MeV/c, μ and π^+ have a much greater range than K^+ and p. Thus, by using the proper amount of brass absorber, K^+ and p will stop in the detector. This is indicated by a large pulse height in the scintillator preceding the stopping layer and no signal in the following layer. The K^+ will then subsequently decay, while the proton will not. Thus, a K^+ should have in addition a small pulse height signal in a scintillator after a suitable delay period. The detector achieved a high efficiency for identifying K^+ with a good rejection for π^+. This method can also be applied to make a trigger for stopping muons.

12.6.4 Planar spark counters

Planar spark counters have been used in applications such as time of flight measurements where it is desirable to obtain the best possible timing resolution. One such device constructed for use at PEP consisted of two parallel electrodes separated by 185 μm with a field $\sim 2\ 10^5$ V/cm maintained across the gap [26]. The anode was constructed from semiconducting glass. A noble gas–organic quencher mixture flowed through the gap at high pressure. The signal from a discharge had a typical risetime of ~ 1 ns and a pulse length of about 5 ns. The counter recovered within a few milliseconds.

The PEP planar spark counters had timing resolutions better than 200 ps under actual experimental conditions [26]. The measured single counter time distribution for Bhabha scattering events are shown in Fig. 12.5, together with the time of flight distribution. The "albedo" events in the histogram are due to electrons and positrons scattered backward from a nearby shower counter. The albedo events occur later and have a

Figure 12.5 Single counter Bhabha scattering time distributions for two spark counters (T1, T2) and the resulting time of flight spectrum (TOF). The dotted curves are due to backward scattered (albedo) events. (After W. Atwood et al., Nuc. Instr. Meth. 206: 99, 1983.)

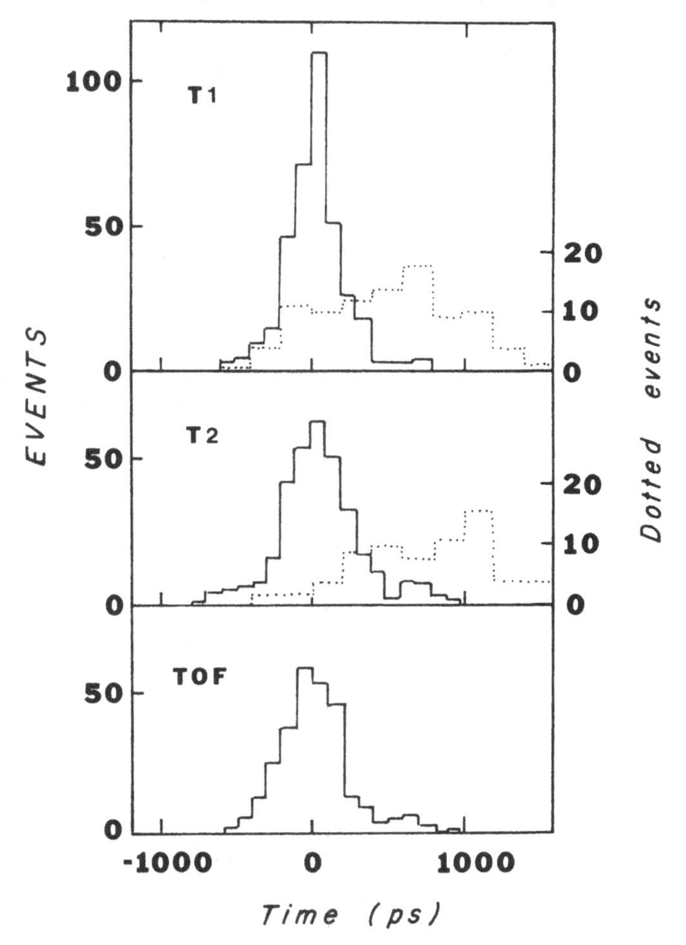

broader distribution than the incident Bhabha particles. The spatial resolution of the counter was 2.6 mm, consistent with the least count of the TDCs used to make the measurement. The pulse height was well correlated with the number of particles striking the counter. There was, however, a degradation in the pulse height and the timing resolution when the counting rate was $\geqslant 300$ Hz.

References

[1] R. Shutt (ed.) *Bubble and Spark Chambers,* New York: Academic, 1967.

[2] T. Dombeck and J. Vanhoy, A study of the bubble formation mechanism in a hydrogen bubble chamber using e^- and e^+ tracks, Nuc. Instr. Meth 177: 347–52, 1980.

[3] J. Ballam and R. Watt, Hybrid bubble chamber systems, Ann. Rev. Nuc. Sci. 27: 75–138, 1977.

[4] J. Benichou, A. Herve, H. Leutz, G. Passardi, W. Seidl, J. Tischhauser, H. Wenninger, and C. Fisher, A rapid cycling hydrogen bubble chamber with high spatial resolution for visualizing charmed particle decays, Nuc. Instr. Meth. 190: 487–502, 1981.

[5] M. Dykes, P. Lecoq, D. Gusewell, A. Herve, H. Wenninger, H. Royer, B. Hahn, E. Hugentobler, E. Ramseyer, and M. Boratav, Holographic photography of bubble chamber tracks: A feasibility test, Nuc. Instr. Meth. 179: 487–93, 1981.

[6] A. Herve, K. Johansson, P. Lecoq, P. Olivier, J. Pothier, L. Veillet, G. Waurick, and S. Tavernier, Performance of the holographic bubble chamber HOBC, Nuc. Instr. Meth. 202: 417–26, 1982.

[7] C. Powell, P. Fowler, and D. Perkins, *The Study of Elementary Particles by the Photographic Method,* New York: Pergamon, 1959.

[8] N. Ushida, T. Kondo, G. Fujioka, H. Fukushima, S. Tatsumi, Y. Takahashi, C. Yokoyama, Y. Homma, Y. Tsuzuki, S. Bahk, C. Kim, J. Park, J. Song, D. Bailey, S. Conetti, J. Fischer, J. Trischuk, H. Fuchi, K. Hoshino, M. Miyanishi, K. Niu, K. Niwa, H. Shibuya, Y. Yanagisawa, S. Errede, M. Gutzwiller, S. Kuramata, N.W. Reay, K. Reibel, T.A. Romanowski, R. Sidwell, N.R. Stanton, K. Moriyama, H. Shibata, T. Hara, O. Kusumoto, Y. Noguchi, M. Teranaka, H. Okabe, J. Yokota, J. Harnois, C. Hebert, J. Hebert, S. Lokanathan, B. McLeod, S. Tasaka, P. Davis, J. Martin, D. Pitman, J.D. Prentice, P. Sinervo, T.S. Yoon, H. Kimura, and Y. Maeda, Measurement of the D^0 lifetime, Phys. Rev. Lett. 45: 1049–52, 1980.

[9] P. Rice-Evans, *Spark, Streamer, Proportional, and Drift Chambers,* London: Richelieu, 1974, Chaps. 3–6.

[10] G. Charpak, Evolution of the automatic spark chambers, Ann. Rev. Nuc. Sci. 20: 195–254, 1970.

[11] K. Foley, W. Love, S. Ozaki, E. Platner, A. Saulys, E. Willen, and S. Lindenbaum, The Brookhaven double vee magnetic spectrometer, Nuc. Instr. Meth. 108: 33–60, 1973.

[12] J. Sandweiss, The high resolution streamer chamber, Phys. Today, Oct. 1978, pp. 40–45.

[13] R. Majka, T. Cardello, S. Dhawan, A. Disco, R. Kellog, T. Ludlam, P. Nemethy, J. Sandweiss, A. Slaughter, L. Tzeng, and I. Winters, Design and performance of a high resolution streamer chamber, Nuc. Instr. Meth. 192: 241–52, 1982.

[14] X. Artru, G. Yodh, and G. Mennessier, Practical theory of the multi-layered transition radiation detector, Phys, Rev. D 12: 1289–1306, 1975.

[15] J. Cobb, C. Fabjan, S. Iwata, C. Kourkoumelis, A. Lankford, G. Moneti, A. Nappi,

R. Palmer, P. Rehak, W. Struczinski, and W. Willis, Transition radiators for electron identification at the CERN ISR, Nuc. Instr. Meth. 140: 413–27, 1977.

[16] M. Deutschmann, W. Struczinski, C. Fabjan, W. Willis, I. Gavrilenko, S. Maiburov, A. Shmeleva, P. Vasiljev, V. Tchernyatin, B. Dolgoshein, V. Kantserov, P. Nevski, and A. Sumarokov, Particle identification using the angular distribution of transition radiation, Nuc. Instr. Meth. 180: 409–12, 1981.

[17] T. Ludlam, E. Platner, V. Polychronakos, M. Deutschmann, W. Struczinski, C. Fabjan, W. Willis, I. Gavrilenko, S. Maiburov, A. Shmeleva, P. Vasiljev, V. Chernyatin, B. Dolgoshein, V. Kantserov, P. Nevski, and A. Sumarokov, Particle identification by electron cluster detection of transition radiation photons, Nuc. Inst. Meth. 180: 413–18, 1981.

[18] C. Fabjan, W. Willis, I. Gavrilenko, S. Maiburov, A. Shmeleva, P. Vasiljev, V. Chernyatin, B. Dolgoshein, V. Kantserov, P. Nevski, and A. Sumarokov, Practical prototype of a cluster counting transition radiation detector, Nuc. Instr. Meth. 185: 119–24, 1981.

[19] G. Dearnalley and D. Northrop, *Semiconductor Counters for Nuclear Radiation*, London: Spon, 1963.

[20] R. Rancoita and A. Seidman, Silicon detectors in high energy physics: physics and applications, Rivista del Nuovo Cimento Vol. 5, No. 7, 1982.

[21] S.R. Amendolia, G. Batignani, E. Bertolucci, L. Bosisio, C. Bradaschia, M. Budinich, F. Fidecaro, L. Foa, E. Focardi, A. Giazotto, M.A. Giorgi, M. Givoletti, P.S. Marrocchesi, A. Menzione, D. Passuello, M. Quaglia, L. Ristori, L. Rolandi, P. Salvadori, A. Scribano, A. Stefanini, and M.L. Vincelli, Construction and performance of a silicon target for the decay path measurement of long lived mesons, Nuc. Instr. Meth. 176: 449–56, 1980.

[22] B. Hyams, U. Koetz, E. Belau, R. Klanner, G. Lutz, E. Neugebauer, A. Wylie, and J. Kemmer, A silicon counter telescope to study short lived particles in high energy hadronic interactions, Nuc. Instr. Meth. 205: 99–105, 1983.

[23] M. Conversi and G. Brosco, Flashtube hodoscope chambers, Ann. Rev. Nuc. Sci. 23: 75–122, 1973.

[24] J. Astbury, D. Binnie, A. Duane, J. Gallivan, J. Jafar, M. Letheren, J. McEwen, D. Miller, D. Owen, V. Steiner, and D. Websdale, A large detector for slow K^+ mesons, Nuc. Instr. Meth. 115: 435–43, 1974.

[25] C. M. Jenkins, J. Albright, R. Diamond, H. Fenker, J.H. Goldman, S. Hagopian, V. Hagopian, W. Morris, L. Kirsch, R. Poster, P. Schmidt, S.U. Chung, R. Fernow, H. Kirk, S. Protopopescu, D. Weygand, B. Meadows, Z. Bar-Yam, J. Dowd, W. Kern, and M. Winik, Existence of Ξ resonances above 2 GeV, Phys. Rev. Lett. 51: 951–4, 1983.

[26] W. Atwood, G. Bowden, G. Bonneaud, D. Klem, A. Ogawa, Y. Pestov, R. Pitthan, and R. Sugahara, A test of planar spark counters at the PEP storage ring, Nuc. Inst. Meth. 206: 99–106, 1983.

Exercises

1. A 10-GeV/*c* pion beam produces 10 bubbles/cm in a liquid hydrogen bubble chamber. What is the maximum energy required for bubble formation? How does this compare with the energy required to develop a grain in an emulsion?

2. What field is necessary for spark formation in a neon spark chamber with a 1-cm gap?

3. What clearing field is necessary to remove all the electrons from the 1-cm gap of a neon spark chamber within 1 μs?

4. Find the total energy lost to transition radiation by a 100-GeV/c proton traversing 500 lithium foils. Ignore interference effects. What is the mean number of emitted X-rays?

13

Triggers

A well-designed trigger is an essential ingredient for a successful particle physics experiment. The trigger must efficiently pass the events under study without permitting the data collection systems to become swamped with similar but uninteresting background events. Since the design of a trigger depends critically on the intent of the experiment and is strongly influenced by the choice of beam parameters, target, geometry, and so forth, it is impossible to give a prescription here on how to set up a trigger for any situation. Instead, we must content ourselves in this chapter with considering some general classes of trigger elements and with examining some specific examples in more detail. It should be mentioned that some experiments do not use a trigger. For example, neutrino experiments sometimes accept any event that occurs within a gate following the acceleration cycle.

13.1 General considerations

A trigger is an electronic signal indicating the occurrence of a desired temporal and spatial correlation in the detector signals. The desired correlation is determined by examining the physical process of interest in order to find some characteristic signature that distinguishes it from other processes that will occur simultaneously. Most triggers involve a time correlation of the form $B \cdot F$, where B is a suitably delayed signal indicating the presence of a beam particle and F is a signal indicating the proper signature in the final state. The time coincidence increases the probability that the particles all come from the same event.

Let us first consider the beam portion of the trigger. Figure 13.1 shows a typical fixed target beam line containing scintillation counters S_1, S_2, and A, threshold Cerenkov counters C_1 and C_2, and PWC P_1. Suppose the experiment requires a K^- beam of a certain momentum at the target. With suitable gases and pressures in C_1 and C_2, the signal $C_1 \cdot \overline{C}_2$ can be used to separate kaons from pions and protons in the beam. The small scintillation counters S_1 and S_2 require that the beam particles traverse the Cerenkov counters near the beam axis. The anticounter (or veto counter) A is a large scintillation counter with a hole approximately the target diameter in the center. The anticounter is placed directly before the target. The signal from this counter vetos any event accompanied by an off-axis (or halo) particle, which could confuse the downstream detectors. The scintillation counter BC is a small counter placed directly in the path of the beam. It can be used to veto events accompanied by a beam track. The PWC P_1 can be used to ensure that only one beam track is present within a certain time interval. A typical kaon beam signal then is

$$B = S_1 \cdot S_2 \cdot C_1 \cdot \overline{C}_2 \cdot (P_1 = 1) \cdot \overline{A} \cdot \overline{BC}$$

The B signal as defined here can be used for devices that require a pretrigger. It is sometimes useful to record events with just the B requirement. This will give events with a minimum amount of bias with respect to the final state of interest. It is also useful for checking the efficiency of downstream detectors.

In a colliding beam machine a beam signal is usually provided each beam–beam crossing. This may result from the coincidence between two scintillator hodoscopes located close to the beam pipes or from pickup electrodes near the interaction point.

The final state portion of the trigger is obviously much more varied since it depends on the particular experiment. The F trigger usually makes use of some aspect of the topology of the desired events. These may include [1]

1. multiplicity,
2. direction,

Figure 13.1 A simple experimental arrangement of detectors. S_1, S_2, S_3, A, BC, CS: scintillation counters; C_1, C_2: threshold Cerenkov counters; P_1, P_2: MWPCs.

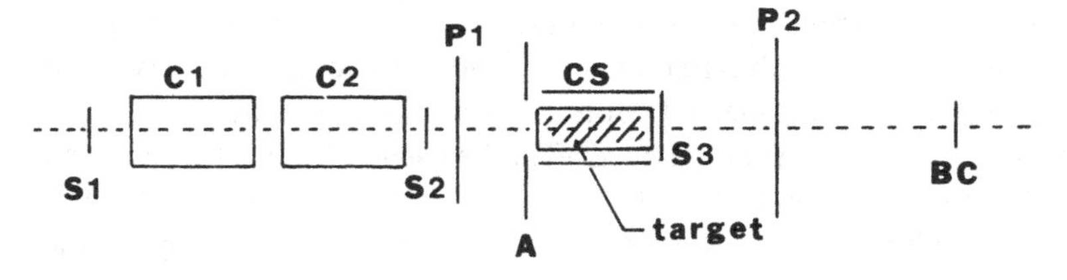

3. deflection,
4. opening angle, and
5. coplanarity.

Multiplicity is frequently used in conjunction with other requirements. Here one demands that a certain number of particles are present at a specified location. Multiplicity is usually determined using a scintillator hodoscope or a MWPC. The multiplicity can also be used to set an upper limit to the number of particles in order to reject unanalyzable "junk" events.

Triggers frequently use a counter telescope or a hodoscope coincidence matrix to require that a particle points back to the target or interaction region. Two separate direction requirements may be combined to demand that a particle undergo a certain deflection or that a particle be produced at a given angle.

Particle kinematics can play an important role in the design of a trigger for reactions involving 2-body final states. This is particularly useful in elastic scattering. The final state particles and the beam must all lie in a plane (coplanarity). In addition, the beam momentum and forward scattering angle determine the forward and recoil momenta and the recoil scattering angle. In high rate reactions one can define narrow solid angles around these directions and require a coincidence between the signals from the two arms. An alternate approach is to use large solid angle detectors, but to use hodoscopes or hard-wired electronics to require the proper correlation.

Figure 13.2 shows the fast electronics for the 150-GeV/c pp elastic scattering experiment of Fidecaro et al. [2]. The INTERACTION coincidence indicates that there was one and only one particle in the forward (F) arm and at least one backward (B) particle in the counters around the target. The trigger also requires one and only one particle in the backward vertical (BV) and backward horizontal (BH) hodoscopes. The correlation between scattering angle and recoil energy is imposed by delaying the FV signal for a time corresponding to the recoil proton time of flight and then forming the coincidence WINDOW. A rough angular correlation between the backward and forward particles is made with the vertical coincidence matrix. Figure 13.3a shows the deviation of the measured forward proton angle in pp elastic scattering from the angle calculated from the recoil particle direction. Figure 13.3b shows the corresponding coplanarity distribution. The narrow distributions show that the trigger was very successful in selecting elastic events.

For the purpose of illustration let us consider a simple example. Suppose we are examining the reaction

$$K^- p \rightarrow \overline{K}^0 n$$

We have already discussed a suitable signal to tell us when a single K^- has come down the beam line. The target could be liquid hydrogen. The final state consists of two neutral particles, one of which will typically decay into two pions after a short distance. A signature in the detector that is indicative of a vee (decaying neutral particle) is no tracks followed some distance away by two tracks, or $0 \rightarrow 2$ for short. Thus, referring to Fig. 13.1, one could look for a forwardly produced $\overline{K}^0$ with the signal $\overline{S}_3 \cdot (P_2 = 2)$. Since the recoil neutron is also neutral, one could attempt to tighten the trigger by also vetoing with the cylindrical hodoscope CS surrounding the target. Thus, a suitable final state signal might be

$$F = \overline{S}_3 \cdot (P_2 = 2) \cdot \overline{CS}$$

Several factors complicate the design of a trigger. The first is the presence

Figure 13.2 Trigger logic for a pp elastic scattering experiment. Inputs are from horizontal (H) and vertical (V) scintillator hodoscopes in the forward (F) and backward (B) spectrometer arms and counters near the target. (G. Fidecaro et al., Nuc. Phys. B 173: 513, 1980.)

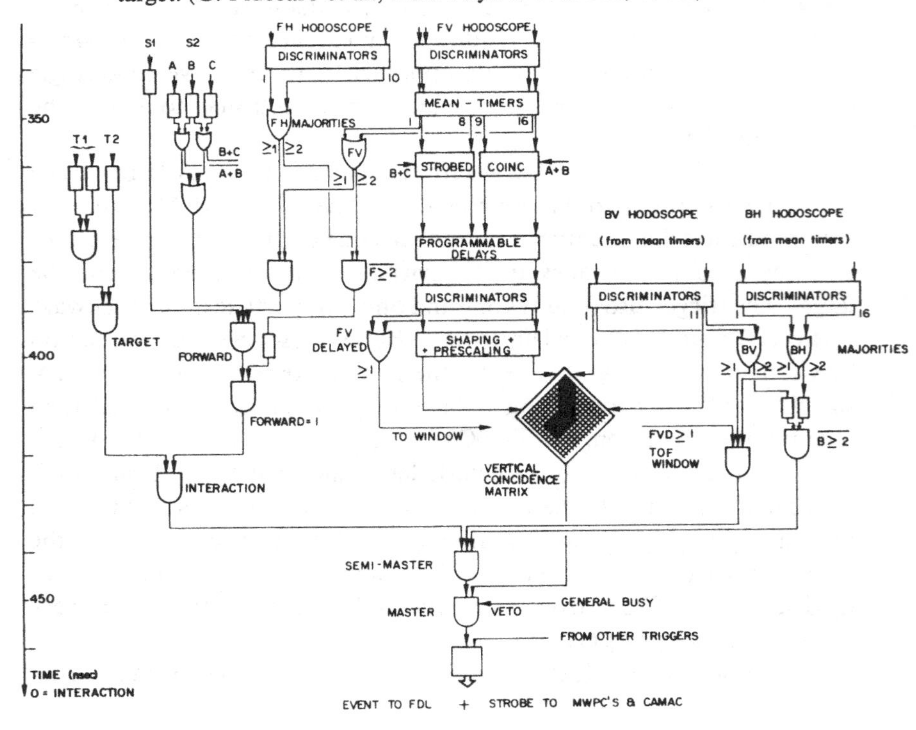

of background reactions, which have similar characteristics to the process under study. In our example $\Lambda\pi^0$ is also an all neutral final state produced in K^-p interactions. The F signal will also be generated by this final state when the Λ is produced forwardly. If the background is significant, one must take further steps, such as surrounding the target with a thin lead cylinder to convert the photons from the π^0 decay.

A second problem is the presence of accidental hits in the detectors. These extra hits could arise from uncorrelated particles traversing the detector or from noise in the detector itself. Accidentals increase the number of hits in a detector and could give false F signals. A competing problem is chamber inefficiency. In this case there would be no hit in a detector plane, even though a charged particle crossed it. This leads to a loss of true F signals.

13.2 Identified particle triggers

Often we wish to trigger on the production of a certain type of particle. In this section we will consider the particles commonly used for

Figure 13.3 (a) Polar angle correlation between the measured angle for forwardly scattered particles and the forward angle calculated from the recoil particle direction for pp elastic scattering. (b) Measurement of the coplanarity of the beam, forward particle, and recoil particle. (G. Fidecaro et al., Nuc. Phys. B 173: 513, 1980.)

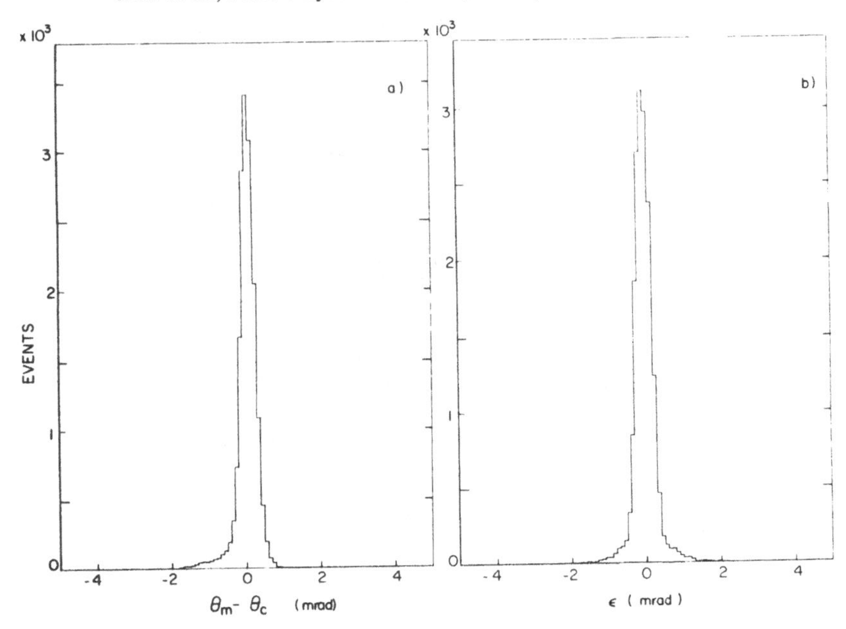

Table 13.1. *Identified particle triggers*

Particle	Method	Example references
γ, π^0	multiplicity	G. Donaldson et al., Phys. Rev D 14: 2839, 1976
	lead-glass	R. Kephart et al., Phys. Rev D 14: 2909, 1976
	NaI	L. O'Neill et al., Phys. Rev D 14: 2878, 1976
	lead-scintillator	R. Baker et al., Nuc. Phys. B 156: 93, 1979
	liquid argon	D. Scharre et al., Phys. Rev D 23: 43, 1981
e	Cerenkov	A. Maki et al., Phys. Lett. 106B: 423, 1981
	lead-scintillator	D. Blockus et al., Nuc. Phys. B 201: 205, 1982
	transition radiator	M. Adams et al., Phys. Rev. D 27: 1977, 1983
	liquid argon	C. Kourkoumelis et al., Phys. Lett. 81B: 405, 1979
μ	penetration	B. Gordon et al., Phys. Rev. D 20: 2645, 1979
	magnetized iron	D. Antreasyan et al., Phys. Rev. Lett 45: 863, 1980
$\pi^\pm$	multiplicity	D. Aston et al., Nuc. Phys. B 166: 1, 1980
	Cerenkov	A. Berglund et al., Nuc. Phys. B 166: 25, 1980
$K^\pm$	Cerenkov	T. Armstrong et al., Nuc. Phys. B 224: 193, 1983
	K^+ detector	C.M. Jenkins et al., Phys. Rev. Lett. 51: 951, 1983
K^0	$n \rightarrow n + 2$	C. Bromberg et al., Phys. Rev. D 22: 1513, 1980
	Cerenkov (π)	J. Wise et al., Phys. Lett. 98B: 123, 1981
	charged particle veto	M. Alston-Garnjost et al., Phys. Rev. D 17: 2226, 1978
p, $\bar{p}$ (fast)	Cerenkov	S.U. Chung et al., Phys. Rev. Lett. 45: 1611, 1980
p (recoil)	multiplicity	C. Daum et al., Nuc. Phys. B 182: 269, 1981
	range	K. Fujii et al., Nuc. Phys. B 187: 53, 1981
	solid state detector	R. Schamberger et al., Phys. Rev. D 17: 1268, 1978
n, $\bar{n}$	plastic scintillator	A. Robertson et al., Phys. Lett. 91B: 465, 1980
	liquid scintillator	R. Baker et al., Nuc. Phys. B 156: 93, 1979
	^{3}He-filled PWC	E. Pasierb et al., Phys. Rev. Lett. 43: 96, 1979

Table 13.1. *(cont.)*

Particle	Method	Example references
	np elastic	K. Egawa et al., Nuc. Phys. B 188: 11, 1981
	charged particle veto	D. Cutts et al., Phys. Rev. D17: 16, 1978
$\Lambda\bar{\Lambda}$	$n \rightarrow n+2$	F. Lomanno et al., Phys. Lett. 96B: 223, 1980
	multiplicity	D. Aston et al., Nuc. Phys. B195: 189, 1982
	fast p	J. Bensinger et al., Phys. Rev. Lett. 50: 313, 1983

triggering and examine the techniques that have been used for their detection. Many of these subtriggers would not make a sufficiently tight trigger themselves but would be satisfactory if used in coincidence with other requirements on the interaction under study. Additional information such as time of flight or pulse heights may be recorded with every event. This can be used in the offline analysis to enhance the fraction of events containing rarer particles, such as K mesons or antiprotons. Table 13.1 contains a summary of identified particle techniques together with example references.

13.2.1 γ, π^0

Photons are usually identified by the large electromagnetic shower they create in matter. Lead sheets can be used to convert the photons. Methods used in detecting the photon include using multiplicity, lead–glass Cerenkov counters, NaI scintillation counters, lead–scintillation shower counters, and liquid argon calorimeters. One can distinguish showers created by electrons by using a thin scintillator veto before the photon detector. Additional scintillator requirements may be needed in the trigger to reduce background.

Triggers requiring a π^0 almost invariably make use of its 2γ decay mode. If the detector is sufficiently fine grained so that the two photons can be resolved, the π^0 events can be separated on the basis of the 2γ effective mass.

The production of "direct" photons is a subject of considerable interest [3]. By direct we mean photons that originate from the fundamental interactions of the hadronic constituents, as opposed to "ordinary" photons, which result from the electromagnetic decays of π^0's or other particles. This type of process is usually studied at large transverse momentum, so that the direct photon yield is enhanced. The π^0 decays can fake a direct photon event if either of the decay γ's is not detected or if the two decay γ's cannot be resolved by the detector. Thus, an experiment that attempts to directly measure the π^0 background requires a large solid angle, highly segmented γ detector.

A second method using thin convertors can be used to separate the classes of events [3]. The shower initiated by an unresolved π^0 is due to two γ's, while the shower from a direct γ event is only due to one. Thus, the conversion efficiencies are different, and the two types of events may be separated statistically. This method requires that any nonlinearities in the detector must be well understood. Both methods require good detection efficiency for low energy γ's in order to resolve asymmetric π^0 decays.

Figure 13.4 (a) Pulse height spectrum for e^+ triggers using an electromagnetic shower counter. The solid line events have an associated Cerenkov counter signal. (b) Pulse height spectrum from the shower counters for events without a Cerenkov counter signal. (c) Measured hadron contamination versus trigger transverse momentum. (D. Drijard et al., Phys. Lett. 108B: 361, 1982.)

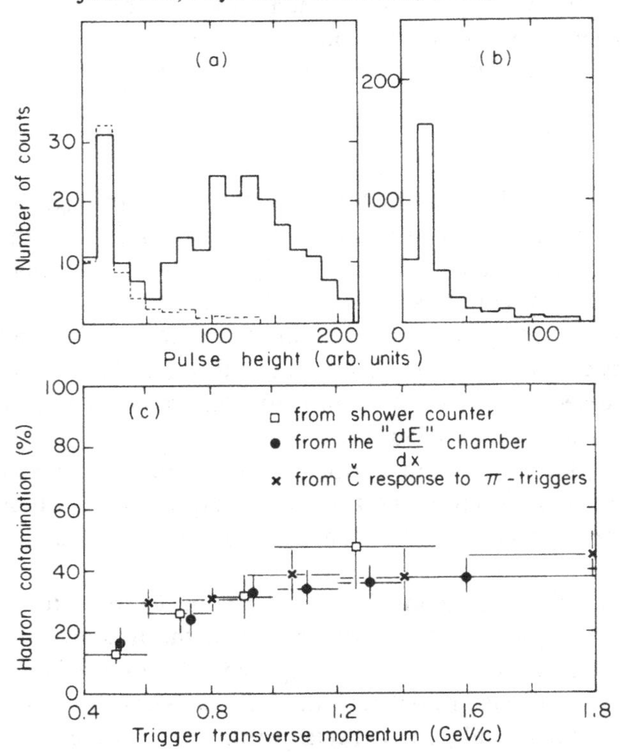

13.2.2 *e*

Electrons and positrons are usually detected by the energy deposited by their electromagnetic shower. In certain momentum intervals a trigger may also be constructed using Cerenkov counters. Specific detection methods include using threshold Cerenkov counters, lead-scintillator shower counters, transition radiators, and liquid argon calorimeters. Major backgrounds include high momentum pions for Cerenkov counters and e^+e^- pairs from photon conversions. Figure 13.4 shows the pulse height spectrum for e^+ triggers from an electromagnetic shower counter [4]. The full-line spectrum in (a) is for triggers with an associated Cerenkov counter signal. The normalized dashed-line spectrum is for triggers with no Cerenkov signal. It is evident that the majority of low pulse height events are due to hadronic background. The Cerenkov and shower counter signals may be combined to produce a more efficient trigger.

13.2.3 μ

Muon triggers usually make use of the muons' ability to penetrate large depths of matter before being absorbed. Thus, the basic experimental arrangement would consist of a massive hadron absorber followed by a MWPC or counter hodoscope. Sometimes the functions of hadron absorber and spectrometer are combined by using magnetized iron toroids between the interaction point and the muon counters. The major background in the trigger comes from the small fraction of hadron "punch through" that manages to penetrate the absorber and from muons resulting from π and K decays.

A useful procedure for estimating the number of muons arising from hadron decays involves adjusting the amount of absorber [5]. By removing absorber, the probability that the hadron will decay is proportionally increased. This background rate is inversely proportional to the absorber density. By extrapolating the measured rate to infinite density, one can estimate the number of muon triggers not arising from hadron decays, as shown in Fig. 13.5.

13.2.4 $\pi^{\pm}$

Since pions are the most copiously produced hadrons, any arbitrary track in strong interactions is likely to be a pion. This is especially true in large multiplicity events ($n \geqslant 4$), which only have a small electromagnetic background. A tighter trigger could require pulses from a series of Cerenkov counters.

13.2.5 $K^{\pm}$

The production rate of charged kaons is roughly 10% of the charged pion rate. Thus, filtering the kaons out of the pion background requires a fairly good trigger. Fast kaons can be selected using a pair of threshold Cerenkov counters. One counter is set to veto pions, while the second gives a positive kaon signal above a certain momentum. Triggers for slow K^+ recoils have been used with a K^- beam to study Ξ resonances. The trigger was derived from a K^+ detector similar to the one discussed in Section 12.6.

13.2.6 K^0

The trigger requirements for a K_S usually include the $n \to n + 2$ multiplicity requirement for a vee in some manner. The background, which includes Λ production and accidentals, can be cleaned up offline by cuts on the invariant mass of the decay tracks and the lifetime distribution. Sometimes the decay pions are required to give Cerenkov counter signals.

13.2.7 $p, \bar{p}$

Fast forward protons or antiprotons can be identified with large Cerenkov counters downstream of the interaction region. A recoil proton subtrigger is quite common as one part of an overall trigger requirement.

Figure 13.5 Muon event rate versus the inverse of the absorber density. Events have at least one muon with momentum greater than 8 GeV/c. (After J. Ritchie et al., Phys. Rev. Lett. 44: 230, 1980.)

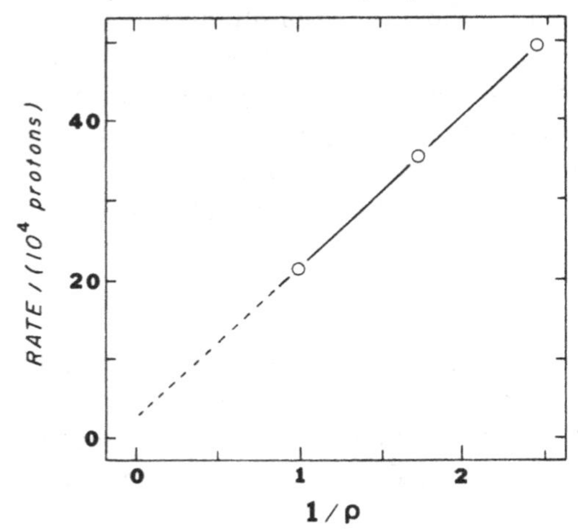

The target can be surrounded by a scintillator hodoscope, and the trigger can make use of a multiplicity of one requirement. The exact requirement should be determined with care since a simple veto on more than one side counter firing could be biased because of delta ray production and interactions in the counters. For low momentum recoil protons a range absorber may be added with a counter behind the absorber used in anticoincidence.

13.2.8 $n, \bar{n}$

Neutron triggers make use of strong interactions of neutrons in matter. A recoil neutron subtrigger is common for interactions in deuterium, heavy nuclei, or in charge exchange scattering. Triggers have been constructed using plastic scintillator, liquid scintillator, ^{3}He filled gas PWC, and the recoil proton from np elastic scattering. When used with a thin veto counter to reject charged particles, the major backgrounds are from γ and K^0's.

13.2.9 $\Lambda, \bar{\Lambda}$

Most Λ triggers make use of the $p\pi^-$ decay mode. A common trigger uses the $n \rightarrow n + 2$ vee requirement. The $K^0 \rightarrow \pi^+\pi^-$ background can be suppressed by using a downstream Cerenkov counter in the area of acceptance of the decay proton. Simple multiplicity requirements on the downstream tracks may be sufficient to pick up Λ's. Analogous $\bar{\Lambda}$ triggers may be devised using its $\bar{p}\pi^+$ decay mode.

The decay proton emerges with most of the Λ momentum. We can give a rough kinematic argument why this is so. The Q (break up energy) for the Λ decay is small, so that in the Λ rest frame the π^- and p have only small momentum. In the Lorentz transformation to the laboratory frame the Λ acquires a velocity β. Since the π^- and p had only a small momentum in the Λ rest frame, they will also acquire a velocity $\approx \beta$ in the transformation to the LAB coordinate system. Thus, in the LAB

$$p_p = m_p \beta\gamma \qquad p_\pi = m_\pi \beta\gamma$$

so that the proton track tends to be much stiffer. The detection of a fast forward proton can be used to construct an efficient Λ trigger with π^- and K^- beams.

13.2.10 $\Sigma, \Xi, \Omega,$

The Σ, Ξ, and Ω hyperons can be used for triggering, particularly at high energies. We have already mentioned in Chapter 4 how DISC Cerenkov counters may be used to identify charged hyperons. The characteristic decay modes of the hyperons have also been used to construct triggers. We will consider these triggers in more detail since they consist of combinations of subtriggers for previously mentioned particles.

Wilkinson et al. [6] constructed a trigger for Σ^+ hyperons from the decay chain

$$\Sigma^+ \rightarrow p\pi^0 \qquad \pi^0 \rightarrow \gamma\gamma$$

They required a positively charged particle in the forward direction plus a minimum energy deposited in a lead–glass counter array. Approximately 6% of the triggers gave a reconstructed Σ^+. The signal was very clean. The major sources of background were accidental coincidences of beam tracks with γ's (3%) and $K^+ \rightarrow \pi^+\pi^0$ decays (2%).

A trigger for Σ^- was constructed from the decay mode [7]

$$\Sigma^- \rightarrow n\pi^-$$

The neutron was detected using a 2-absorption length steel–scintillator calorimeter. The front of the calorimeter was covered by a charged particle veto and 10 cm of lead–glass to veto photons. The trigger required a neutron signal in coincidence with a negative particle in the MWPCs. The trigger rate was 150 Σ^- triggers in the 800-ms beam spill per 5×10^8 protons incident on the production target.

Negative Ξ hyperons have been detected using a trigger based on the decay mode [8]

$$\Xi^- \rightarrow \Lambda\pi^- \qquad \Lambda \rightarrow \pi^- p$$

The trigger required a single charged particle before the decay region and at least one particle in both the right and left parts of a MWPC which followed the decay region and a bending magnet. The Ξ^- trigger was enhanced by requiring a stiff positive particle.

Cox et al. [9] have constructed a trigger for Ξ^0 using the decay chain

$$\Xi^0 \rightarrow \Lambda\pi^0 \qquad \pi^0 \rightarrow \gamma\gamma \qquad \Lambda \rightarrow \pi^- p$$

A neutral beam was defined using collimators, a sweeping magnet, and a scintillator veto. The Λ portion of the trigger required charged tracks in the positive and negative regions of a set of chambers following a bending magnet. Approximately 50% of the raw triggers contained a Λ. The photons from the π^0 decay were detected in either a lead-glass counter array or in a lead-scintillator shower detector. The Ξ^0/Λ ratio was 2%.

The lifetime of the Ω^- was measured by Bourquin et al. [10] using the CERN SPS charged hyperon beam. The beam was passed through a DISC Cerenkov counter. They triggered on the

$$\Omega^- \rightarrow \Lambda K^- \qquad \Lambda \rightarrow \pi^- p$$

decay mode by requiring a coincidence between the DISC counter, a multiplicity counter ($n > 1$), and a small downstream (proton) counter that was inaccessible to negative particles. The trigger rate was 22 events per 10^6 beam particles. About 1% of the raw triggers contained an Ω^-. Backgrounds were mainly due to multiparticle events and Ξ^- decays. This experiment is discussed in more detail in Chapter 15.

13.3 Deposited energy triggers

Photons, electrons, π^0's, and jets are typically detected with a calorimeter. The simplest requirement for a trigger is that the total deposited energy in the detector exceed some threshold. With finely divided calorimeters one can sharpen the trigger by looking at the energy deposited in individual channels or clusters of channels or by looking at correlations between channels. Subdivision of the calorimeter elements gives the freedom to optimize the acceptance of the desired type of particle, while rejecting other types of particles or particles not originating from the region of interaction.

We will now consider several examples of this type of trigger. Amaldi et al. [11] used an array of lead-glass Cerenkov counters to trigger on single photons at the ISR. The counters were arranged in a 9×15 matrix and mounted on rails so that the distance to the beam intersection point could be adjusted. Several triggers were used. In the first the counters were divided into fifteen 3×3 submatrices, and the nine counters in each submatrix were connected in parallel. A trigger was generated when at least one of the submatrices showed a deposited energy greater than the electronic threshold of 1.25 GeV. A second trigger required that one of the 5×3 central counters show a deposited energy greater than an electronic threshold of 0.5 GeV. The major backgrounds for the direct photon search were signals from π^0 or η, where one of the decay photons was not detected or was unresolved, and from the annihilation of $\bar{n}$ in the lead-glass producing additional π^0's.

Gollin et al. [12] used a calorimeter as part of a two-muon trigger at Fermilab. The calorimeter was constructed from 15 modules, each consisting of five alternating layers of steel plates and scintillation counters. The steel plates were magnetized and served as targets for the beam muons. The counters in consecutive modules were grouped into overlapping clusters of 10. The trigger required that at least 5 of the 10 counters in any cluster exceed some threshold. This requirement selected hadrons preferentially, since electromagnetic showers have a shorter longitudinal development. The trigger efficiency curve was quite broad, rising from 10 to 90% over about 50GeV in shower energy.

Cobb et al. [13] used a finely divided liquid argon calorimeter at the ISR to trigger on the production of single particles and jets with large transverse momentum. The trigger required a total energy deposit above threshold in the calorimeter. This requirement is not sufficient to efficiently trigger on large p_t single particles, since it is also satisfied by events with a large number of low energy particles. For this reason, they also imposed a localized energy deposit requirement on the trigger. Overlapping clusters of four or five calorimeter elements were summed together in each coordinate and the sums were taken in coincidence.

Bromberg et al. [14] also used a calorimeter to provide a trigger signal proportional to the transverse momentum of single particles or jets. The experiment had four calorimeter modules on each side of the beam. Each calorimeter module had an electromagnetic front end consisting of lead and steel sheets interleaved with plastic scintillators. This was followed by a hadronic section consisting of alternating iron and scintillator layers. The scintillator signals in the electromagnetic section of each module were summed together, as were the hadronic signals. The signals from the four PMTs associated with each module were summed together and attenuated appropriately so that the resulting signal was proportional to the transverse momentum. The single particle trigger required that the signal in one of the eight modules exceed some threshold. For the jet trigger the signals from the four modules on each side of the beam were combined and the sum required to exceed a threshold. Figure 13.6 shows the transverse momentum distribution of jet-triggered events for two different threshold conditions. The plotted p_t value was calculated from ADC information from each of the PMTs. The ADC signal was proportional to the area under the pulse signal, while the trigger hardware was only sensitive to the height of the signal. For this reason, the distribution shown in Fig. 13.6 does not have a sharp threshold.

13.4 Higher level triggering

In some experiments involving rare processes, a second or third level of triggering is required after an event passes the fast triggers we have discussed up to now. The purpose of the high level triggers is to enrich the sample of recorded events or to increase the sensitivity of the experiment [15]. By rejecting noninteresting events at the trigger stage, the number of data tapes is reduced, thereby simplifying bookkeeping and handling and reducing the time necessary to analyze the data. In addition, if a decision regarding the noninteresting events can be made in a time shorter than the data readout time, the deadtime of the experiment can be reduced and the sensitivity to interesting events can be increased. The higher level triggers can enrich the data sample by (1) eliminating events with insufficient information that will eventually be discarded by the offline analysis programs; (2) demanding the presence of at least one track with a specific charge, angle, or momentum; or (3) requiring the overall event to satisfy a specific kinematic constraint.

The trigger generally uses a programmable device such as (1) a look-up table, (2) a hard-wired processor, (3) a microprogrammable processor, or (4) an emulator [15]. A look-up table is generally a random-access memory (RAM), which is addressed by signals from the detectors of the experiment. The look-up table may be used in a logical mode, whereby the incoming detector signals form a bit pattern, which corresponds to an address in RAM. The addressed location need only contain a single bit specifying whether that bit pattern was acceptable or not. When used in an arithmetic mode, the address can be formed by an operand to evaluate the value of a function. Hard-wired processors can execute a particular

Figure 13.6 Transverse momentum distribution of triggered particles in a calorimeter. Data are shown for two values of the detector threshold. (After C. Bromberg et al., Nuc. Phys. B 171: 1, 1980.)

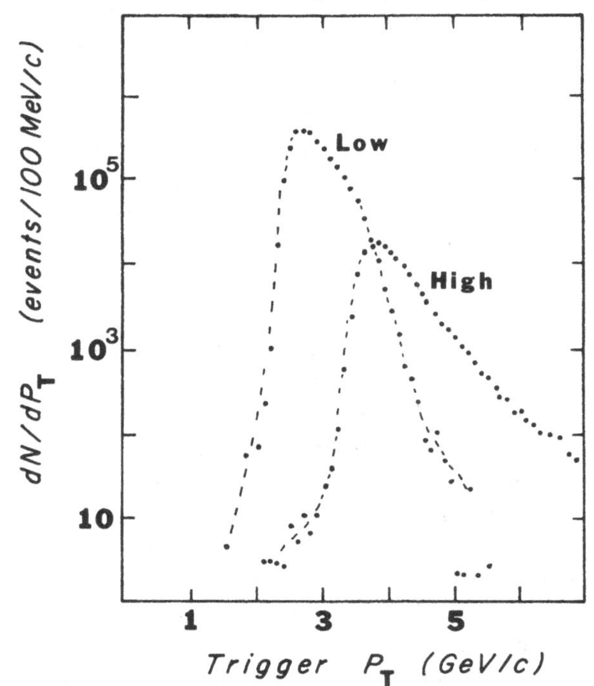

algorithm very quickly, but it is difficult to change the algorithm on short notice. Greater flexibility is available in a microprogrammable processor. However, these devices must be programmed in microcode. Emulators are microprocessors that can execute the instruction set of a high level machine. They are the easiest of these devices to program but are also slower.

An example of eliminating events that will eventually be rejected is a "loose track requirement." Eventually an offline analysis program will be used to determine a particle track from the pattern of chamber hits (accidental or real) in the tracking chambers. The program will require some minimum number of hits along a path before it can reliably recognize a track. A typical drift chamber uses a double layer of drift cells with the layers offset by half a cell width to help resolve the left–right ambiguity. A microprocessor can be used to determine the number of these double planes that have hits in adjacent cells and to veto events that do not have a sufficient number.

A second category of higher level trigger selects events where at least one track has some property. Pizer et al. [16] have developed a programmable track selector (PTS) using a look-up memory table that can make track validity decisions in less than 1 μs. The system uses two MWPCs as inputs. Each of the 1024 wires in the first chamber addresses a word in memory. Each memory location contains the first and last wire numbers in the second chamber, which form a valid combination with that wire in the first chamber. This device can be used to make rough momentum selections or to test for angular or coplanarity correlations in elastic scattering. Fidecaro et al. [2] have used the PTS as part of a pp elastic scattering trigger. One PTS was used to demand a positive recoil particle coming from the target. A second PTS required a high momentum forwardly produced positive particle.

A similar device has been used at the Omega spectrometer at CERN to enhance the trigger fraction for K^+ production [17]. Three planes of scintillator hodoscopes and two Cerenkov counters were used as inputs to the programmable trigger. Because of the presence of a magnetic field, a correlation exists between the hits in any two planes for particles of a given charge and in a restricted momentum interval. Signals from some of the planes were used as inputs to an ECL RAM module called a bit assigner. For each input the bit assigner contained a pattern of bits that should be present for another hodoscope if the signals from the event satisfy the correlation. The output of the bit assigner was then sent to a unit that performed a bit by bit AND with the actual signals from the second plane.

Two-input trigger matrices have been used for triggering the European Muon Spectrometer at CERN [18]. A simplified experimental layout and logic diagram for a single muon trigger are shown in Fig. 13.7. Horizontal and vertical scintillator hodoscope were used downstream of the bending magnet. The hodoscope information was input into two trigger matrices (M1 and M5) to determine if the track pointed back toward the target. The shaded areas on M1 and M5 in Fig. 13.7 represent coincidences corresponding to good tracks. The outputs of the target-pointing matrices were used as inputs to additional trigger matrices, which selected final state muons whose scattering angle (M2) and momentum (M3) were larger than preselected values. A NIM output was provided from the trigger matrix for those channels where the coincidence requirement was satisfied. The input–output delay for obtaining a good signal was only 25 ns. The matrix could be manually set or programmed using CAMAC.

If a third plane is added to the correlation requirement, the momentum

Figure 13.7 Trigger logic diagram for single muon events at the European Muon Spectrometer. (O. Allkofer et al., Nuc. Instr. Meth. 179: 445, 1981.)

THE EMC TRIGGER SYSTEM

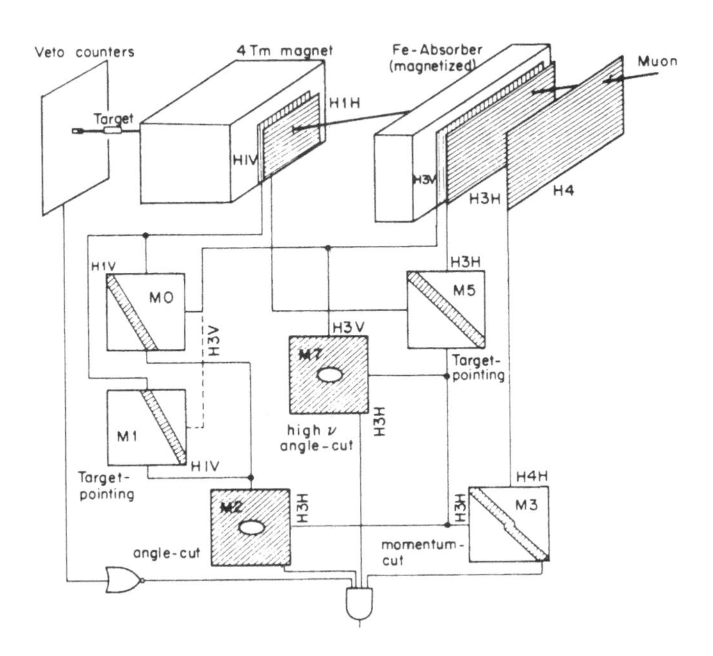

of the particle can be tightly specified in addition to its charge. Such a device (RAM) has been built at the Multi Particle Spectrometer at BNL [19]. The RAM is a 3-dimensional array of programmable AND gates. Each dimension of the RAM has a corresponding fine-grained detector array as input. The detector output is strobed into a fast register and used to address a particular location in memory. The memory is loaded with a pattern of logical 0's and 1's before the experiment on the basis of Monte Carlo studies of the predicted behavior of the trigger particles. If a logical 1 is present in the addressed memory location, the RAM will pass a higher level trigger signal.

The RAM is a high resolution device that can determine if the 3-dimensional correlation is satisfied within 250 ns. It is easily adaptable to different experiments by simply reprogramming the pattern of 0's and 1's. The pattern can be checked during an experiment by comparing the memory pattern with a disc file between beam spills. The device responds accurately in the presence of additional particles. In addition, it can be used to demand the presence of more than one particle of the required type.

Lastly we will describe a mass dependent trigger used in a muon pair production experiment at Fermilab [20]. The experiment was designed to study the properties of short-lived, high mass states that decay into muon pairs. The effective mass of a state can be written

$$M^2 \approx 2p_1p_2(1 - \cos\theta) \tag{13.1}$$

when $p_1 \gg m_\mu$ and $p_2 \gg m_\mu$. Unfortunately, the production of low mass muon pairs exceeds by many orders of magnitude the production of the more interesting higher mass pairs. In order to improve the efficiency of data collection, the experimenters constructed a mass dependent trigger.

Figure 13.8 Detector arrangement for use with a two-muon effective mass trigger. (M) bending magnet; (J_x, J_y, F_x) scintillator hodoscopes. (After G. Hogan, Nuc. Instr. Meth. 165: 7, 1979.)

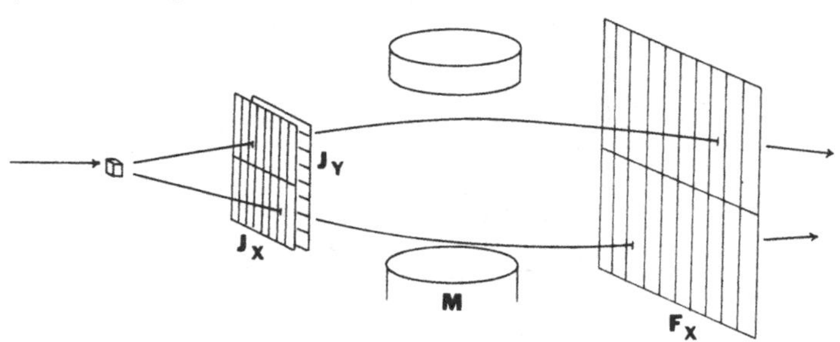

Figure 13.8 shows a schematic of the experimental arrangement. There is a double layer hodoscope (J_x and J_y) between the target and the bending magnet and an x-measuring hodoscope after the magnet. The basic idea of the trigger is to use the correlations in the hodoscope planes to determine p_1, p_2, and θ and then to use a hard-wired digital processor to determine M. Consider the logarithm of Eq. 13.1,

$$\ln \tfrac{1}{2} M^2 = \ln p_1 + \ln p_2 + \ln(1 - \cos \theta) \qquad (13.2)$$

If we assume the muons are produced in the target, the separation of the hits in J_x and J_y will determine the opening angle θ, and the positions of the two hits in J_x and F determine the momenta p_1 and p_2. The hodoscope signals were input into gate matrices constructed so that the output was $\ln p_1$, $\ln p_2$, and $\ln(1 - \cos \theta)$. These signals then entered a digital adder, which generated a signal that monotonically increased with muon pair mass. This signal was compared with a switch-selected, mass-cut value to determine whether or not to accept an event.

The trigger was able to select high mass muon pairs in less than 300 ns. The mass resolution is shown in Fig. 13.9. It can be seen that one can choose a mass cut to study the production of a high mass state such as the J/ψ and at the same time eliminate much of the background from low mass states, such as the ρ.

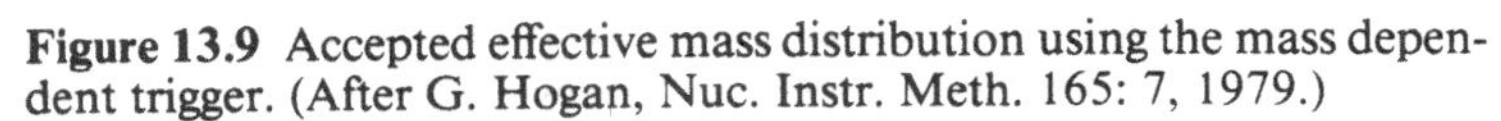
Figure 13.9 Accepted effective mass distribution using the mass dependent trigger. (After G. Hogan, Nuc. Instr. Meth. 165: 7, 1979.)

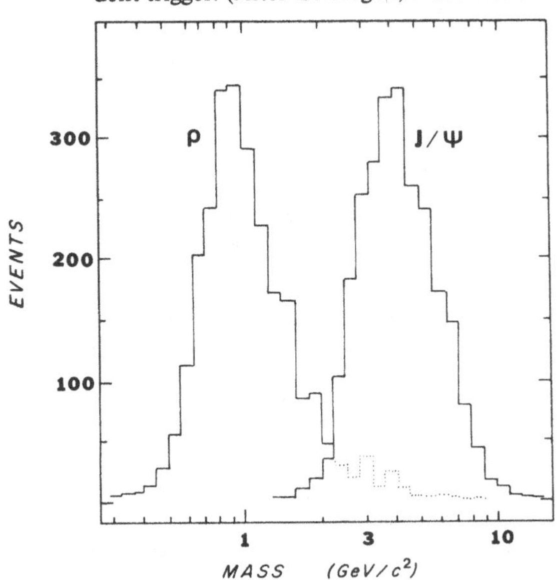

References

The references given in this chapter should aid the reader in finding a more complete description of the hardware used in the triggers, rates, backgrounds, etc., and are of course a good source of additional references. Another excellent source of references concerning experiments with specified beams, final state particles, and momentum regions is the Particle Data Group, An indexed Compilation of Experimental High Energy Physics Literature, LBL-90, September 1978.

[1] M. Turala, Application of wire chambers for triggering purposes, Nuc. Instr. Meth. 176: 51–60, 1980.

[2] G. Fidecaro, M. Fidecaro, L. Lanceri, S. Nurushev, L. Piemontese, Ch. Poyer, V. Solovianov, A. Vascotto, F. Gasparini, A. Meneguzzo, M. Posocco, C. Voci, R. Birsa, F. Bradamante, M. Giorgi, A. Penzo, P. Schiavon, A. Villari, W. Bartl, R. Fruhwirth, Ch. Gottfried, G. Leder, W. Majerotto, G. Neuhofer, M. Pernicka, M. Regler, M. Steuer and H. Stradner, Measurement of the polarization parameter in pp elastic scattering at 150 GeV/*c*, Nuc. Phys. B173: 513–45, 1980.

[3] T. Ferbel and W. Molzon, Direct photon production in high energy collisions, Rev. Mod. Phys. 56: 181–221, 1984.

[4] D. Drijard, H.G. Fischer, H. Frehse, W. Geist, P.G. Innocenti, D.W. Lamsa, W. T. Meyer, A. Norton, O. Ullaland, H.D. Wahl, G. Fontaine, C. Ghesquiere, G. Sajot, W. Hofmann, M. Panter, K. Rauschnable, J. Spengler, D. Wegener, P. Hanke, M. Heiden, E.E. Kluge, T. Nakada, A. Putzer, M. Della Negra, D. Linglin, R. Gokieli, and R. Sosnowski, Further investigation of beauty baryon production at the ISR, Phys. Lett. 108B: 361–6, 1982.

[5] J. Ritchie, A. Bodek, R. Coleman, W. Marsh, S. Olsen, B. Barish, R. Messner, M. Shaevitz, E.Siskind, H. Fisk, Y. Fukushima, G. Donaldson, F. Merritt, and S. Wojcicki, Prompt muon production at small x_F and P_T in 350 GeV p-Fe collisions, Phys. Rev. Lett. 44: 230–3, 1980.

[6] C. Wilkinson, R. Handler, B. Lundberg, L. Pondrom, M. Sheaff, P. Cox, C. Dukes, J. Dworkin, O. Overseth, A. Beretvas, L. Deck, T. Devlin, K. Luk, R. Rameika, R. Whitman, and K. Heller, Polarization of Σ^+ hyperons produced by 400 GeV protons, Phys. Rev. Lett. 46: 803–6, 1981.

[7] L. Deck, A. Beretvas, T. Devlin, K. Luk, R. Rameika, R. Whitman, R. Handler, B. Lundberg, L. Pondrom, M. Sheaff, C. Wilkinson, P. Cox, C. Dukes, J. Dworkin, O. Overseth, and K. Heller, Polarization and magnetic moment of the Σ^- hyperon, Phys. Rev. D 28: 1–20, 1983.

[8] R. Handler, R. Grobel, B. Lundberg, L. Pondrom, M. Sheaff, C. Wilkinson, A. Beretvas, L. Deck, T. Devlin, B. Luk, R. Rameika, P. Cox, C. Dukes, J. Dworkin, O. Overseth, and K. Heller, Magnetic moments of charged hyperons, in G. Bunce (Ed.), *High Energy Spin Physics — 1982,* AIP Conf. Proc. No. 95, New York: AIP, 1983, pp. 58–63.

[9] P. Cox, J. Dworkin, O. Overseth, R. Handler, R. Grobel, L. Pondrom, M. Sheaff, C. Wilkinson, L. Deck, T. Devlin, K. Luk, R. Rameika, P. Skubic, K. Heller, and G. Bunce, Precise measurement of the Ξ^0 magnetic moment, Phys. Rev. Lett. 46: 877–80, 1981.

[10] M. Bourquin, R.M. Brown, J. Chollet, A. Degre, D. Froidevaux, J. Gaillard, C. Gee, J. Gerber, W. Gibson, P. Igo-Kemenes, P. Jeffreys, M. Jung, B. Merkel, R. Morand, H. Plothow-Besch, J. Repellin, J. Riester, B. Saunders, G. Sauvage, B. Schiby, H. Siebert, V. Smith, K. Streit, R. Strub, J. Thresher, and S. Tovey, Measurement of Ω^- decay properties in the CERN SPS hyperon beam, Nuc. Phys. B241: 1–47, 1984.

[11] E. Amaldi, M. Beneventano, B. Borgia, A. Capone, F. DeNotaristefani, U. Dore, F.

Ferroni, E. Longo, L. Luminari, P. Pistilli, I. Sestilli, G.F. Dell, L.C.L. Yuan, G. Kantardjian, and J. Dooher, Single direct photon production in pp collisions at $\sqrt{s} = 53.2$ GeV in the P_t interval 2.3 to 5.7 GeV/c, Nuc. Phys. B150: 326–44, 1979.

[12] G.D. Gollin, F.C. Shoemaker, P. Surko, A.R. Clark, K.J. Johnson, L.T. Kerth, S.C. Loken, T.W. Markiewicz, P.D. Meyers, W.H. Smith, M. Strovink, W.A. Wenzel, R.P. Johnson, C. Moore, M. Mugge, and R.E. Shafer, Charm production by muons and its role in scale non-invariance, Phys. Rev. D 24: 559–89, 1981.

[13] J. Cobb, S. Iwata, D. Rahm, P. Rehak, I. Stumer, C. Fabjan, M. Harris, J. Lindsay, I. Mannelli, K. Nakamura, A. Nappi, W. Struczinski, W. Willis, C. Kourkoumelis, and A. Lankford, A large liquid argon shower detector for an ISR experiment, Nuc. Instr. Meth. 158: 93–110, 1979.

[14] C. Bromberg, G. Fox, R. Gomez, J. Pine, J. Rohlf, S. Stampke, K. Yung, S. Erhan, E. Lorenz, M. Medinnis, P. Schlein, V. Ashford, H. Haggerty, R. Juhala, E. Malamud, S. Mori, R. Abrams, R. Delzenero, H. Goldberg, S. Margulies, D. McLeod, J. Solomon, R. Stanek, A. Dzierba, and W. Kropac, Jet production in high energy hadron-proton collisions, Nuc. Phys. B 171: 1–37, 1980.

[15] C. Verkerk, Use of intelligent devices in high energy physics experiments, in C. Verkerk (ed.), Proc. of the 1980 CERN School of Computing, CERN Report 81-03, pp. 282–324.

[16] I. Pizer, J. Lindsay, and G. Delavallade, Programmable track selector for nuclear physics experiments, Nuc. Inst. Meth. 156: 335–8, 1978.

[17] T. Armstrong, W. Beusch, A. Burns, I. Bloodworth, E. Calligarich, G. Cecchet, R.Dolfini, G. Liguori, L. Mandelli, M. Mazzanti, F. Navach, A. Palano, V. Picciarelli, L. Perini, Y. Pons, M. Worsell, and R. Zitoun, Using MBNIM electronics for the trigger in a high energy physics experiment, Nuc. Inst. Meth. 175: 543–7, 1980.

[18] O.C. Allkofer, J.J. Aubert, G. Bassompierre, K.H. Becks, Y. Bertsch, C. Besson, C. Best, E. Bohm, D.R. Botterill, F.W. Brasse, C. Broll, J. Carr, B. Charles, R.W. Clifft, J.H. Cobb, G. Coignet, F. Combley, J.M. Crespo, P.F. Dalpiaz, P. Dalpiaz, W.D. Dau, J.K. Davies, Y. Declais, R.W. Dobinson, J. Drees, A. Edwards, M. Edwards, J. Favier, M.I. Ferrero, J.H. Field, W. Flauger, E. Gabathuler, R. Gamet, J. Gayler, P. Ghez, C. Gossling, J. Haas, U. Hahn, K. Hamacher, P. Hayman, M. Henckes, H. Jokisch, J. Kadyk, V. Korbel, M. Maire, L. Massonnet, A. Melissinos, W. Mohr, H.E. Montgomery, K. Moser, R.P. Mount, M. Moynot, P.R. Norton, A.M. Osborne, P. Payre, C. Peroni, H. Pessard, U. Pietrzyk, K. Rith, M.D. Rousseau, E. Schlosser, M. Schneegans, T. Sloan, M. Sproston, W. Stockhausen, H.E. Stier, J.M. Thenard, J.C. Thompson, L. Urban, M. Vivargent, G. von Holtey, H. Wahlen, E. Watson, V.A. White, D. Williams, W.S.C. Williams and S.J. Wimpenny, A large magnetic spectrometer system for high energy muons, Nuc. Inst. Meth. 179: 445–66, 1981.

[19] E. Platner, A. Etkin, K. Foley, J.H. Goldman, W. Love, T. Morris, S. Ozaki, A. Saulys, C. Wheeler, E. Willen, S. Lindenbaum, J. Bensinger, and M. Kramer, Programmable combinational logic trigger system for high energy particle physics experiments, Nuc. Instr. Meth. 140: 549–52, 1977.

[20] G. Hogan, Design of a fast mass dependent trigger, Nuc. Instr. Meth. 165: 7–13, 1979.

Exercises

1. Design a subtrigger for a fixed target accelerator experiment that produces a signal OPEN whenever there are two downstream

tracks with an opening angle greater than 1°. Include a description of the detectors, their arrangement, and electronics.

2. The dominant decay modes of the η meson are

$$\eta \rightarrow \gamma\gamma \quad (39\%) \rightarrow 3\pi^0 \quad (32\%)$$

Design a trigger to study the 2γ decay mode. What can you do to minimize the background from $3\pi^0$?

3. The charmed D^+ mesons have a $c\tau = 0.028$ cm and a significant branching ratio into a K^- plus pions. Design a trigger to study D^+ decays.

4. Show that the fraction F of potential events that can be recorded by a data acquisition system is

$$F = (1 + R\tau)^{-1}$$

where R is the trigger rate and τ is the time needed to read out an event. What trigger rate can be tolerated if the deadtime fraction should be less than 10% and if the readout time is 2 μs?

5. Define the sensitivity S of a trigger to be the rate at which good events are recorded. Show that the sensitivity of a two-level trigger is

$$S = \frac{pfR}{1 + R(t + f\tau)}$$

where p is the purity of the event sample selected by the second-level trigger, R is the first-level trigger rate, t is the second-level decision time, f is the fraction of events that pass the second-level trigger, and τ is the readout time. Examine the quantities in this equation to determine under what conditions a two-level trigger can make a significant improvement in sensitivity.

6. Suppose that a calorimeter experiment defines a good trigger event to be any one in which the sum (p_t) of the magnitudes of the deposited transverse momenta equals or exceeds a threshold value (q). Assume that the probability of getting an event with transverse momentum p is $\propto \exp(-\beta p)$ and that the calorimeter can only resolve p_t to within $\pm\delta p_t$, where $0 \leqslant \delta \leqslant 1$. Show that the ratio r of the probability of getting a trigger in which the actual p_t is less than q to the probability of a good trigger is

$$r = \exp(\delta\beta q) - 1$$

14

Detector systems

Particle physics experiments are conceived to study some aspect of particle properties or of the interactions between particles. The first step in designing an experiment is to find an efficient trigger for the desired experiment. Then one must select suitable detectors to gather the necessary data. Often the choice of detectors and their configuration involves compromises between competing goals. In this chapter we will examine some important considerations involved in designing a detector system.

14.1 Magnetic spectrometers

A spectrometer, as the word is generally used, is a system for measuring the momentum of a particle. The spectrometer consists of a magnet and tracking detectors that determine the momentum through a measurement of the particle's curvature. The motion of a particle with charge q in a magnetic field B is governed by the equation

$$d\mathbf{p}/dt = q/c\ \mathbf{v} \times \mathbf{B} \tag{14.1}$$

If the field is uniform and we neglect the presence of matter in the particle's path, the magnitude of the momentum remains constant with time and the particle will follow a helical trajectory. The actual trajectory will differ from an exact helix due to spatial inhomogeneities in the magnetic field and radiation energy losses. If matter is present, there will be additional changes in the trajectory due to ionization energy loss and multiple scattering.

The momentum is related to B and the radius of curvature ρ by

$$p = 0.2998 B\rho \qquad \text{GeV}/c,\ \text{T, m} \tag{14.2}$$

The radius of curvature, chord l, and sagitta s of the circular segment shown in Fig. 14.1 are related by

$$\rho = \frac{l^2}{8s} + \frac{s}{2}$$
$$\simeq \frac{l^2}{8s} \tag{14.3}$$

The accuracy in the determination of p is related by Eq. 14.2 to the accuracy in the measurement of ρ. This is in turn related by Eq. 14.3 to the accuracy in determining the sagitta of the arc. Thus we have

$$p = 0.3\,\frac{Bl^2}{8s} \tag{14.4}$$

$$\left|\frac{\delta p}{p}\right| = \left|\frac{\delta s}{s}\right| \tag{14.5}$$

The quantity $\delta p/p$ is referred to as the momentum resolution. We see that the momentum resolution is equal to the relative error in the sagitta determination. In addition to the deviations of the trajectory from a perfect helix mentioned earlier, the resolution of the tracking detectors (setting error) and distortions in the measuring system contribute to δs.

Gluckstern [1] has shown that the error in the curvature ($k = 1/\rho$) due to measurement error is

$$\delta k = (\epsilon/L^2)\sqrt{A_N} \tag{14.6}$$

where ϵ is the rms measurement error, L is the projected track length in the magnetic field, and A_N is a parameter that depends on the number N of points that are measured. For large N

$$A_N = \frac{720}{N+5} \tag{14.7}$$

The contribution to the momentum resolution from measurement error then is

$$\left|\frac{\delta p}{p}\right| = \sqrt{A_N}\,\frac{\epsilon}{L^2}\,\frac{p}{0.3B} \tag{14.8}$$

We see that the resolution is improved faster by increasing L than by increasing B or decreasing ϵ.

Figure 14.1 The sagitta (s), chord (l), and radius (ρ) of a circular arc.

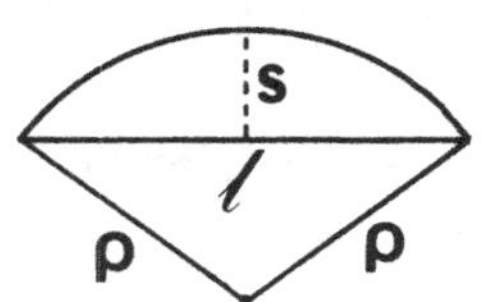

The error in the curvature due to multiple scattering is

$$\delta k_{\mathrm{ms}} = \sqrt{\xi C_N / L} \tag{14.9}$$

where ξ is the rms projected angle per unit thickness due to multiple scattering and C_N is a parameter that equals 1.43 for large N. The contribution to the momentum resolution is

$$\left|\frac{\delta p}{p}\right| = \frac{p}{0.3B}\sqrt{\frac{\xi C_N}{L}} \tag{14.10}$$

The trajectory of a particle before and after the magnet can be determined using wire chambers or scintillator telescopes. Assume we have a uniform magnetic field that ends abruptly near the magnet edges. The entrance and exit angles are related to the bending in the field by the relation

$$\sin\beta_{\mathrm{i}} + \sin\beta_{\mathrm{o}} = \frac{\int B\,dl}{3.33p} \qquad \mathrm{GeV}/c,\ \mathrm{T},\ \mathrm{m} \tag{14.11}$$

as shown in Fig. 14.2. This equation can be used to determine p once the magnetic field has been accurately measured. The momentum resolution is related to the error in determining the total bend angle by

$$\frac{\delta p}{p} \simeq \frac{\delta\alpha}{\alpha} \simeq \frac{3.33\,\Delta x\,p}{r\int B\,dl} \tag{14.12}$$

where r is the distance to the counter, and

$$\delta\alpha = \Delta x / r \tag{14.13}$$

Figure 14.2 Geometry of a particle orbit through a bending magnet.

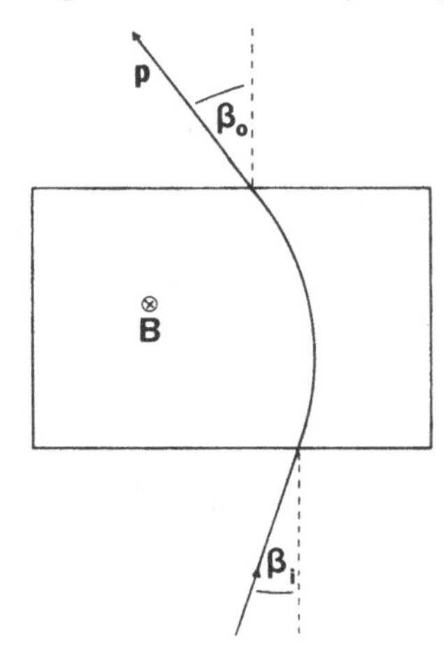

A number of methods have been used to map out the magnetic field in spectrometer magnets. The field may be quite nonuniform since the magnets tend to have large gaps to maximize the solid angle available to the experiment. Common methods for measuring the field include rotating coils, nuclear magnetic resonance, or Hall probes. Usually the three components of the field are measured for many points over a large mesh. The data is sometimes fit to a high-order polynomial for calculational convenience. Amako et al. [2] have described an automated Hall probe system used for field mapping at KEK. The temperature-compensated Hall probes were mounted in a copper block and positioned on a movable table using computer-controlled stepping motors. Different field components were measured by rotating the probe orientation. The uncertainty in the field was less than 32 G everywhere in the magnet volume for a 12.8-kG central field.

Now consider as an example the double arm spectrometer shown in Fig. 14.3. The spectrometer was used in an experiment to study pp elastic scattering at 90° in the center of mass between 5 and 13 GeV/c [3]. Each arm consisted of two bending magnets and a three-counter scintillator telescope to measure the exit angle. For 90° CM scattering both final state protons emerge at the same angle in the laboratory.

The solid angle acceptance of this type of spectrometer is given by

$$\Delta\Omega = A/r^2 \tag{14.14}$$

where A is the area of the defining counter in the telescope and r is its distance from the target. For the 7 × 5-in.^{2} counter located about 100 ft

Figure 14.3 A simple two-arm spectrometer. The beam enters from the left. (C. Akerlof et al., Phys. Rev. 159: 1138, 1967.)

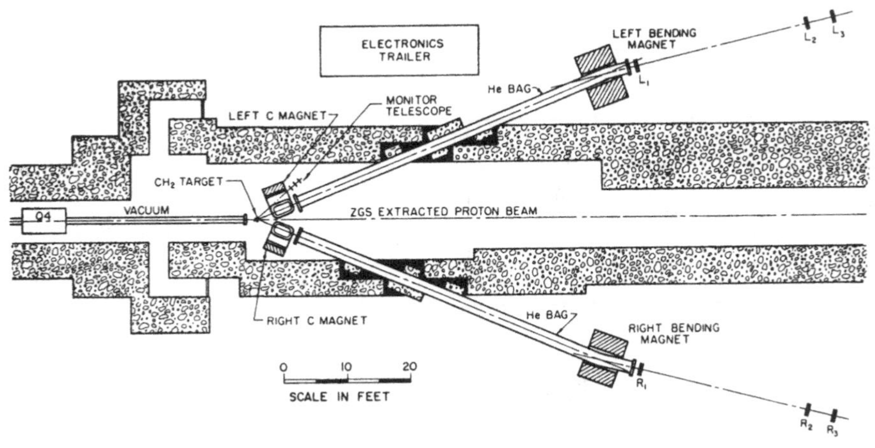

from the target this corresponds to a solid angle of 24 μsr. There is a correction to the actual acceptance due to focusing at the edges of the dipole magnets.

In principle, the R counter should be identical to the L counter. However, in this case the finite target size, beam divergence, multiple scattering, and field integral error could lead to a loss of true coincidences by allowing the companion of a proton detected in the L arm to escape from the R arm. For this reason, the R counter was slightly larger than the corresponding L counter. The C magnets were used to ensure that the protons always entered the bending magnets at the same angle and approximately the same location. This permitted the study of elastic scattering at a large number of incident momenta without moving the bending magnets or counter telescopes.

The momentum resolution is related to the spatial and angular divergences of the beam by

$$\Delta x = \alpha_1 \, \Delta\theta_0 + \alpha_2 \, \Delta p_0 / p_0 \tag{14.15}$$

The quantity $\Delta\theta_0$ can be related to $\Delta p_0/p_0$ by the kinematics of elastic scattering. The α_i are beam transport matrices. Thus, the momentum resolution will be determined by the counter width Δx.

The normalized number of left–right coincidences is shown in Fig. 14.4 as a function of the current in the right bending magnet. The number of events peak at the calculated value for elastic scattering. The plateau around the elastic point is due to the overmatching of the R counters. As the current deviates from the value for elastic scattering, the event rate decreases rapidly.

14.2 Design considerations

The design of a detector system is ultimately determined by the physics goals that are being pursued [4]. There is no universal spectrometer capable of measuring all high energy processes. Each system must emphasize certain features important to the physical processes under study to the detriment of others. "Known" physics questions can be addressed using smaller, specialized detectors, but searches for "new" physics require a general purpose detector system.

The first obvious consideration is whether the experiment will be performed at a fixed target or colliding beam accelerator. Fixed target spectrometers can benefit from the Lorentz transformation, which folds the full CM angular range into the forward direction in the LAB. These spectrometers tend to be long and of small transverse extent. Colliding beam spectrometers, on the other hand, derive no such solid angle benefit and must completely surround the interaction region to obtain full CM coverage.

A second consideration is the nature of the beam particles. In e^+e^- colliding beams with customary luminosities, the event rates are small and every event can be recorded. In proton machines the fact that the particles interact strongly and that the luminosities tend to be higher results in much higher event rates. Here an efficient trigger system is essential to reduce the data rate to manageable levels.

Electrons in linear accelerators are accelerated in very short pulses with a very high repetition rate. This beam structure makes it very difficult to use time coincidence techniques to identify events [5]. Electron interactions are also associated with intense electromagnetic showers in the forward direction, and the detector design must allow for this.

There are five major areas of concern in most detector systems: magnetic field, tracking chambers, calorimetry, particle identification, and

Figure 14.4 Left–right coincidences in the spectrometer of Fig. 14.3 as a function of the current in the right bending magnet. (C. Akerlof et al., Phys. Rev. 159: 1138, 1967.)

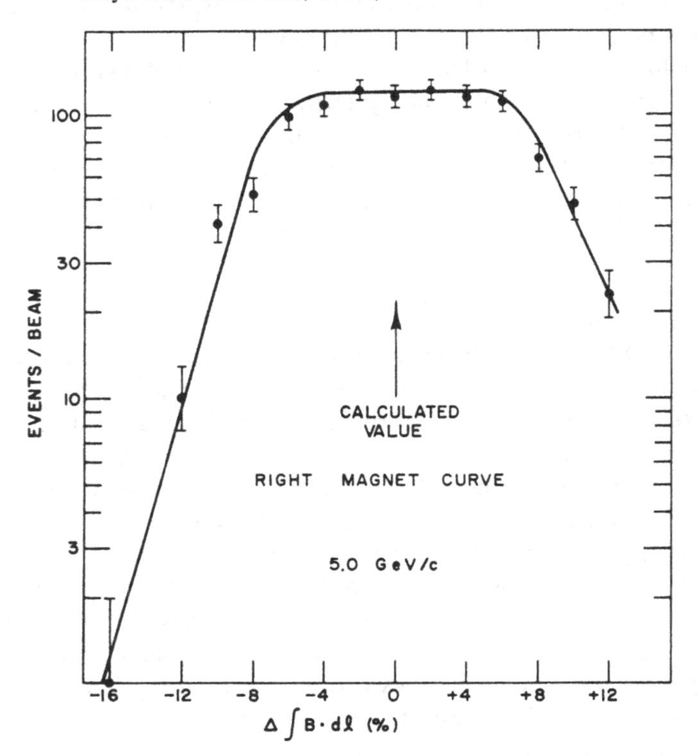

triggering. The physics goals will determine the relative importance of each. Most spectrometers employ a magnetic field and tracking chambers to measure the charge and momentum of any charged particles. However, there are detector systems, such as the Crystal Ball described in Section 14.4, that do not use a magnetic field.

The choice of magnetic field configuration puts severe constraints on the design of the spectrometer [6, 7]. Figure 14.5 depicts five magnetic field configurations for use with colliding beams, while Table 14.1 compares some of their characteristics. The magnetic field and the tracking chambers are chosen to obtain good momentum resolution for tracks in the region of most importance. The field cannot be too high or else large numbers of small momentum tracks will spiral through the chambers.

Dipole magnets typically have the field perpendicular to the incident beam. They are used with proton beams, but usually not in e^+e^- machines

Figure 14.5 Magnetic field configurations used with colliding beams.

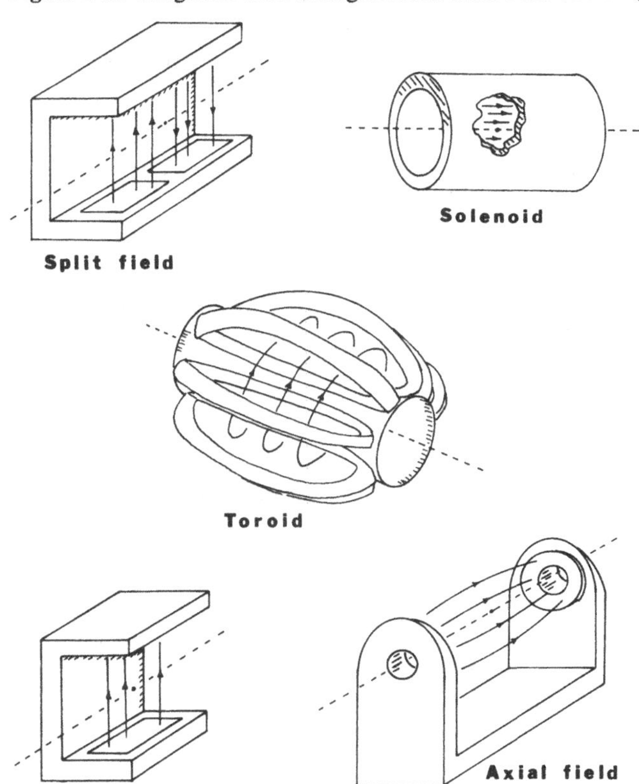

because of synchrotron radiation. Dipoles are good for analyzing forwardly produced particles, which makes them well suited for fixed target accelerators. In addition, they are quite flexible for changing external devices. For colliding beam machines the deflection of the beams must be compensated for by another magnet. This can be avoided in a split field magnet, in which the direction of the field reverses in the magnet center.

Since the field in solenoid magnets is approximately parallel to the beam direction, only a weak compensating magnet may be required. The combination of a solenoid magnet and cylindrical tracking chambers has been widely adopted at e^+e^- colliding beam machines. Solenoids provide uniform azimuthal acceptance and do not disturb the transverse momenta of particles. However, only a limited transverse space is available before entering the coils. A solenoidal field with open access around polar angles of 90° is provided by an axial field magnet (AFM). A spectrometer built around this type of magnet is described in Section 14.4.

Lastly, we mention the toroidal field configuration. Here a large p_t particle must cross the coils and magnet structural elements before entering the magnetic field. The field itself is nonuniform, varying as

$$B(r) = \frac{B_i r_i}{r} \tag{14.16}$$

where B_i is the field at the inner radius r_i. The design is most useful in a muon spectrometer. In this case the fact that the coils are in the particle path is no problem, since the detector invariably has a massive hadronic filter. Multiple scattering limits the ultimate momentum resolution. One interesting feature of the toroid is that the field lines form circles around

Table 14.1. *Comparison of magnetic field configurations*

	Dipole	Split field magnet	Solenoid	Axial field magnet	Toroid
Return yoke	yes	yes	yes	yes	no
Compensating magnet	yes	no	small	small	no
e^+e^- beams	no	no	yes	yes	yes
Coils before field region	no	no	no	no	yes
High p_t measurement	good	good	poor	good	poor
Forward particle measurement	good	good	poor	poor	poor

Source: W. Willis, Phys. Today, Oct. 1978, p. 32; T. Taylor, Physica Scripta 23: 459, 1981.

the beam and no iron return yoke is necessary. The field can be reversed to check systematic effects without affecting the beam.

Particle tracking in colliding beam detectors is usually done in a central detector that surrounds the intersection point. It is the job of the central detector to measure the momenta, directions, and multiplicity of charged tracks coming from the interaction. The central detector usually consists of drift chambers or MWPCs but may include a high resolution vertex detector or a TPC.

Some important "typical" characteristics of MWPCs, drift chambers, and other detectors used for tracking are given in Table 14.2. The desired momentum resolution, ability to distinguish tracks in high multiplicity events, and expected rates determine the solid angle coverage, wire spacing, and rate capabilities of the chambers.

Calorimeters may be desirable to measure the influence of neutral particles and the production of electrons and photons. We have discussed the characteristics of sampling calorimeters in Chapter 11. Table 14.3 contains a summary of some important properties of "continuous" calorimeters. The best resolution has been obtained with NaI crystals, whose resolution typically improves with energy according to the relation, $\sigma_E/E \sim 2\%/E^{1/4}$ with E in GeV. BaF_2 is a promising material for calorimeter applications [8]. It scintillates in the ultraviolet portion of the spectrum with a decay time of only 0.6 ns. It is nonhygroscopic and cheaper than BGO. A calorimeter may be used in the trigger to achieve large event rate reductions in proton machines.

Some experiments look for the production of specific types of particles.

Table 14.2. *Typical properties of tracking detectors*

	Spatial resolution (μm)	Response time (ns)	Recovery time (ns)
Scintillator hodoscope	5000	10	10
MWPC	500	100	[a]
Drift chamber	150	100–1000	[a]
Proportional drift tubes	5000	400	100
Bubble chamber	100	10^6	10^7
Spark chamber	300	1000	10^6
Streamer chamber	200	1000	10^6
Flashtube hodoscope	5000	1000	10^6

[a] Individual wire dead for ~ 100 ns. Other wires are still sensitive.

Table 14.3. *Continuous calorimeters*

	Statistical contribution to energy resolution $cE^{-1/2}$ c[%]	Typical integration time [ns]	Radiation length (cm)	Absorption length (cm)	Relative light output	Radiation for 10% loss (rad)
Lead glass (SF5)	4.5	40	2.5	42	1	2500
Scintillation glass (SCG1-C)	1.1	100	4.35	45	5.1	8.5×10^4
BGO	1	300	1.12	23	10^3	10^4-10^5
NaI	1[a]	250	2.5	41	10^4	10^5
BaF_2	1.7	—	2.1	—	700	—

[a] Energy resolution ~ 1% $E^{-1/4}$.

Source: B. Pope, Calorimetry: working group summary report, in Proc. of the 1983 DPF workshop on collider detectors, LBL-15973, UC-37, CONF-830224, p. 49.

In this case some type of particle identification must be incorporated into the design. We have discussed a number of techniques for accomplishing this in the preceding chapters. The detector space available for this purpose may be limited, particularly in colliding beam machines.

For most identification methods the accuracy of the measurement improves with length. Thus we can make a useful comparison of techniques by plotting the length required to achieve a given accuracy as a function of the energy [6]. Figure 14.6 shows that the required length grows very rapidly. At low energies time of flight and Cerenkov counter systems can be used. For large γ the intensity from transition radiation is sufficient for e/π/K separation and synchrotron radiation may be used for electron identification. There is no "natural" system for the intermediate region. One typically uses the relativistic rise of the ionization energy loss. However, because of the non-Gaussian statistical nature of the energy loss, a very large number of samples must be taken. In general, we note that π/K separation is more difficult than π/e.

Figure 14.6 The length necessary for particle identification using various techniques is shown as a function of the Lorentz factor γ. (TOF) time of flight, (C) Cerenkov counters, (ION) relativistic rise of ionization energy loss, and (TR) transition radiation. (After W. Willis, Phys. Today, Oct. 1978, p. 32.)

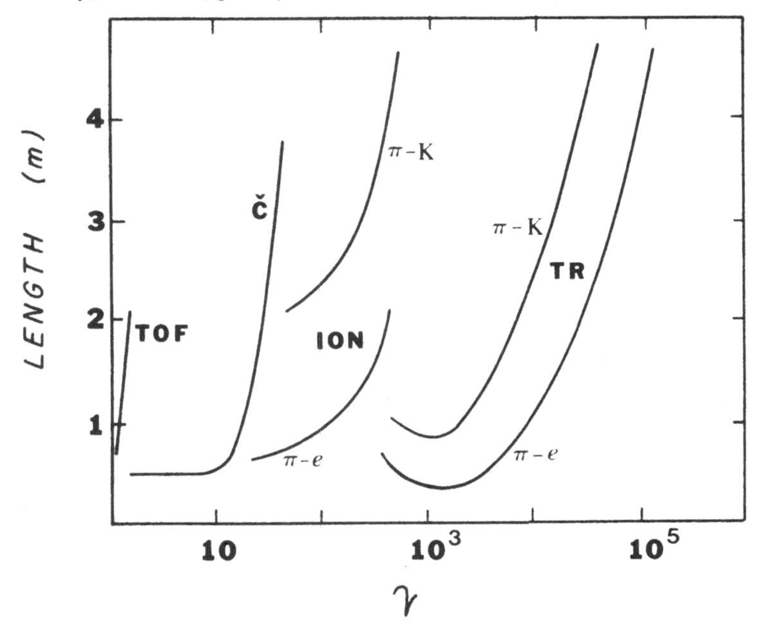

Lastly, some means of triggering must be provided and integrated into the detector system. Scintillation counters, Cerenkov counters, MWPCs, and calorimeters are typical trigger elements. The trigger timing and rate must be carefully incorporated into the overall data acquisition system.

Special care must be taken in designing a detector for use in a high luminosity environment. The major problems are the pileup of unrelated events within the detector's resolving time, rate limitations on detector operation, and the effects of radiation damage. Specialized experiments can handle the problem by using a limited solid angle acceptance. A general detector must use finer spatial segmentation of detector elements. The analog signals must be very fast, and the readout time must be as short as possible. A calorimeter can be used to provide a selective trigger.

Summaries of the important features and diagrams of many spectrometers can be found in the literature [9]. Darriulat [10] has given a summary of high p_t spectrometers, while Fabjan and Ludlam [11] have tabulated properties of many spectrometers that use calorimeters. The characteristics of the PETRA detectors have been reviewed by Wu [12].

14.3 Fixed target spectrometers

In this section we will consider the design of several spectrometers used at fixed target accelerators. Table 14.4 lists some examples of fixed target spectrometers and summarizes some of their important characteristics. The word spectrometer is used here in the more general sense of an integrated system of detectors. The spectrometers listed in Table 14.4 divide naturally into two classes. The neutrino spectrometers emphasize calorimetry and muon detection, while the charged beam spectrometers are likely to give more weight to tracking and particle identification. It should be pointed out that many of these detectors are really facilities, so that experiment dependent equipment may be used to augment the basic spectrometer.

As an example consider the European Muon Collaboration spectrometer at CERN [13]. The spectrometer operates in a secondary muon beamline at the SPS. Its primary objective is the study of "deep" inelastic scattering. In this process the beam muon interacts in the target and produces a final state muon together with other unspecified particles. The spectrometer is designed to have a large acceptance for detecting the final state muon and to make a precise measurement of its production angle and momentum.

The layout of the spectrometer is given in Fig. 14.7. The incoming beam is defined with the beam hodoscope (BH). The spray of muons traveling off the beam axis (beam halo) is eliminated by using the hodoscopes (V) as veto counters in the trigger. Liquid hydrogen, liquid deuterium, and active iron-scintillator calorimeter targets have been used. A polarized target was also used to provide spin information. The target was placed close to the magnet to provide good acceptance.

The momentum determination was made using a large aperture window frame type dipole magnet (FSM) with a bending power of 5 T-m. The entrance and exit angles are measured with drift chambers (W) before and after the magnet. The chambers have a spatial resolution of ~0.3 mm. Tracking inside the magnetic field is done using MWPCs (P1 – P3). These chambers have a short resolving time (75 ns) to help eliminate out of time tracks, such as knock-on electrons from the target. The momentum resolution was ~ 1% for a 100-GeV/c secondary track from liquid hydrogen.

A secondary muon emerging from an interaction will pass through H3 and H4. The muon trajectory after the absorber is measured using the drift chambers W6 and W7. The single muon trigger discussed in Chapter 13 requires a track after the absorber and the absence of a beam veto. In addition, correlations are required between hodoscope planes such that (1) the track points at the target in x and y, (2) the momentum exceeds some minimum requirement, and (3) the scattering angle exceeds a preselected value.

Pions and kaons can be identified over certain momentum ranges using the gas threshold Cerenkov counter (C2) downstream of the target. The calorimeter (H2) consists of a front section of 20 radiation lengths of lead and scintillator to detect photons and electrons and 5 interaction lengths of iron and scintillator to detect hadrons. This is followed by 10 interaction lengths of magnetized iron to absorb the remaining hadrons.

A partial view of another large fixed target detector system is shown in Fig. 14.8. The CHARM detector was used at the SPS to investigate the neutral current interactions of high energy neutrinos [14]. Recall that neutral-current neutrino interactions have an outgoing neutrino plus hadrons in the final state. This type of interaction must be distinguished from charged-current neutrino interactions, which have a muon in the final state. Thus, two essential parts of the overall detector are a calorimeter for hadron detection and a muon detector.

The detector consisted of 78 calorimeter subunits surrounded by a frame of magnetized steel and followed by four toroidal iron magnets. Each calorimeter subunit consists of a 3 m × 3 m × 8 cm marble plate, a layer of 20 scintillators 15 cm wide and 3 m long, and a layer of 128 proportional drift tubes 3 cm wide and 4 m long. The plate material was

Table 14.4. *Examples of fixed target spectrometers*

Spectrometer	Beam	Vertex field	Tracking	Calorimetry	Particle identification	Muon detector	Recoil detector
Neutrino (AGS)	ν	none	PDT	liquid Sc	dE/dx	dipole, PDT	—
CCFR (FNAL)	ν	none	DC	Fe-liquid Sc	—	toroid DC, Sc	—
CDHS (SPS)	ν	none	DC	Fe-Sc	—	1.7 T toroid	—
CHARM (SPS)	ν	none	PDT	marble-PDT streamer tubes, Sc	—	toroid PDT	—
Neutrino (FNAL)	ν	none	PDT flash chambers	PDT flash chambers	—	toroid DC	—
Tagged photon (FNAL)	γ	none	DC, 0.7 T-m, 1.4 T-m dipoles	Pb-Sc, liquid Sc, Fe-Sc	Cer, dE/dx	SC	cyl PC, Sc, liquid Sc

EMC (SPS)	μ	4 T-m	DC, PC 5.2 T-m dipole	Pb-Sc, Fe-Sc	Cer, TOF	—	Streamer chamber, PC, PDT
Multimuon (FNAL)	μ	2 T	DC MWPC	Fe-Sc	—	—	—
E605 (FNAL)	p	none	MWPC, DC dipoles	Pb-Sc, Fe-Sc	RICH	Zn, concrete PDT	—
MPS (FNAL)	p	none	MWPC, DC 1.7 T dipole	Pb-Sc, Fe-Sc	—	1.8 T toroid, MWPC	Pb-glass
LASS (SLAC)	Hadron	2.2 T solenoid	MWPC, Sc 3 T-m dipole	none	Cer, TOF	—	cyl MWPC
MPS II (AGS)	Hadron	4.6 T-m dipole	DC, MWPC Sc	none	Cer, TOF	—	Sc
OMEGA (SPS)	Hadron	1.8 T	MWPC DC, Sc	none	Cer	—	MWPC

Source: Particle Data Group, Major detectors in elementary particle physics, LBL-91, UC-37, 1983.

Figure 14.7 The European Muon Spectrometer at the SPS. (BH) Beam hodoscope, (V) veto hodoscopes, (P) multiwire proportional chambers, (W) drift chambers, (C) Cerenkov counters, (H1,3,4) scintillator hodoscopes, and H2 calorimeter. (O. Allkofer et al., Nuc. Instr. Meth. 179: 445, 1981.)

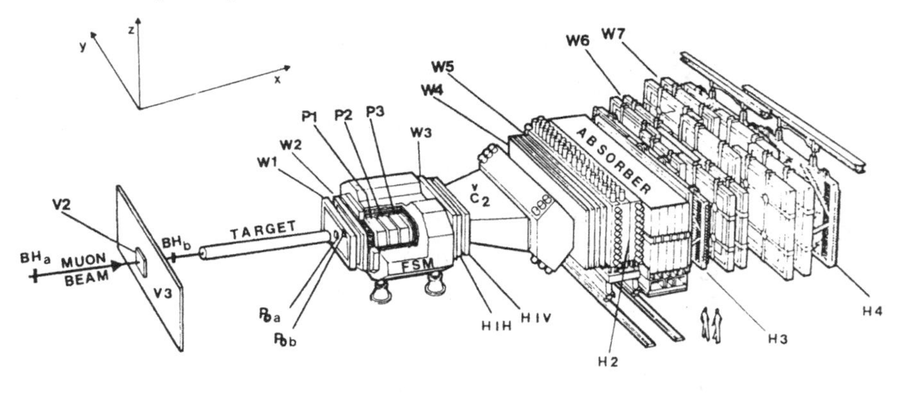

Figure 14.8 The CHARM neutrino detector at the SPS. (A. Diddens et al., Nuc. Instr. Meth. 178: 27, 1980.)

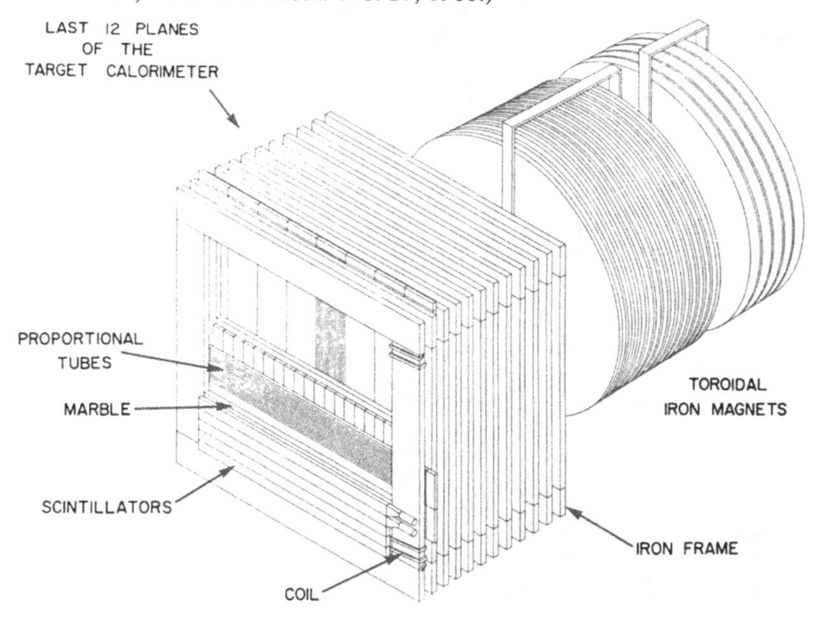

selected in order to equalize the depth of electromagnetic and hadronic showers. The scintillation counters were required to give a precise measurement of single particle ionization, to sample the shower energy deposition, and to serve as an active target for the neutrino beam. The photomultiplier tube output was split between a discriminator system and ADCs for pulse height measurement. The proportional drift tubes had a cross section of 29×29 mm^2. They could locate muon tracks with an accuracy of 1 mm (rms) and also sampled the shower energy deposition. The readout electronics recorded the charge collected at the sense wire and the drift time for particles traversing the tube.

The iron frames contained a toroidal magnetic field of 15 kG. Forward going muons were momentum analyzed in four toroidal iron discs interspersed with proportional drift tubes. The momentum resolution for 80-GeV/c muons was 25% in the frame magnets and 16% for the forward spectrometer.

All triggers were formed from combinations of scintillation counter signals. The inclusive neutrino trigger required no hits in the first scintillator layer and at least one hit in at least four planes in the rest of the calorimeter. This gave a trigger rate of 0.2 events/10^{13} primary protons on target when using the SPS narrow band neutrino beam.

14.4 Colliding beam spectrometers

In this section we will consider several spectrometers that have been developed for use at colliding beam accelerators. In this case the detectors surround the interaction points of the two beams. Colliding beam spectrometers divide naturally into those used at e^+e^- colliders and at hadron colliders. The major features of some representative e^+e^- detectors are summarized in Table 14.5, while some examples of hadron beam spectrometers are listed in Table 14.6.

We first consider the Mark-J spectrometer at PETRA [15]. The detector was designed to give good measurements of the energy and production angles of electrons, photons, and muons with almost 4π acceptance. A cross-sectional view looking down the beam pipe is shown in Fig. 14.9. The electron and positron beams interact in the area (b). The beam pipe is surrounded by a layer of drift tubes (D), which have a spatial resolution of 300 μm and can determine the longitudinal position of the vertex to ~ 2 mm. Next come three layers of electromagnetic shower counters (S). These contain 18 radiation lengths of interleaved lead and scintillator sheets. The scintillators have PMTs connected to both ends. The relative pulse heights and signal timing from the two PMTs allow an independent

Table 14.5. *Examples of e^+e^- colliding beam spectrometers*

Spectrometer	Vertex field	Tracking	Calorimetry	Particle identification	Muon detector	Forward detector
ARGUS (DORIS)	0.8 T solenoid	cyl DC	Pb-Sc	dE/dx, TOF	PDT	Pb-Sc
CELLO (PETRA)	1.3 T solenoid	cyl MWPC, cyl DC	Pb-liquid Ar	—	MWPC	Pb-glass Sc, MWPC
CLEO (CESR)	1.0 T solenoid	MWPC, DC	Pb-PDT	TOF, dE/dx	DC	Pb-PDT, Sc
Crystal Ball (DORIS)	none	PDT	NaI	—	—	NaI Sc
CUSB (CESR)	none	MWPC	NaI, Pb-glass, MWPC	—	1.5 T toroid Sc, DC	Pb-Sc, NaI
DELCO (PEP)	0.3 T open solenoid	cyl DC, DC	Pb-Sc	Cer, TOF	—	Pb-Sc
HRS (PEP)	1.6 T solenoid	cyl DC, PDT	Pb-Sc, MWPC	Cer, TOF	—	Pb-glass, Pb-Sc, MWPC

JADE (PETRA)	0.5 T solenoid	cyl DC	Pb-glass	dE/dx, TOF	DC Sc	Pb-glass, Pb-Sc
MAC (PEP)	0.6 T solenoid	cyl DC	Pb-PWC, Fe-PWC	dE/dx, TOF	toroid DC	Fe-PWC, Sc
MARK II (PEP)	0.5 T solenoid	cyl DC	Pb-liquid Ar	TOF	PDT	Pb-PDT, DC, Pb-Sc
MARK III (SPEAR)	0.4 T solenoid	cyl DC	Pb-PWC	dE/dx, TOF	PDT	Pb-PWC, Sc
MARK-J (PETRA)	none	PDT, DC	Pb-Sc, Fe-Sc	—	1.7 T toroid DC	DC, Sc
MD-1 (VEPP-4)	1.2 T dipole	MWPC	Fe-MWPC	Cer, TOF	MWPC	NaI, Sc
TASSO (PETRA)	0.5 T solenoid	cyl DC, cyl PWC, DC	Pb-liquid Ar, Pb-Sc	TOF, Cer, dE/dx	PDT	Pb-liquid Ar Sc, PDT, Pb-Sc
TPC (PEP)	1.5 T solenoid	cyl DC, TPC	Pb-PWC	dE/dx	PDT	PDT, Pb-PWC

Source: Particle Data Group, Major detectors in elementary particle physics, LBL-91, UC-37, 1983.

Table 14.6. *Examples of colliding hadron beam spectrometers*

Spectrometer	Vertex field	Tracking	Calorimetry	Particle identification	Muon detector	Forward detector
AFS (ISR)	0.5 T open axial	cyl DC	NaI, U, Cu-Sc	dE/dx, Cer	MWPC	Sc
CDF (Tevatron)	1.5 T solenoid	cyl DC	Pb-Sc, Fe-Sc	—	PDT toroid-DC	Pb-PDT, Fe-PDT
SFM (ISR)	1.0 T split field dipole	MWPC	none	dE/dx, Cer, TOF	—	MWPC
UA1 (SPS)	0.7 T dipole	cyl DC	Pb-Sc, Fe-Sc	dE/dx	PDT	Pb-Sc, DC, Fe-Sc
UA2 (SPS)	none	cyl MWPC, cyl DC, 0.4 T-m toroid + DC	Pb-Sc, Fe-Sc	—	—	Sc

Source: Particle Data Group, Major detectors in elementary particle physics, LBL-91, UC-37, 1983.

measurement of the longitudinal position of a track. The shower counters signal the presence of an electron or photon.

The electromagnetic calorimeter is followed by a layer of drift chambers (W). These large chambers have 10-cm drift cells arranged in double layers. The layers are grouped in pairs and displaced by half a wire spacing to remove the left–right ambiguity. The chamber gas is argon/isobutane. The chambers have a spatial resolution of 0.6 mm and are used to measure the trajectory of muons through magnetized iron.

Next there is a hadronic calorimeter (C) consisting of iron and scintillator plates, a trigger counter (T), and a magnetized iron toroid (M). The total bending power is 1.7 T-m. The minimum momentum needed to penetrate the iron is 1.3 GeV/c at normal incidence.

Muons produced at small angles along the beam pipe are measured in end cap chambers. The luminosity is measured by detecting small angle e^+e^- elastic scattering (Bhabha) events in lead-glass counter arrays surrounding the beam pipe. The single muon, fast trigger requires signals from the scintillators in the shower counters, a hit in the counter T, and a beam crossing signal. Events that pass the fast trigger must then satisfy a loose track requirement. A microprocessor examines the drift chamber hit information and requires that at least three inner detector double planes have adjacent hits.

Figure 14.9 The MARK-J detector at PETRA. (After D. Barber et al., Phys. Rep. 63: 337, 1980.)

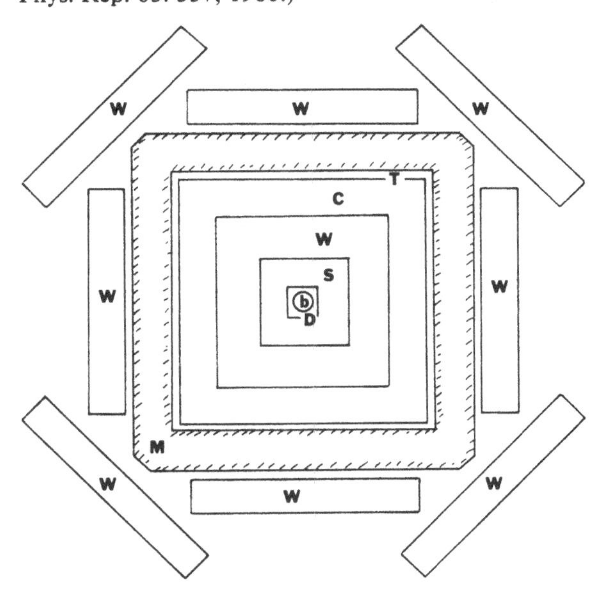

A second example of a colliding beam spectrometer is the Axial Field Spectrometer (AFS) at the ISR at CERN [16]. The primary objective of this facility was to study phenomena at large transverse momentum. In particular, the experimenters sought to measure the inclusive production of hadrons, electrons, photons, and hadronic jets. Since the ISR circulated beams of hadrons, the interaction rate in the AFS was much higher than that experienced in e^+e^- detectors. Particle fluxes could exceed 10^7 particles per second.

The highest priority in the design of the detector was to provide high resolution electromagnetic and hadronic calorimetry over a wide solid angle. The layout of the detector is shown in Fig. 14.10. As we saw in Section 14.2, the axial field magnet design allows unobstructed access near 90°. The magnetic field integral was 0.5 T-m. Tracking was provided by cylindrical drift chambers near the interaction vertex and MWPCs outside the magnetic field. The spectrometer provided a momentum resolution $\Delta p/p \sim 0.01p$, where p is measured in GeV/c. The azimuthal coordinate was provided by the drift times with a resolution of 200 μm. The coordinate along the wire was given by charge division with a resolution of 1.5 cm.

One arm of the spectrometer was provided with three Cerenkov counters for unambiguous particle identification in the range 2–12 GeV/c. The one nearest the interaction region was a silica-aerogel counter

Figure 14.10 Schematic diagram of the Axial Field Spectrometer at the ISR. Not to scale. (b) Beam pipe, (W) cylindrical drift chambers, (M) axial field magnet, (LA) liquid argon calorimeter, (U) uranium calorimeter, (C1) aerogel Cerenkov counter, (P) MWPC, (C2) high pressure gas Cerenkov counter, (C3) atmospheric gas Cerenkov counter, and (Cu) copper calorimeter. (After H. Gordon et al., Nuc. Instr. Meth. 196: 303, 1982.)

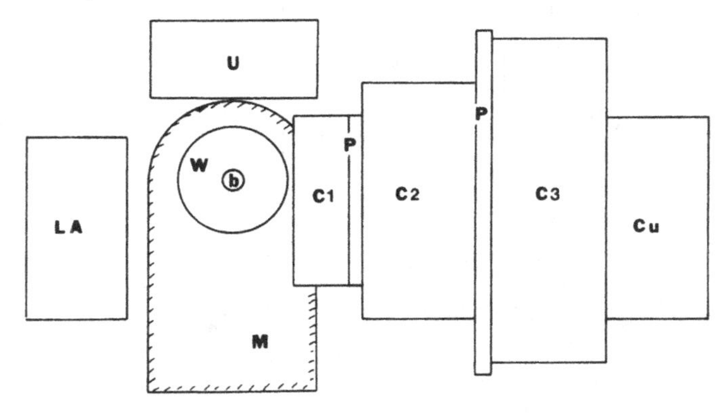

with a refractive index of 1.05. The light from this counter must be wavelength-shifted and transmitted for 2 m to keep the tubes in a low field region. The other two counters used Freon-12 at pressures of 4 and 1 atm.

The calorimeter modules covered 8 sr solid angle around the interaction region. They consisted of an electromagnetic front end followed by a hadronic section. The electromagnetic part consisted of 5 radiation lengths of uranium and scintillator sheets. The scintillator light was wavelength-shifted and transmitted to a PMT at the end of the module. The hadronic part of the calorimeter had 3.6 absorption lengths of uranium, copper, and scintillator sheets. The uranium improved the energy resolution through fission compensation, as discussed in Chapter 11. The energy resolution was $15\%/\sqrt{E}$ for electrons and $34\%/\sqrt{E}$ for pions.

Some specialized detector systems do not use a magnetic field. Figure 14.11 shows a schematic drawing of the Crystal Ball detector, as it was used in experiments at SLAC. The detector has since been moved to an interaction area at PETRA. The detector surrounded one of the e^+e^- interaction regions at the SPEAR colliding beam storage rings [17, 18]. The Crystal Ball was designed to study the spectroscopy of high mass, short-lived particle states by detecting the photons emitted in the transitions between levels. The required large acceptance and high resolution for photon detection was achieved by surrounding the intersection point with a ball of NaI scintillation counters.

The regions closest to the interaction point contained a cylindrical magnetostrictive spark chamber, a cylindrical MWPC for triggering and particle identification, and another cylindrical spark chamber. The crys-

Figure 14.11 The Crystal Ball detector as it was used at SPEAR. (M. Oreglia et al., Phys. Rev. D 25: 2259, 1982.)

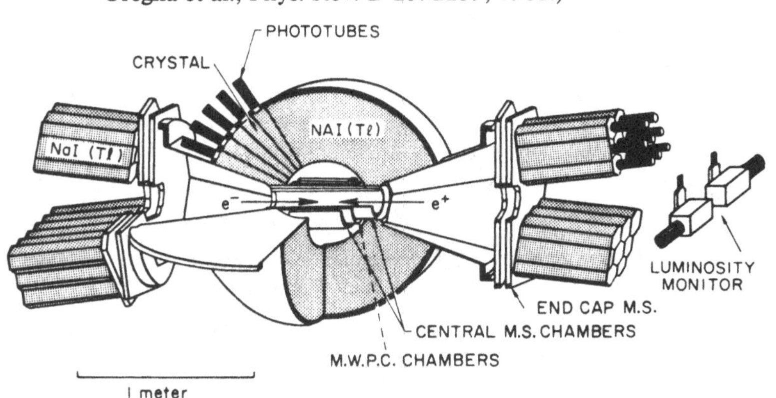

tal ball itself consisted of 672 NaI (Tl) crystals. Each face of a crystal was fitted with white reflecting paper and a sheath of aluminized material for optical isolation. Each crystal was about 16 in. long, corresponding to about 15.7 radiation lengths. The crystals were in the shape of prisms with a small end dimension of 2 in. and a large end dimension of about 5 in. Each counter was connected to a 2-in. photomultiplier tube. The crystals were arranged in the form of a 20-sided regular polyhedron, broken into two separate hemispheres. Some crystals around the beam pipe had to be removed, resulting in 94% of 4π solid angle coverage. End caps, consisting of additional NaI crystals and planar spark chambers, were located approximately 1 m from the interaction point. The end caps increased the NaI solid angle coverage to 98% of 4π. Figure 14.12 shows the pattern of energy deposits in the NaI counters for a $e^+e^-\gamma\gamma$ final state.

The standard deviation on the measurement of the energy of electromagnetic showers was

$$\sigma_E \sim (0.0255 \pm 0.0013)E^{3/4} \qquad \text{GeV}$$

Figure 14.12 Map of the energy deposit in the NaI counters of the Crystal Ball for an $e^+e^-\gamma\gamma$ final state. (M. Oreglia et al., Phys. Rev. D 25: 2259, 1982.)

$\psi' \rightarrow \pi^\circ J/\psi \rightarrow \gamma\gamma e^+ e^-$

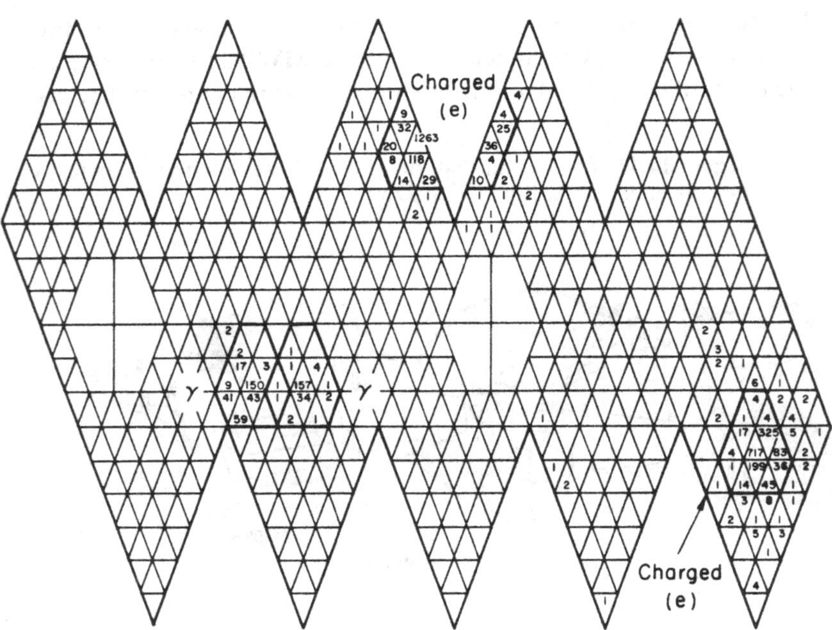

This good resolution allowed the experimenters to extract the natural widths of high mass states from the inclusive photon spectrum and to identify certain reactions through constrained kinematic fitting. The direction of a photon could be determined to within 2° of the true polar angle.

The system was triggered using (1) the total energy deposited in the ball, (2) separate energy deposited in two hemispheres, (3) charged particle multiplicity, or (4) back to back energy loss. All triggers were required to come within ±20 ns of the beam crossing signal from SPEAR. The trigger rate was 3–4 Hz at maximum luminosity.

14.5 Nucleon decay spectrometers

A third major class of large detector systems are the nucleon decay spectrometers [19, 20]. These devices differ from those in the previous two sections, since no accelerator beams are involved. Instead a large mass of material is continuously monitored by some type of sensitive detector, either on the surface of the material or interspersed throughout its volume. The detectors generally define a fiducial volume in the center of the monitored material. The signature of a nucleon decay would be an event totally inside the fiducial volume with an energy release of about 940 MeV and with a net momentum of zero.

Table 14.7 summarizes the characteristics of some of the currently operating nucleon decay detectors. The detectors can be separated into three main classes. The first uses water as the source of decaying protons and looks for the Cerenkov light emitted by the charged decay particles in reactions such as $p \rightarrow e^+\pi^0$ or μ^+K^0. The pulse height and arrival time of the PMT signals can be recorded and used to reconstruct the Cerenkov cone. This enables the direction of the particle to be determined. The detector can be triggered by requiring a minimum coincidence of PMT signals.

The second large class of detectors uses a sampling calorimeter made up of some type of gaseous chamber interspersed with plates of material. The monitored material is usually iron plate. Active detectors can be gas proportional chambers or gas detectors operated in a saturated mode. The detectors can be triggered by demanding signals from a minimum number of contiguous planes. The third class of detectors use liquid scintillator. This type of detector has also been used to study solar neutrinos.

The major background in nucleon decay experiments arises from neutrinos and muons produced in the atmosphere by cosmic rays. The muons can be efficiently attenuated by locating the detector far below the surface of the earth. For example, the muon flux at the FREJUS detector

Table 14.7. *Examples of nucleon decay detectors*

Detector	Location	Depth (MWe)[a]	Monitored material	Sensitive elements
FREJUS	France	4400	1000 tons iron plates	flashtubes, Geiger tubes
HPW	United States	1500	780 tons water	PMT
IMB	United States	1570	7000 tons water	PMT
Kamiokande	Japan	2700	3000 tons water	PMT
NUSEX	Italy–France	$\geqslant$5000	150 tons iron plates	streamer tubes
Kolar	India	7600	140 tons iron plates	gas PC
Baksam	USSR	850	80 tons liquid Sc	PMT
Soudan	United States	1800	30 tons concrete and iron	gas PC
Mt. Blanc	Italy–France	4270	30 tons iron and liquid Sc	PMT, streamer chamber

[a] MWe = Meters of water equivalent.

Source: Particle Data Group, Major detectors in elementary particle physics, LBL-91, UC-37, 1983; S. Weinberg, Sci. Amer., June 1981, p. 64; J. M. Lo Secco, F. Reines, and D. Sinclair, Sci. Amer., June 1985, p. 54.

is 6.2 ± 1.2 muons/day-m^2, a reduction of about 10^6 of the flux at sea level [19]. The outer layer of the detector is used to veto muons coming in from outside the detector. Good nucleon decay events must originate in the fiducial volume of the detector. On the other hand, the neutrino flux is largely unaffected by attenuation in the earth and enters the detector from all directions, regardless of depth. A neutrino interaction inside the fiducial volume may look like a nucleon decay, and this limits the ultimate sensitivity of the experiments.

14.6 Data acquisition

A large detector system is capable of generating immense amounts of digital and analog signals. It is the function of the data acquisition system to control the flow of this information. In general, it is desirable to at least monitor the performance of the experiment as it is happening (online) and to save some of the data for subsequent (offline) analysis. A minimal system for a simple experiment would likely use CAMAC modules as the interface between the experimental equipment and a small computer.

Most experiments make use of a minicomputer to control the online monitoring and data collection for the experiment [21]. Figure 14.13 shows a typical online data acquisition system for a more complicated experiment. The data acquisition system is intimately connected with the trigger. Two constraints may require a highly selective trigger. First, with present technology no more than 10–100 events/sec can be handled and stored on permanent memory. Second, the offline data analysis may require ~ 1 sec/event so that the cost and availability of computer time may restrict the total number of events that can be processed.

The fast trigger detectors include devices such as scintillation counters, Cerenkov counters, and MWPCs. The signals are fed into the trigger logic, where discriminators standardize the pulses, and coincidence units indicate temporal correlations of signals from different detectors. If a good event has occurred, a fast trigger pulse is generated within a few hundred nanoseconds of the event. This causes other triggered devices such as drift chambers, ADCs, and TDCs to be read out.

All of the data is fed into some device that prepares the data in a recognizable fashion and channels it onto a pathway to the computer. Some systems may have a higher level trigger that can make a more complicated decision based on hodoscope matrix correlations or coarse track finding within about 1 μs. Still more complicated decisions may be made using hardware processors or dedicated minicomputers. Sometimes these special purpose devices have been designed to emulate the instruction set of a larger, more powerful computer [22]. Then the online filtering program can first be thoroughly tested using the large computer before being used for an experiment. The emulator is slower than the large computer but is much cheaper.

Since the events occur randomly, it may be necessary to provide some temporary storage media or buffer that can store the data from one event while the detector and trigger logic are examining another one. The buffer holds the data until the online computer finishes processing an event and makes the data available for writing onto magnetic tape, at which point it is ready to accept the data from another event. If the trigger rate, computer processing rate, and buffer capabilities are not carefully balanced, many events may occur while the data acquisition system is busy processing a preceding event.

A more sophisticated data acquisition system for a large experiment is outlined in Table 14.8. At the LEP storage ring 25 μs will be available

Figure 14.13 A data acquisition system.

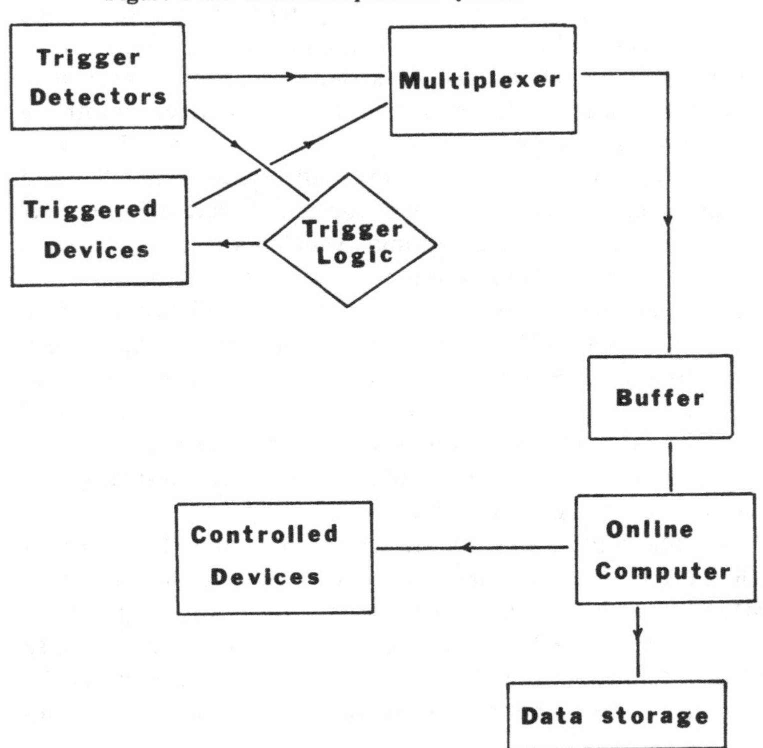

Table 14.8. *Data acquisition flow*

Level		Characteristic time	Equipment	Data treatment
1	fast trigger	0.1–25 μs	NIM electronics; fast correlation matrices	trigger detector signals, rough topology
2	detector readout	0.025–5 ms	programmable processors using integer arithmetic	drift chambers, analog signals; suppression of empty channels, clusters
3	transfer to computer	5–25 ms	floating point processors, emulator	calculate track vectors, particle reconstruction
4	data recording	0.025–1 sec	minicomputer	filtering based on full event reconstruction

Source: R. Dobinson, Physica Scripta 23: 487, 1981.

between bunch crossings. Assuming that there is only a small probability for more than one interaction per bunch crossing, trigger decisions that can be made in less than 25 μs will not introduce any deadtime. At each of the higher levels the event selection can be refined. Only the data for events that reach level 4 are recorded [23].

The online computer can perform various auxiliary functions. Analog measuring devices must be calibrated [24]. This generally requires operating the detector in a test beam or under special conditions or injecting a known test signal and measuring the detector's response.

Monitoring of the detectors is a second important auxiliary function of the online computer [24]. The monitoring can be passive, in the sense that the software merely histograms channel hit frequencies and calculates distributions at the same time that the experiment is taking data. Generally this works adequately, so long as sufficient statistics are accumulated for all channels in a reasonable length of time. Otherwise an active monitoring system may be required. An active system injects a test signal as early as possible into the detector elements and then checks that the detectors return the correct response.

Some examples of specific tasks for the monitoring software include checking the high voltages on power supplies, currents in magnets, and scaler contents. This information can be used as part of a feedback loop to adjust the settings of tuneable devices.

References

[1] R. Gluckstern, Uncertainties in track momentum and direction due to multiple scattering and measurement errors, Nuc. Instr. Meth. 24: 381–9, 1963.
[2] K. Amako, K. Kawano, S. Sugimoto, and T. Matsui, High precision field measuring system for a large aperture spectrometer magnet, Nuc. Inst. Meth. 197: 325–30, 1982.
[3] C. Akerlof, R. Hieber, A. Krisch, K. Edwards, L. Ratner, and K. Ruddick, Elastic proton-proton scattering at 90° and structure within the proton, Phys. Rev. 159: 1138–49, 1967.
[4] Some examples of selecting detectors for specific types of e^+e^- physics studies are given in K. Winter, Some detector arrangements at LEP, Physica Scripta 23: 569–78, 1981.
[5] W. Panofsky, The evolution of SLAC and its program, Phys. Today, October, 1983, pp. 34–41.
[6] W. Willis, The large spectrometers, Phys. Today, October 1978, pp. 32–9.
[7] T. Taylor, The choice of spectrometer magnets for LEP, Physica Scripta 23: 459–64, 1981.
[8] D. Anderson, G. Charpak, Ch. Von Gagern, and S. Majewski, Recent developments in BaF_2 scintillator coupled to a low pressure wire chamber, Nuc. Inst. Meth. 225: 8–12, 1984.
[9] Particle Data Group, Major detectors in elementary particle physics, LBL-91, Supplement, UC-37, March 1983.

[10] P. Darriulat, Large transverse momentum hadronic processes, Ann. Rev. Nuc. Part. Sci. 30: 159-210, 1980.
[11] C. Fabjan and T. Ludlam, Calorimetry in high energy physics, Ann. Rev. Nuc. Part. Sci. 32: 335-89, 1982.
[12] S.L. Wu, e^+e^- Physics at PETRA — The first five years, Phys. Rep. 107: 59-324, 1984.
[13] O.C. Allkofer, J.J. Aubert, G. Bassompierre, K.H. Becks, Y. Bertsch, C. Besson, C. Best, E. Bohm, D.R. Botterill, F.W. Brasse, C. Broll, J. Carr, B. Charles, R.W. Clifft, J.H. Cobb, G. Coignet, F. Combley, J.M. Crespo, P.F. Dalpiaz, P. Dalpiaz, W.D. Dau, J.K. Davies, Y. Declais, R.W. Dobinson, J. Drees, A. Edwards, M. Edwards, J. Favier, M.I. Ferrero, J.H. Field, W. Flauger, E. Gabathuler, R. Gamet, J. Gayler, P. Ghez, C. Gossling, J. Hass, U. Hahn, K. Hamacher, P. Hayman, M. Henckes, H. Jokisch, J. Kadyk, V. Korbel, M. Maire, L. Massonnet, A. Melissinos, W. Mohr, H.E. Montgomery, K. Moser, R.P. Mount, M. Moynot, P.R. Norton, A.M. Osborne, P. Payre, C. Peroni, H. Pessard, U. Pietrzyk, K. Rith, M.D. Rousseau, E. Schlosser, M. Schneegans, T. Sloan, M. Sproston, W. Stockhausen, H.E. Stier, J.M. Thenard, J.C. Thompson, L. Urban, M. Vivargent, G. von Holtey, H. Wahlen, E. Watson, V.A. White, D. Williams, W.S.C. Williams, and S.J. Wimpenny, A large magnetic spectrometer system for high energy muons, Nuc. Inst. Meth. 179: 445-66, 1981.
[14] A.N. Diddens, M. Jonker, J. Panman, F. Udo, J.V. Allaby, U. Amaldi, G. Barbiellini, A. Baroncelli, V. Blobel, G. Cocconi, W. Flegel, W. Kozanecki, E. Longo, K.H. Mess, M. Metcalf, J. Meyer, R.S. Orr, F. Schneider, A.M. Wetherell, K. Winter, F.W. Busser, P.D. Gall, H. Grote, P. Heine, B. Kroger, F. Niebergall, K.H. Ranitzsch, P. Stahelin, V. Gemanov, E. Grigoriev, V. Kaftanov, V. Khovansky, A. Rosanov, R. Biancastelliok, B. Borgia, C. Bosio, A. Capone, F. Ferroni, P. Monacelli, F. DeNotaristefani, P. Pistilli, C. Santoni, and V. Valente, A detector for neutral current interactions of high energy neutrinos, Nuc. Instr. Meth. 178: 27-48, 1980.
[15] D.P. Barber, U. Becker, H. Benda, A. Bohm, J.G. Branson, J. Bron, D. Buikman, J.D. Burger, C.C. Chang, H.S. Chen, M. Chen, C.P. Cheng, Y.S. Chu, R. Clare, P. Duinker, G.Y. Fang, H. Fesefeldt, D. Fong, M. Fukushima, J.C. Guo, A. Hariri, G. Herten, M.C. Ho, H.K. Hsu, R.W. Kadel, W. Krenz, J. Li, Q.Z. Li, M. Lu, D. Luckey, C.M. Ma, D.A. Ma, G.G.G. Massaro, T. Matsuda, H. Newman, M. Pohl, F.P. Poschmann, J.P. Revol, M. Rohde, H. Rykaczewski, K. Sinram, H.W. Tang, L.G. Tang, Samuel C.C. Ting, K.L. Tung, F. Vannucci, X.R. Wang, P.S. Wei, M. White, G.H. Wu, T.W. Wu, J.P. Xi, P.C. Yang, C.C. Yu, X.H. Yu, N.L. Zhang, and R.Y. Zhu, Physics with high energy electron-position colliding beams with the Mark J. detector, Phys. Rep. 63: 337-91, 1980.
[16] H. Gordon, R. Hogue, T. Killian, T. Ludlam, M. Winik, C. Woody, D. Burckhart, V. Burkert, O. Botner, D. Cockerill, W.M. Evans, C.W. Fabjan, T. Ferbel, P. Frandsen, A. Hallgren, B. Heck, M. Harris, J.H. Hilke, H. Hofmann, P. Jeffreys, G. Kantardjian, G. Kesseler, J. Lindsay, H.J. Lubatti, W. Molzon, B.S. Nielsen, P. Queru, L. Rosselet, E. Rosso, A. Rudge, R.H. Schindler, T. Taylor, J. v.d. Lans, D.W. Wang, Ch. Wang, W.J. Willis, W. Witzeling, H. Boggild, E. Dahl-Jensen, I. Dahl-Jensen, Ph. Dam, G. Damgaard, K.H. Hansen, J. Hooper, R. Moller, S.O. Nielsen, L.H. Olsen, B. Schistad, T. Akesson, S. Almehed, G. von Dardel, G. Jarlskog, B. Lorstad, A. Melin, U. Mjornmark, A. Nilsson, M.G. Albrow, N.A. McCubbin, M.D. Gibson, J. Hiddleston, O. Benary, S. Dagan, D. Lissauer, and Y. Oren, The Axial Field Spectrometer at the CERN ISR, Nuc. Instr. Meth. 196: 303-13, 1982.
[17] E. Bloom and C. Peck, Physics with the Crystal Ball detector, Ann. Rev. Nuc. Part. Sci. 33: 143-97, 1983.

[18] M. Oreglia, E. Bloom, F. Bulos, R. Chestnut, J. Gaiser, G. Godfrey, C. Kiesling, W. Lockman, D.L. Scharre, R. Partridge, C. Peck, F.C. Porter, D. Antreasyan, Yi-Fan Gu, W. Kollmann, M. Richardson, K. Strauch, K. Wacker, D. Aschmann, T. Burnett, C. Newman, M. Cavalli-Sforza, D. Coyne, H. Sadrozinski, R. Hofstadter, I. Kirkbride, H. Kolanoski, K. Konigsmann, A. Liberman, J. O'Reilly, B. Pollock, and J. Tompkins, Study of the reaction $\Psi' \rightarrow \gamma\gamma J/\Psi$, Phys. Rev. D 25: 2259–77, 1982.
[19] M. Nieto, W. Haxton, C. Hoffman, E. Kolb, V. Sandberg, and J. Toevs (eds.), *Science Underground,* AIP Conf. Proc. No. 96, New York: AIP, 1983.
[20] D.H. Perkins, Proton decay experiments, Ann. Rev. Nuc. Part. Sci. 34: 1–52, 1984.
[21] F. Kirsten, Computer interfacing for high energy physics experiments, Ann. Rev. Nuc. Sci. 25: 509–54, 1975.
[22] P. Kunz, Use of emulating processors in high energy physics, Physica Scripta 23: 492–98, 1981.
[23] Many of the things that were considered in the design of the data acquisition system for the European Muon Spectrometer are described in R. Dobinson, Practical Data Acquisition Problems in Large High Energy Physics Experiments, Proc. 1980, CERN School of Computing, CERN Report 81-03, 1981, pp. 325–61.
[24] M. Breidenbach, Calibration, monitoring, and control of complex detector systems, Physica Scripta 23: 508–11, 1981.

Exercises

1. Tracking is to be done with a set of 10 chambers equally spaced over 5 m in a 1-T magnetic field. What is the resolution on the measurement of a 20-GeV/c particle if the chambers have 200 μm spatial resolution? Estimate the contribution of multiple scattering to the momentum resolution if each of the chambers contains 0.05 radiation lengths of material.

2. Consider a rectangular dipole as shown in Fig. 14.2. Prove Eq. 14.11. Calculate the displacement Δx in the direction along the face of the magnet between the particle's entry and exit points.

3. Estimate the deflection angle of a 50-GeV muon that passes radially through a 2-m-wide toroidal field. Assume the toroid has a field of 0.5 T at the inner toroid radius of 1 m.

4. Design a fixed target spectrometer to measure inclusive hadron production from a hadron beam. Specify the locations and characteristics of any counters, tracking chambers, magnets, particle identification detectors, or calorimeters that are used. What trigger would you use?

5. Design a colliding beam spectrometer that can be used to search for the existence of the hypothetical Higgs particle H using the reaction

$$e^+e^- \rightarrow \mu^+\mu^-H^0$$

6. An experiment is to be done at a colliding beam accelerator with bunched beams and a luminosity per bunch collision L (i.e., L is the instantaneous luminosity $\mathscr{L}$ integrated over the bunch collision time). Suppose that it is not possible to interpret events unambiguously if more than one interaction occurs per bunch collision.
   a. If the cross section for interesting processes is σ and the total cross section is σ_t, then find the mean number of interesting events per bunch collision.
   b. What is the probability of getting one and only one interesting event per bunch collision?
   c. What is the mean number of background events per bunch collision?
   d. What is the probability of getting zero background events per bunch collision?
   e. What is probability of getting one and only one interesting event per bunch collision and nothing else?

15
Some fundamental measurements

This final chapter has three goals. First, we want to show how experimentalists have measured properties of subatomic particles. This section takes the form of a survey of some of the applicable techniques. Second, we want to discuss some of the considerations involved in measurements of particle interactions, such as total cross sections, elastic differential cross sections, polarization experiments, and new particle searches. Finally, we want to illustrate these measurements with examples of actual particle physics experiments.

15.1 Particle properties

In this section we will describe some of the methods used to measure the basic properties of the elementary particles. As mentioned in Chapter 1, these properties include charge, mass, spin, magnetic moment, lifetime, and branching ratios. Many specialized techniques have been developed for measuring some of these properties, particularly for the electron and nucleons. We will not attempt to survey all the applicable procedures for each particle, since many of the methods use techniques from atomic and molecular physics that fall outside the scope of this book. Instead we will follow the philosophy of the preceding chapters and discuss selected examples in more detail.

15.1.1 Charge

The sign of a particle's charge may be inferred from the direction of its deflection (if any) in a magnetic field of known orientation. The magnitude of the charge can be determined if the momentum of the particle and the strength of the magnetic field are known. Deflection in an electric field could also be used, although this is not so useful for high energy particles.

An alternative method for measuring the magnitude of a particle's charge is to measure a charge dependent quantity, such as the ionization energy loss or the Rutherford scattering cross section. This method has been used in a free quark search to measure the masses of particles produced in e^+e^- collisions [1]. The energy loss was sampled in selected scintillation counters, whereas the time of flight of the particle was measured in some others. Combining the value of β from the time of flight counters with the mean value of dE/dx allows the charge to be calculated from the Bethe–Bloch equation. The system calibration was provided by cosmic rays.

15.1.2 Mass

Many specialized techniques have been used to determine the masses of particles. For example, the π^- and K^- masses have been determined from the photons emitted in pionic and kaonic atoms [2]. The muon mass has been measured in the "g-2" spin procession experiments [3, 4].

A common method is to measure two independent kinematic quantities involving the mass of the particle. One of these quantities is usually the momentum, since it can easily be measured using a spectrometer. The other quantity could be the ionization, range, or velocity. The velocity could in turn be calculated from the results of a dE/dx, Cerenkov, or time of flight measurement. The laws of conservation of energy and momentum could then be used to calculate the mass of the particle. Källén [4] has given a nice summary of particle mass measurements.

As an example of a mass measurement consider the experiment of Daum et al. [5], which placed an upper limit on the mass of the muon neutrino from a study of $\pi^+ \rightarrow \mu^+ \nu_\mu$ decays at rest. Energy and momentum conservation give ($c = 1$)

$$m_\pi = \sqrt{p^2 + m_\mu^2} + \sqrt{p^2 + m_\nu^2}$$

where p is the momentum of the decay muon. The ν_μ mass can then be calculated from

$$m_\nu^2 = m_\pi^2 + m_\mu^2 - 2m_\pi\,(p^2 + m_\mu^2)^{1/2}$$

The error in the squared mass arising from an error in the momentum measurement is

$$\delta(m_\nu^2) = \frac{2m_\pi p}{\sqrt{p^2 + m_\mu^2}}\,\delta p \approx 76\;\delta p$$

since p is known to be around 30 MeV. The errors due to the uncertainties in the π^+ and μ^+ masses contribute similar amounts. Thus, a measurement of the muon momentum with an accuracy of about 0.001 MeV can be combined with prior measurements of the π^+ and μ^+ masses to determine the ν_μ mass with an error of ~ 0.1 MeV^2 on the mass squared.

A π^+ beam was passed through a lucite light guide with a thin 1-mm-thick scintillation counter embedded in the downstream side, as shown in Fig. 15.1. The scintillation counter was in a vacuum chamber located between the gaps of a bending magnet. The magnetic field was such that μ^+ from π^+ decays at rest near the downstream edge of the scintillator could follow a trajectory through a set of collimators to a solid state detector on the other side.

Stopping π^+ deposited 3.6 MeV in the scintillator. A gate signal was generated whenever the PMT signal exceeded a discriminator threshold of 3 MeV. The magnetic field was kept uniform to ± 0.15 G over the region of allowed μ^+ trajectories by shimming the pole faces. The field was

Figure 15.1 Magnetic spectrometer for muon momentum measurement. Some of the important features include (4) MWPC for monitoring the beam profile, (10) pion stop scintillator, (12) accepted muon trajectories, (15) NMR probe, (16) silicon surface barrier detector, and (20) ^{241}Am calibration source. (R. Abela et al., Phys. Lett. 146B: 431, 1984.)

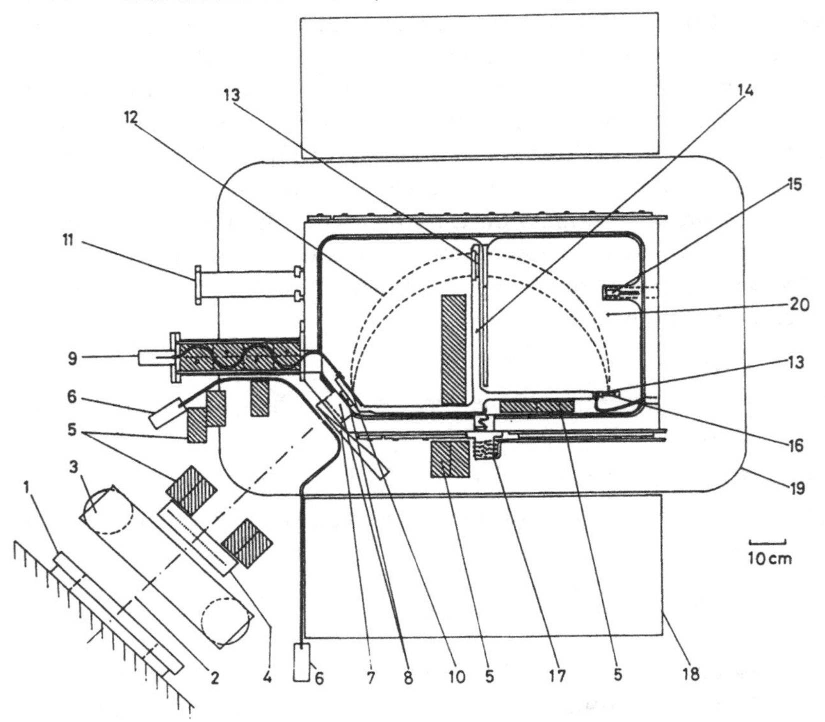

prevented from varying with time using a fixed NMR probe in a feedback circuit with a set of correction coils. The silicon surface detector signal was sent to a pulse height analyzer, which was gated by the stopping π^+ signal from the small scintillation counter. Figure 15.2 shows the gated pulse height spectrum. The peak is due to the μ^+ from stopping π^+. Note the excellent energy resolution obtained with the silicon detector. The shoulder to the right of the peak is due to additional energy losses in the silicon detector from the e^+ in $\mu^+ \rightarrow e^+ \nu_e \bar{\nu}_\mu$ decays.

The good event rate had a sharp cutoff above a certain magnetic field. This effect was used to determine the muon momentum from the very precise measurement of B. The measured value was $p = 29.7914 \pm 0.0008$ MeV/c. This result together with the π^+ and μ^+ masses gives the value $m_\nu^2 = -0.163 \pm 0.080$ MeV^2. This corresponds to an upper limit for the ν_μ mass of 0.25 MeV at the 90% confidence level.

15.1.3 Spin

Many particles have a spin degree of freedom associated with them. According to the laws of quantum mechanics, the magnitude and component of the spin along one fixed direction are quantized. The spin

Figure 15.2 Gated pulse height spectra from the silicon detector. (R. Abela et al., Phys. Lett. 146B: 431, 1984.)

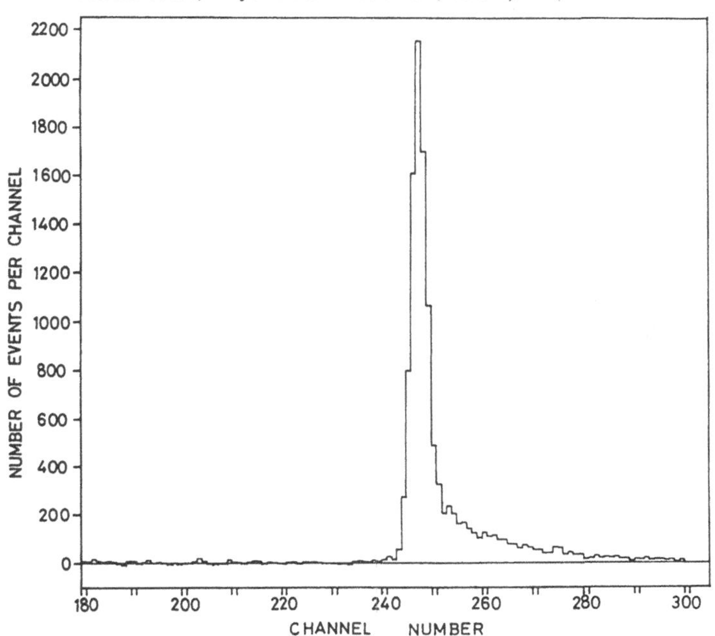

of the particle manifests itself in the decay angular distributions of unstable particles and in the production angular distributions of particle interactions. We will discuss the spin dependence of high energy interactions in more detail in Section 4.

The spins of the electron and nucleons have been determined from the hyperfine structure in optical spectroscopy, from atomic and molecular beam experiments, or from experiments in bulk matter using magnetic resonance techniques [6]. The spins of the charged pions have been determined using the principle of detailed balance, which relates the cross section for a reaction with the cross section for the inverse reaction [7].

15.1.4 Magnetic moment

Closely associated with the spin of a particle is the magnetic moment μ. However, whereas the magnitude of the spin is quantized, the magnitude of the magnetic moment is not. The two quantities are related through the equation

$$\mu = g\mu_B \mathbf{S} \tag{15.1}$$

where μ_B is the Bohr magneton and g is an experimental quantity that must be determined for each particle. The Dirac theory predicts that $g = 2$ for pointlike particles.

Any particle possessing a magnetic moment is subject to the force

$$\mathbf{F} = \mu \nabla B \tag{15.2}$$

in an inhomogeneous magnetic field. Note that this force is the dominant electromagnetic effect on the motion of a neutral particle. In addition, the magnetic moments of the particles are subject to a torque in a uniform magnetic field given by

$$\mathbf{N} = \mu \times \mathbf{B} \tag{15.3}$$

This torque causes the moment (and spin) to precess in the magnetic field with an angular frequency Ω_s given in the LAB frame by the Bargmann–Michel–Telegdi equation [8]

$$\mathbf{\Omega}_s = \frac{e}{m\gamma}[(1 + G)\mathbf{B}_\parallel + (1 + G\gamma)\mathbf{B}_\perp] \tag{15.4}$$

where $\mathbf{B}_\parallel$ ($\mathbf{B}_\perp$) is the component of $\mathbf{B}$ along (perpendicular to) the particle's velocity, and

$$G = \tfrac{1}{2}(g - 2) \tag{15.5}$$

is the g factor anomaly representing any deviation from pointlike behavior. It follows then that the total precession angle of the moment for a neutral particle traveling perpendicular to a uniform magnetic field is

$$\phi_{p} = \frac{e}{m\beta\gamma}(1 + G\gamma)\int B_{\perp}\,dl \tag{15.6}$$

The momentum of a charged particle will also bend in the magnetic field at the cyclotron frequency

$$\Omega_{c} = \frac{e}{m\gamma}B_{\perp} \tag{15.7}$$

Thus, for a charged particle the net angle between the momentum and moment after traversing the field is

$$\phi_{net} = \frac{e}{m\beta}G\int B_{\perp}\,dl \tag{15.8}$$

We see that for a pointlike charged particle with $G = 0$, the spin precesses at the same rate as the momentum.

The magnetic moments of the electron and nucleons have been determined using atomic physics techniques [6]. The g factor of the muon has been measured in a classic set of storage ring experiments at CERN [3, 9]. A by-product of this effort has been a precise demonstration of the validity of relativistic time dilation.

Schachinger et al. [10] have made a precise measurement of the magnetic moment of the Λ hyperon. The Λ, which were produced inclusively from interactions of a 300-GeV proton beam on a beryllium target at FNAL, had a polarization of ~8%. This made it possible to send the Λ through a bending magnet and measure the precession of the net polarization. The amount of precession gives G from Eq. 15.6 and then μ from Eq. 15.1. Good precision was possible since (1) the Λ were produced inclusively with a large cross section, (2) the detector had a large acceptance, and (3) the high energy gives the Λ a long mean decay length.

The experimental arrangement is shown in Fig. 15.3. The proton beam struck the target at a vertical angle. By parity conservation the Λ polarization must be perpendicular to the plane formed by the beam and Λ momenta. Thus, the accepted Λ will have the initial polarization along the -x direction in the figure. The vertical magnetic field causes the direction of the net polarization to precess. A veto counter following the magnet defines the beginning of the Λ decay region. The angle of the polarization after the magnet was determined by measuring the angular distribution of the proton from $\Lambda \rightarrow \pi^{-}p$ decays. In the Λ rest frame this is given by

$$\frac{1}{N}\frac{dN}{d\Omega} = \frac{1}{4\pi}(1 + \alpha\mathbf{P}\cdot\mathbf{k}) \tag{15.9}$$

where **P** is the Λ polarization vector, **k** is a unit vector along the proton momentum, and $\alpha = 0.647$ is the Λ decay asymmetry parameter.

The precession field was carefully mapped and monitored online using an NMR probe. Many checks were made to minimize any biases. The production angle, precession field, and spectrometer field were all periodically reversed. The precession angle is shown as a function of the integral field in Fig. 15.4. The small background consisted mainly of K^0's, Λ's produced by neutrons interacting in the collimators, and Λ's from Ξ^0 decays. The result for the magnetic moment was $\mu = (-0.6138 \pm 0.0047)\mu_N$.

15.1.5 Lifetime

The relativistic time dilation effect makes it possible to measure the lifetime of an unstable particle by measuring the distribution of its decay points at high energies. We have seen that in the LAB frame this

Figure 15.3 Experimental arrangement for the Λ magnetic moment measurement.

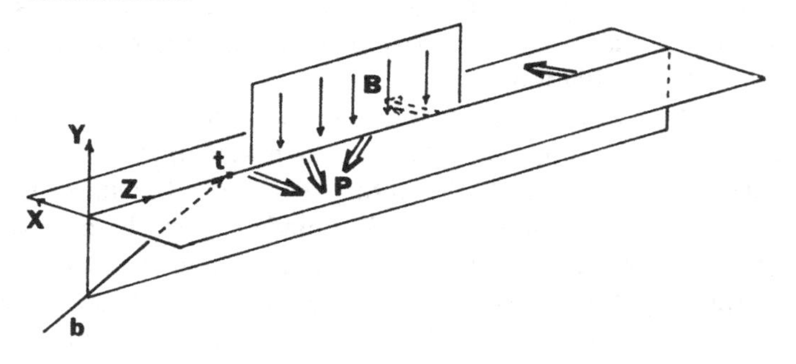

Figure 15.4 Precession angle of the Λ versus the measured field integral. (After L. Schachinger et al., Phys. Rev. Lett. 41: 1348, 1978.)

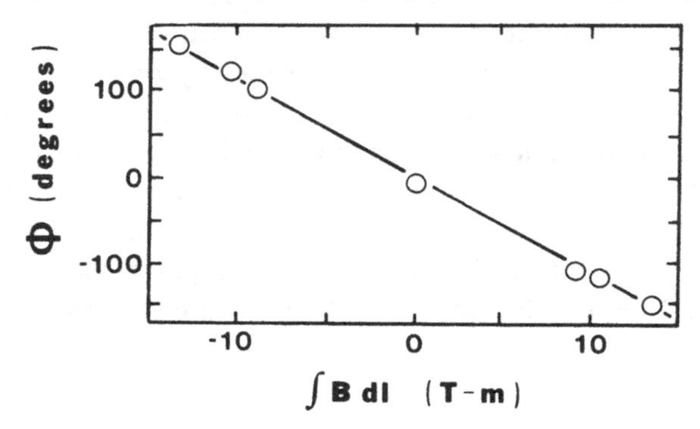

distribution is exponential with a slope that depends on the particle's lifetime. Figure 15.5 shows the experimental arrangement used by Bourquin et al. [11] at the CERN SPS to measure the lifetime of the Ω^- hyperon. The Ω^- were produced by interactions of the primary proton beam in a small target. The trigger for Ω^- in the beam used a DISC Cerenkov counter and was described in Chapter 13. Decays of the type $\Omega^- \rightarrow \Lambda K^-$, $\Lambda \rightarrow \pi^- p$, among others were reconstructed from tracks in the drift chambers. They first selected events containing a Λ by assigning the proton mass to positive tracks and the pion mass to negative tracks and looking for combinations where the effective mass was within 10 MeV of the Λ mass. They then assigned the kaon mass to the other negative track and calculated the ΛK^- effective mass. These events contained a large background of $\Xi^- \rightarrow \Lambda\pi^-$ decays. The Ξ^- decays could be removed by requiring that the effective mass $m(\Lambda\pi^-)$, obtained by interpreting the negative track as a π^-, be larger than 1.350 GeV. Since the Ξ^- mass is 1.321 GeV, this cut removes practically all the true Ξ^- decays, but only about 14% of true Ω^- events. Contamination from $\Omega^- \rightarrow \Xi^0\pi^-$ events was removed by requiring that the sum of the Λ and K^- momenta measured in the spectrometer match within errors the incident Ω^- momentum. The final sample of about 12,000 $\Omega^- \rightarrow \Lambda K^-$ decays had a very clean, narrow peak at the Ω^- mass. Samples of Ξ decays were also collected as a monitor on the experiment.

The Ω^- decay point was reconstructed from the intersection of the Λ and K^- momenta. The decay point distribution for Ω^- events is shown in Fig. 15.6. The solid line represents Monte Carlo calculations that took

Figure 15.5 Experimental apparatus for the Ω^- lifetime measurement. (M. Bourquin et al., Nuc. Phys. B 241: 1, 1984.)

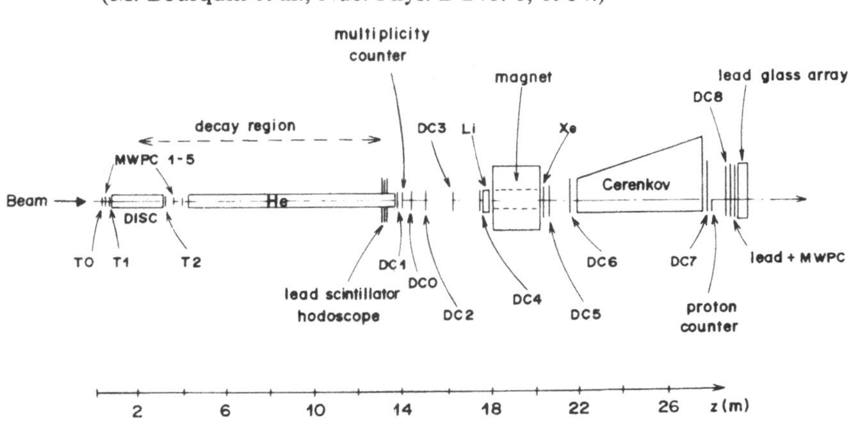

into account various systematic effects. The Ω^- lifetime extracted from the slope was $(0.823 \pm 0.013) \times 10^{-10}$ sec.

15.1.6 Branching ratios

Unstable particles generally decay into any combination of other particles that are allowed by energy conservation and are not forbidden by some selection rule. The fraction of the time a particle decays into a specific channel is referred to as its branching ratio. Measurements of branching ratios involve finding the number of events with the desired decay mode, correcting for background and detection efficiencies, and normalizing to the total number of decaying particles, or to one of its principal decay modes. As an example, a measurement of the branching ratio for $\pi^0 \rightarrow \gamma e^+e^-$ decays is described by Schardt et al. [12].

Figure 15.6 Decay point distribution for $\Omega^- \rightarrow \Lambda K^-$ events. (M. Bourquin et al., Nuc. Phys. B 241: 1, 1984.)

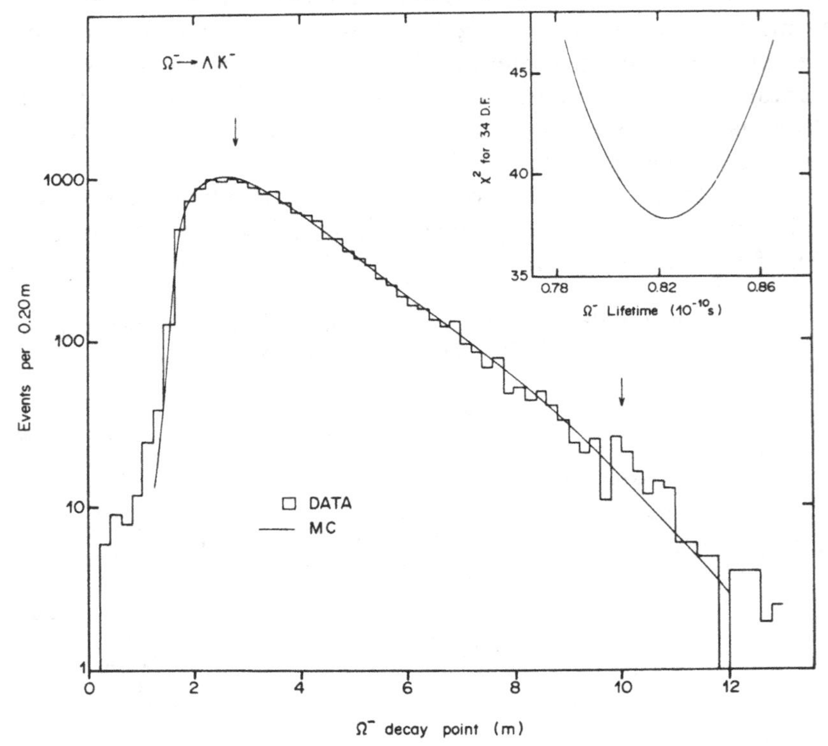

15.2 Total cross section

Two methods are commonly used to directly measure total cross sections. In the first, one attempts to record every interaction that takes place. This requires a detector with $\sim 4\pi$ acceptance. This method is often used for colliding beam experiments. The total cross section is given by

$$\sigma_T = R/\mathscr{L} \tag{15.10}$$

where R is the measured total interaction rate and $\mathscr{L}$ is the measured incident luminosity. A similar procedure is used for bubble chamber experiments. In that case one measures the total number of interactions in some volume of the chamber and normalizes to the number of incident particles.

The second direct method is the transmission experiment. Here one measures the intensity of particles before and after the target and infers the total cross section from the relation

$$I = I_0 \exp(-N\sigma_T) \tag{15.11}$$

where I (I_0) is the transmitted (incident) intensity and N is the number of target nuclei/cm^2. This method is frequently employed at fixed target accelerators. We will also mention several methods of extracting the total cross section from the measurements of small angle elastic scattering in Section 15.3.

As an example of the first method, we consider a measurement of the pp total cross section at the CERN ISR [13]. The apparatus shown in Fig. 15.7 consists of two large scintillator hodoscopes surrounding each of the beam pipes leaving the intersection region. The intersection point itself was sandwiched above and below by two large streamer chambers. Lead oxide plates in the chambers aided in converting photons.

Most interactions were detected by coincidences between the hodoscopes in the two arms. The signals from the near and far hodoscopes were ORed to produce an ARM_i signal. The distribution of time differences between the arms showed a narrow central peak, corresponding mostly to true beam–beam events, plus smaller side peaks 18 ns before and after the central peak. These side peaks were due to coincidences with stray beam particles and were removed by placing a ± 10-ns time window around the central peak.

The observed coincidence rates were corrected for accidentals, beam–gas collisions, and collisions with material near the intersection point. The streamer chambers were used to measure corrections necessary for events that triggered only one or neither of the hodoscope arms. Finally, for some events, such as small angle elastic or single diffraction scattering, the particles remain in the beam pipe and are not detected. The rates for these processes had to be calculated from other measurements.

The luminosity was measured using the "Van de Meer method." This involves measuring the interaction rate as a function of the vertical displacement of the two beams relative to each other. The results of this experiment confirmed the rising value of the total cross section in the region $\sqrt{s} = 24-63$ GeV.

The second method of measuring the total cross section is the transmission experiment [14, 15]. A schematic of an idealized transmission experiment is given in Fig. 15.8. The incoming beam passes first through the

Figure 15.7 Apparatus for a total cross section measurement at the ISR. (H) Scintillator hodoscope and (SC) streamer chamber. (After K. Eggert et al., Nuc. Phys. B 98: 93, 1975.)

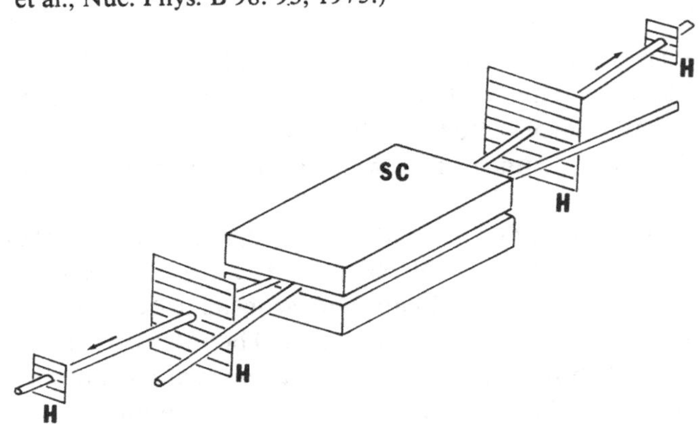

Figure 15.8 The "good-geometry" transmission arrangement for measuring the total cross section. (B) Incident beam counters, (T) target, and (F) transmission counters.

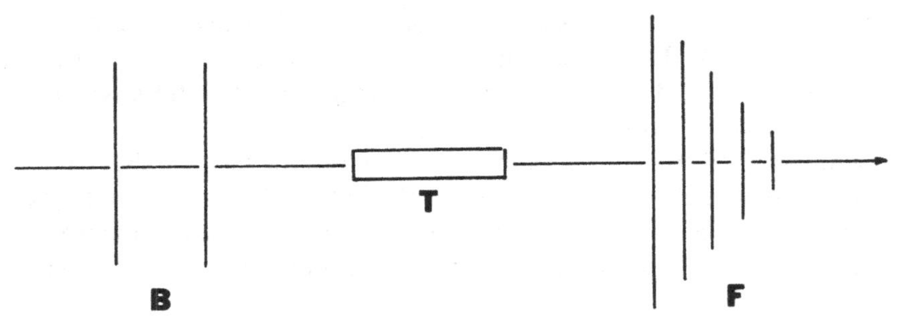

counters B and then the target T. Some of the incident particles will have a nuclear interaction in the target, while others will not. The total cross section is related to the total rate I for particles that do not interact by

$$\sigma_T = \frac{1}{N} \ln\left(\frac{I_0}{I}\right) \tag{15.12}$$

Ideally one would like to have an infinitesimal detector that can measure every noninteracting particle. However, even if this were possible, it would lead to errors, since the beam has a finite size, and since particles that traverse the target without a nuclear interaction may still undergo multiple scattering in the target material.

The finite size of the transmission counters leads to a new difficulty. Particles may undergo small angle nuclear elastic or diffraction scattering and yet remain inside the acceptance of the transmission counter. For this reason, a series of transmission counters are used, each subtending a different solid angle. The effects of beam size and multiple scattering are only important at small solid angles. Away from this region the transmitted intensity has the form

$$I = I_0 e^{-N\sigma_T}\left[1 + N \int_0^\Omega \frac{d\sigma}{d\Omega}\, d\Omega\right] \tag{15.13}$$

Thus an extrapolation of the finite solid angle data to zero solid angle gives the cross section with the small angle scattering properly accounted for.

As an example, consider the experiment of Citron et al. [16], which used the transmission method at the AGS to measure the $\pi^{\pm}$p total cross sections. The incoming beam intensity was measured using coincidences between three scintillation counters and a threshold Cerenkov counter. Data were taken with a liquid hydrogen target in the beam path. Empty target corrections were made by mounting an identical chamber on the same frame. The frame rolled on rails, so that either target could easily be placed in the beamline.

Seven circular transmission counters were used, subtending solid angles between 2 and 30 msr. The counters were arranged so that the radii decreased along the beam path. This arrangement permitted a determination of the statistical error on the difference between two adjacent cross section measurements and made the reduction in efficiency of later counters due to absorption the same for the target empty and target full runs.

Additional logic was used to measure the rate of accidental and circuit deadtime. Muon contamination in the pion beam was vetoed using an iron absorber and scintillator behind the transmission counters. Final corrections were necessary to remove contributions to the measured cross sections from electromagnetic processes.

15.3 Elastic scattering

Elastic scattering has been used for many years as an aid in understanding the effective size or any internal structure of particles. The measured elastic cross section is actually the sum of nuclear scattering events, electromagnetic Coulomb scattering, and scattering due to the interference of the electromagnetic and nuclear amplitudes

$$\frac{d\sigma}{dt} = \frac{d\sigma_{\text{nuc}}}{dt} + \frac{d\sigma_{\text{Coul}}}{dt} + \frac{d\sigma_{\text{int}}}{dt} \tag{15.14}$$

For small angles the nuclear differential cross section has an approximately exponential dependence on t

$$\frac{d\sigma_{\text{nuc}}}{dt} = A \exp(bt) \tag{15.15}$$

where A is the nuclear differential cross section at $t = 0$ and b is the slope parameter. Sometimes it is more interesting to examine the local slope, defined by

$$b(t) = \frac{d}{dt}\left(\ln \frac{d\sigma}{dt}\right) \tag{15.16}$$

The total cross section may be determined from elastic scattering measurements by using the optical theorem [14, 17]

$$\sigma_{\text{T}}^2 = \frac{16\pi A}{1 + \rho^2} \tag{15.17}$$

where ρ is the ratio of the real to imaginary part of the forward nuclear elastic scattering amplitude.

The Coulomb scattering cross section is the square of the amplitude [14]

$$C = \frac{2\alpha}{|t|} G^2(t) e^{i\alpha\phi(t)} \tag{15.18}$$

where $G(t)$ is the particle form factor, $\phi(t)$ is the phase of the Coulomb amplitude, and α is the fine structure constant. The form factor is related to the electromagnetic structure of the particle. The interference cross section is

$$\frac{d\sigma_{\text{int}}}{dt} = -2[\rho + \alpha\phi(t)] \frac{2\alpha}{|t|} G^2(t) \sqrt{\frac{A}{1 + \rho^2}}\, e^{1/2(bt)} \tag{15.19}$$

It is important to note that near $t = 0$ the Coulomb cross section goes like $1/t^2$, while the interference cross section goes like $1/t$.

We consider as an example the high energy elastic scattering measurement of Schiz et al. [18] at Fermilab. The experimenters measured the πp and pp elastic differential cross sections at 200 GeV/c for the t interval $0.02 < t < 0.66$ GeV^2. Figure 15.9 shows the experimental arrangement. The incoming beam particles were identified using four Cerenkov counters. Two groups of high resolution PWCs were used on each side of the liquid hydrogen target to measure the scattering angle. The forwardly scattered beam was focused onto a plane downstream of the spectrometer magnet. A scintillation counter located there vetoed unscattered tracks. Another set of PWCs following the magnet measured the bend angle.

The experiment used a two-level trigger. The first level, which came from the scintillators, essentially required a good beam track, together with a signal at the end of the forward arm, and no veto. This trigger was dominated by beam halo particles, so a higher level was also employed using information from the PWCs. A hardware device required that the trajectory intercept a preset window in the veto plane and that the projected scattering angle exceed $-t = 0.01$ GeV^2. Special "beam" triggers were also collected for alignment and normalization, and some first-level triggers were taken to study the efficiency and biases of the hardware trigger device.

A number of cuts were applied to the raw data to eliminate unusable tracks and to help extract the elastic signal. The data was grouped into bins of $q = \sqrt{-t} \simeq p_b\theta$, where p_b is the beam momentum and θ is the scattering angle. The number of scatters in each bin $N_s(q)$ was then determined by

Figure 15.9 Experimental arrangement for measurement of high energy elastic scattering. (C) Cerenkov counter, (P) MWPC, (t) target, (M) bending magnets, (V) veto counter, and (S) trigger scintillator. (After A. Schiz et al., Phys. Rev. D 24: 26, 1981.)

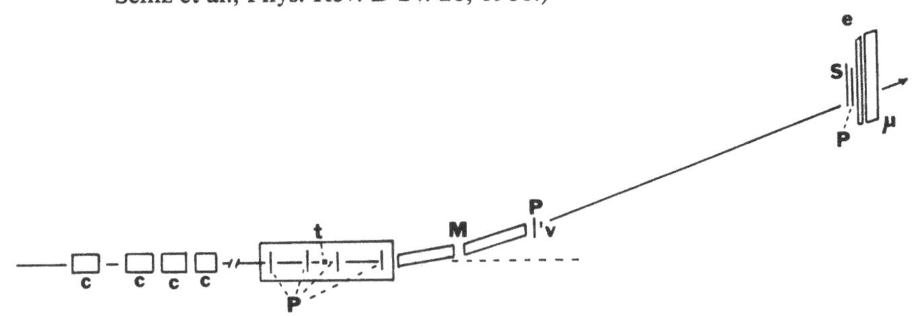

making a normalized, target empty subtraction from the measured events.

The data had to be corrected for several effects. First, radiation of photons by the scattered particles may cause a good elastic event to be missed. Thus a correction factor C_{rad} was calculated that was largest (8%) for large q pions. Second, the measured data still contained some inelastic contamination. A factor C_{in} was determined by fitting the recoil mass distribution to a peak at the proton mass plus background as shown in Fig. 15.10. The largest corrections (9%) occurred for high q proton events. A third correction C_{pl} was necessary to account for events with two nuclear scatters in the target. These events incorrectly enhance the large q bins.

The total elastic cross section may be determined by integrating the measured $d\sigma/dt$ over t. The data must be extrapolated for the regions of t not measured. There is a check on the extrapolation at $t = 0$ from the optical theorem. The measured pp elastic cross section at 200 GeV/c accounts for almost 18% of the total cross section.

15.4 Polarization experiments

It has become increasingly clear that spin dependent interactions remain important up to the highest available accelerator energies [19]. For reactions where one or both of the initial or final state particles has nonzero spin, there are many spin dependent observables in addition to

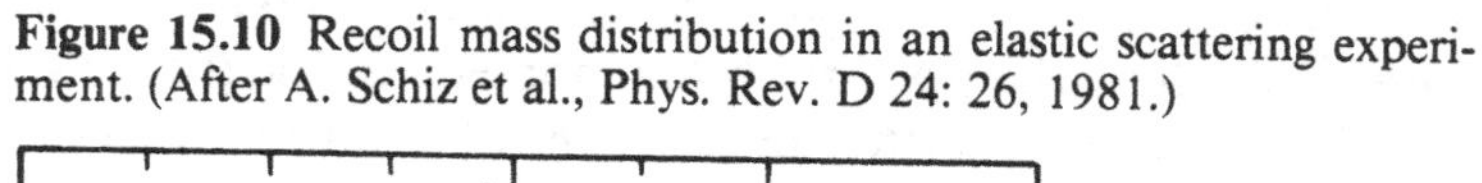

Figure 15.10 Recoil mass distribution in an elastic scattering experiment. (After A. Schiz et al., Phys. Rev. D 24: 26, 1981.)

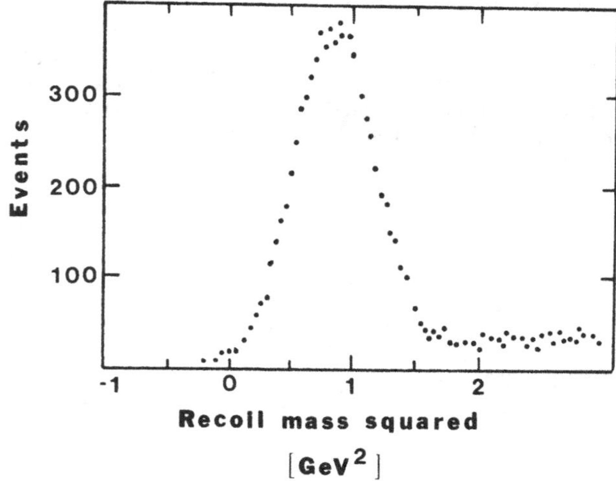

the ordinary spin-averaged differential cross section. Remarkable structure has been observed in some of these observables, for example, the two particle asymmetries in pp elastic scattering.

The simplest of the spin dependent observables is the analyzing power $\mathscr{A}$. Consider the elastic scattering of a spin 0 beam particle from a spin $\frac{1}{2}$ target particle. If we denote by N_L (N_R) the fraction of the events in which the fast forward particle scatters at an angle θ to the left (right), then the analyzing power is defined by

$$\mathscr{A} = \frac{1}{P_T}\frac{N_L - N_R}{N_L + N_R} \tag{15.20}$$

where P_T is the average polarization of the target particles. Note that $\mathscr{A}$ is a function of the angle θ and of the particle's momentum. To measure $\mathscr{A}$ in this way would require two forward spectrometers. However, the same quantity $\mathscr{A}$ can be obtained by using one spectrometer at an angle θ, to the left say, and instead taking data with the target polarization in opposite directions. Then $\mathscr{A}$ is given by

$$\mathscr{A} = \frac{1}{P_T}\frac{N_+ - N_-}{N_+ + N_-} \tag{15.21}$$

where N_+ (N_-) is the fraction of events with the target polarized up (down). Because of the expense of building a second spectrometer and the ease of reversing the target polarization, most high energy experiments are performed this way. The analyzing power is sometimes referred to as the asymmetry or polarization parameter.

Measurements of observables that only involve initial state polarizations are readily performed if a polarized beam and a polarized target are available. The polarization of final state particles can be measured by scattering them off a secondary target, such as carbon. However, the requirement for double scattering makes measurements of observables that involve final state polarizations much less precise than those that only involve the initial state. Table 15.1 lists some of the observables that have been measured in spin 0 – spin $\frac{1}{2}$ and spin $\frac{1}{2}$ – spin $\frac{1}{2}$ 2-body scattering. The table uses the 2-body to 2-body scattering notation of Chapter 1. The various observables are classified depending on whether the initial state particles have been prepared with a net polarization and whether the polarization of final state particles is analyzed along some direction. The polarization of the fast forward, final state particles (1) is usually not analyzed because this requires rescattering the fast particles from a secondary target, and the value of the analyzing power of most substances at high energies is quite small. Many of the components of the observables in

Table 15.1 are either 0 or are related to each other because of symmetry principles.

The analysis of polarized target data is complicated by the fact that only the protons in hydrogen atoms in the target material are polarized. The remaining ~90% of the target nuclei contained mostly in carbon and oxygen atoms are not polarized. This nonpolarized background is reduced for the case of elastic scattering. The nuclei in the heavier atoms have a Fermi momentum distribution, which makes it more difficult for events from those atoms to satisfy the elastic correlations. To determine the background under the elastic signal coming from these unpolarized nuclei, special runs can be taken using carbon or Teflon targets.

Consider as an example the measurements by the SLAC–Yale group [20, 21] of polarized electron-polarized proton scattering at the SLAC linac. The experiment was designed to explore the internal spin structure of the proton using a polarized electron probe, in analogy with the classic series of experiments that used (unpolarized) electrons to study the spatial structure of the proton.

Table 15.1. *Summary of spin 0–spin ½ and spin ½–spin ½ observables*

	Particle[a]			
Observable	b	t	1	2
Spin 0–spin ½				
σ_0	np	np	na	na
A_i	np	p	na	na
P_i	np	np	na	a
D_{ij}	np	p	na	a
Spin ½–spin ½				
σ_0	np	np	na	na
A_i	p	np	na	na
A_i	np	p	na	na
A_{ij}	p	p	na	na
P_i	np	np	na	a
C_{ij}	np	np	a	a
D_{ij}	np	p	na	a
K_{ij}	p	np	na	a
H_{ijk}	p	p	na	a

[a]Abbreviations: p, initial state particle is polarized along one of three orthogonal directions; np, initial state particle is not polarized; a, component of polarization of final state particle is analyzed along one of three orthogonal directions; na, polarization of final state particle is not analyzed.

The overall layout of the experiment is shown in Fig. 15.11. A longitudinally polarized electron beam strikes a longitudinally polarized proton target, and any outgoing electrons are identified and measured inclusively in the forward spectrometer. In order to obtain the polarized beam, a polarized electron source had to be installed on the linac. The source used ultraviolet light to photoionize a beam of ^{6}Li atoms in which the electrons were strongly polarized. The direction of the polarization was either parallel or antiparallel to the electron momentum, depending on the direction of the current in the polarizing coils. The polarization was frequently reversed to minimize systematic errors. The polarization of the beam leaving the source was measured by double Mott scattering of a portion of the beam on a gold foil. The polarization of the accelerated high energy beam was measured by elastic (Moller) scattering from a magnetized iron foil. The beam had an average intensity of 5×10^8 electrons/pulse and a polarization of 80%.

The incoming electron beam interacted in a polarized proton target located in the 5-T field of a superconducting solenoid. The actual target material consisted of frozen beads of butanol doped with the organic free radical porphyrexide. The target was maintained at a temperature of 1 K using a ^{4}He evaporation cryostat. The polarization of the free protons in the target averaged 60%. This was measured by surrounding the target with the coil of a NMR circuit. Radiation damage from the electron beam caused serious depolarization in the target. In order to minimize this effect and to ensure uniform polarization, the beam was rastered quickly across the target. In addition, techniques had to be developed to anneal away some of target damage and to quickly change the target material.

The forward spectrometer used two large dipole bending magnets and a

Figure 15.11 Schematic of the SLAC polarized electron-polarized proton scattering experiments. (PeB) Polarized electron beam, (PPT) polarized proton target, (M) bending magnet, (C) Cerenkov counter, (W) MWPC, (H) hodoscope, and (LG) lead-glass shower counter.

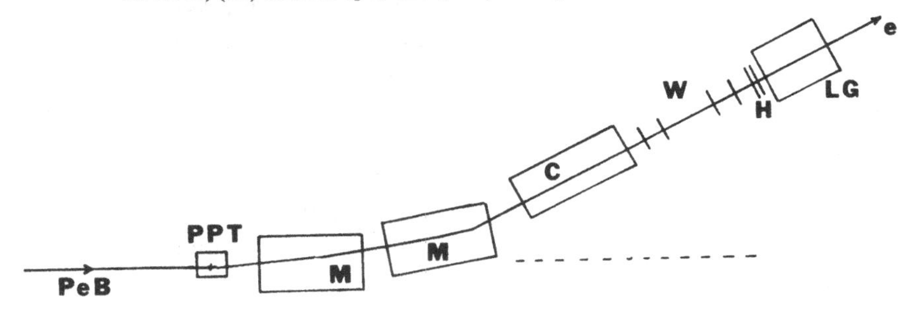

4000-wire set of PWCs for particle tracking. The outgoing electrons were identified using a 4-m-long gas threshold Cerenkov counter and a segmented lead-glass shower counter.

The measured asymmetry averaged over the target polarization is

$$\Delta = \frac{N_+ - N_-}{N_+ + N_-}$$

where N_+ and N_- are the normalized event rates when the electron has positive and negative helicity, respectively. The measured asymmetry was very small ($\sim 1\%$). For this reason careful attention had to be paid to minimizing possible systematic errors due to variations in the intensity, position, or direction of the beam. The measured asymmetry is related to the intrinsic asymmetry A associated with polarized ep scattering by

$$\Delta = P_e P_p f A \tag{15.22}$$

where P_e and P_p are the electron and proton polarization and f is the fraction of free protons (hydrogen atoms) in the target.

The instrinsic asymmetry is a function of the variable $x = Q^2/2M\nu$, where Q^2 is the relativistic 4-momentum transfer to the proton, ν is the energy loss of the electron, and M is the mass of the proton. At large x the asymmetry becomes very large and positive. This is interpreted by QCD models to mean that the probability is large that a quark that carries a large fraction of the nucleon's momentum will also carry its spin.

15.5 New particle searches

Among the most satisfying and exciting experiments in physics are those that discover the existence of new particles. Sometimes the discovery was hoped for, yet still unexpected, as in the case of the J/ψ. This particle was discovered almost simultaneously in 1974 by a production experiment at the AGS at BNL and by a formation experiment at the SPEAR storage ring at SLAC. Other times fairly precise predictions exist for a particle state, but some combination of very high mass, small production cross section, or large background conspire to make the experimental measurement very difficult. This was the case for the W gauge bosons, which were discovered in 1983 by two groups at the CERN SPS $\bar{p}p$ collider.

First let us consider the classic experiment of Samuel Ting and collaborators, who discovered the J half of the J/ψ particle [22, 23]. Ting was interested in searching for possible high mass vector mesons. These strongly interacting particles have the same quantum numbers as the photon. Any such particle should decay into e^+e^- pairs. However, the branching ratio for e^+e^-, which comes from electromagnetic processes, should be much smaller than that for pairs of hadrons. In addition, the angular distribution of electromagnetic processes fall off very sharply. Thus, the experiment required a high resolution detector with good acceptance for e^+e^- pairs that could handle very high rates of beam particles and that had a very high rejection capability for hadron pairs.

A schematic of the experiment is shown in Fig. 15.12. A slow extracted beam from the AGS was focused onto a beryllium target. The incident beam intensity was $\sim 10^{12}$ protons/pulse. The spectrometer consisted of two identical arms. Each arm contained three bending magnets $M_0 - M_2$, two threshold Cerenkov counters C_0 and C_e, and a lead-glass shower counter S for particle identification, MWPCs for tracking, and scintillation counter hodoscopes. The downstream portion of the arms were elevated by 10.33° in order to decouple the measurements of the particles' momenta and their production angles. The magnetic spectrometer had a mass resolution of ± 5 MeV and a mass acceptance of 2 GeV, making it ideally suited for searching over a large mass range for a narrow resonance.

The combination of Cerenkov and shower counters gave a rejection against hadron pairs by a factor $> 10^8$. The Cerenkov counter was filled with H_2 gas to minimize the production of knock-on electrons. The counters were separated by a magnetic field so that the electrons produced in the first counter would not be detected by the second.

Figure 15.12 Double arm spectrometer for high mass electron pairs. (T) Target, (M_0, M_1, M_2) bending magnets, (C_0, C_e) Cerenkov counters, (S) lead-glass and shower counters. The C_B Cerenkov counter was located below M_0 and is not shown. (After S.C.C. Ting, Rev. Mod. Phys. 49: 235, 1977.)

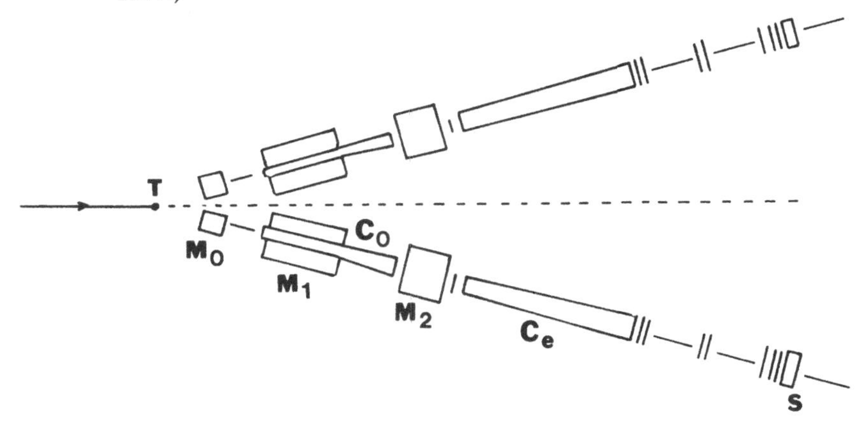

The system was calibrated using electrons from the decay $\pi^0 \rightarrow \gamma e^+e^-$. One of the e^+ or e^- particles was tagged with a highly directional Cerenkov counter (not shown) close to the target, while the other particle went through the spectrometer. The experimental e^+e^- mass spectrum near 3 GeV is shown in Fig. 15.13. The J/ψ resonance stands out clearly above a small background of nonresonant and misidentified events. The dotted histogram represents data taken at a lower magnet current. It is one of

Figure 15.13 Mass spectrum for electron pairs. The bin width is 25 MeV. The dotted spectrum was collected with the spectrometer magnet running at a reduced current. (After S.C.C. Ting, Rev. Mod. Phys. 49: 235, 1977.)

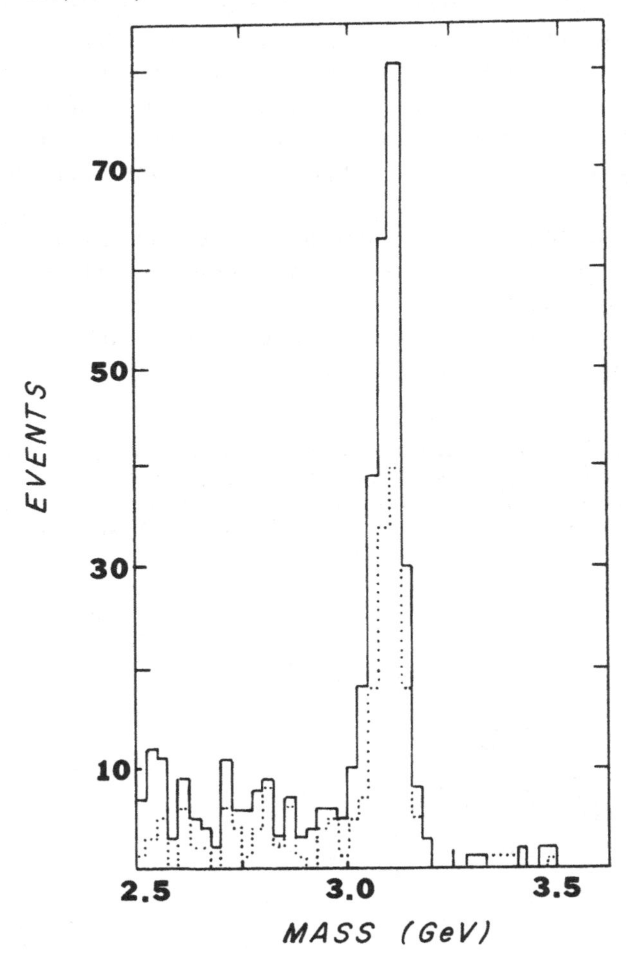

many checks that the peak was not due to an instrumental effect. Lowering the current causes the particle to traverse different portions of the spectrometer and would presumably shift any effect caused by the apparatus itself.

Lastly, let us mention the discovery of the charged W vector bosons by a group headed by Carlo Rubbia at CERN [24]. The highly successful electroweak theory had predicted that the mass of the W should be around 82 GeV. The W was predicted to decay via

$$W^{\pm} \rightarrow e^{\pm}\nu$$

with a significant branching ratio, so the experiment was designed to search for the production of electrons and neutrinos. In order to reach the very high CM energies where such massive particles could be produced, a large project was initiated at CERN to convert the SPS to a $\bar{p}p$ collider. An elaborate $\bar{p}$ cooling scheme similar to that discussed in Chapter 4 was devised by Van de Meer and coworkers in order to get the luminosity up to useful levels.

The UA1 detector at the SPS is shown in Fig. 15.14. The interaction point is surrounded by a central detector, which consists of a cylindrical drift chamber 5.8 m long and 2.3 m in diameter. The central detector sits in a 0.7-T dipole field. The typical momentum resolution is 20% for a 1-m-long, 40-GeV/c particle. The central detector was used to determine the overall topology of the event and to measure the charge and momentum of electron candidates. Electrons were identified by a large energy deposit in the lead–scintillator electromagnetic calorimeter and by the lack of penetration into the iron–scintillator hadron calorimeter. The electromagnetic shower counters were 27 radiation lengths deep and had an energy resolution of $15\%/\sqrt{E}$ (GeV). The electromagnetic calorimeters extended over 99% of 4π, so that the neutrino could be identified through an unbalance in the visible energy flow transverse to the beam axis. Since muons were capable of carrying substantial amounts of energy outside of the calorimeters, the detector was surrounded by eight layers of proportional drift tubes.

The trigger required at least 10 GeV of transverse energy in two electromagnetic calorimeter elements. The trigger rate was 0.2 events/sec at a luminosity of 5×10^{28} cm^{-2} s^{-1}. Two parallel analyses were used on the final sample of high quality events with associated vertices in the central detector. The first analysis examined charged tracks for clean, isolated electrons. A second independent analysis looked at the calorimeter information for events with missing transverse energy. With a few exceptions

Figure 15.14 The UA1 detector at CERN. (Courtesy of the UA1 collaboration and the Particle Data Group.)

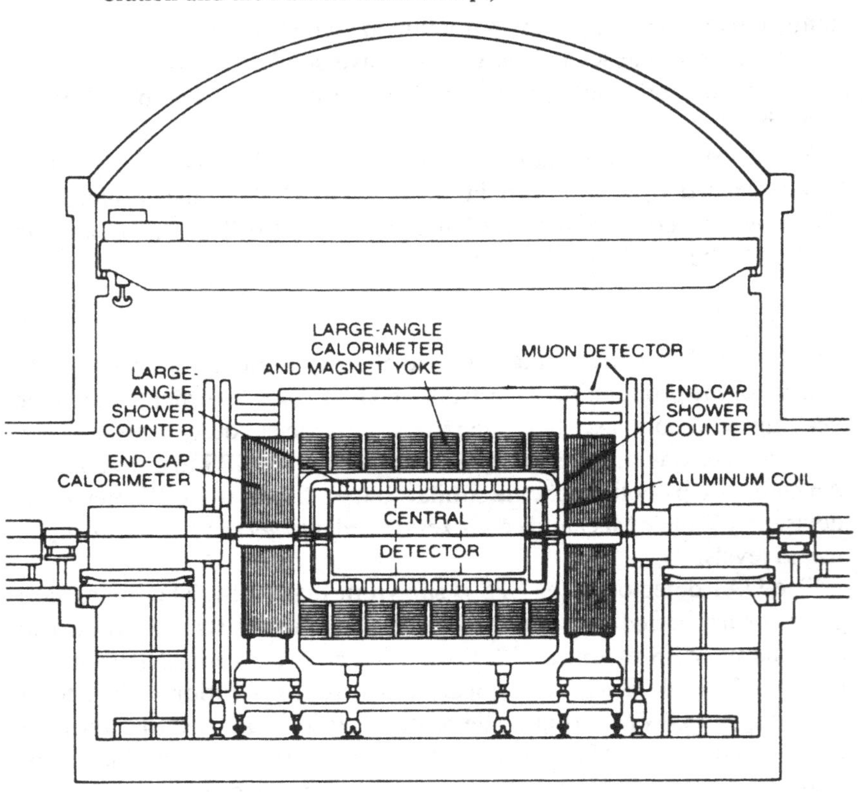

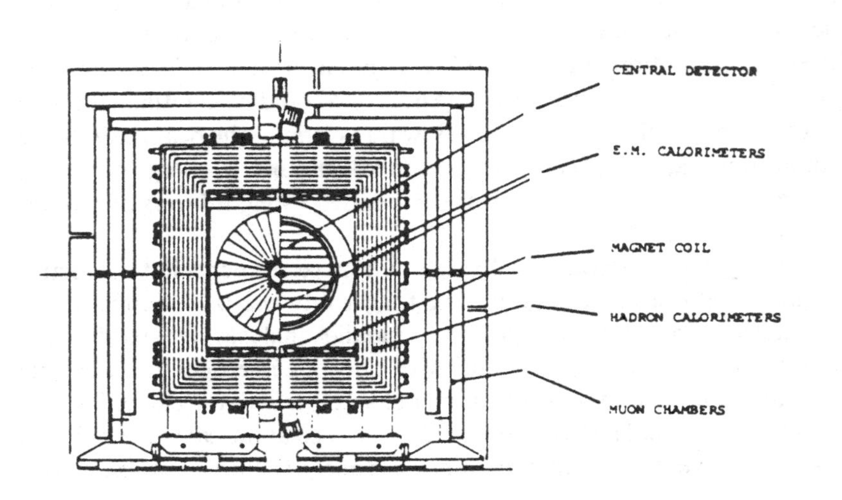

both analyses led to the same set of final events. Figure 15.15 shows the association of the missing transverse momentum for the electron candidate events. The missing energy lies antiparallel to the electron direction, suggesting a 2-body decay. Assuming that these events are the decay products of the W, and taking the W decay kinematics and transverse momentum of the W at production into account, the data gave $m_W = 80.9 \pm 1.5$ GeV, in excellent agreement with the theoretical predictions.

The UA1 experiment illustrates two important points concerning experiments at the multi-TeV accelerators now under consideration. First is the increasing difficulty in using magnetic spectrometers. The high energy events are characterized by very large multiplicities of charged particles. It becomes more and more difficult to pattern-recognize the tracks of all these particles and to accurately measure their momenta. The second feature is the increasing importance of finely segmented, good resolution calorimeters. The measurement of vector energy flow is a very powerful tool for examining the new physics of jets and high mass particle states.

Figure 15.15 Components of the missing transverse energy are plotted relative to the electron direction. (After G. Arnison et al., Phys. Lett. 129B: 273, 1983.)

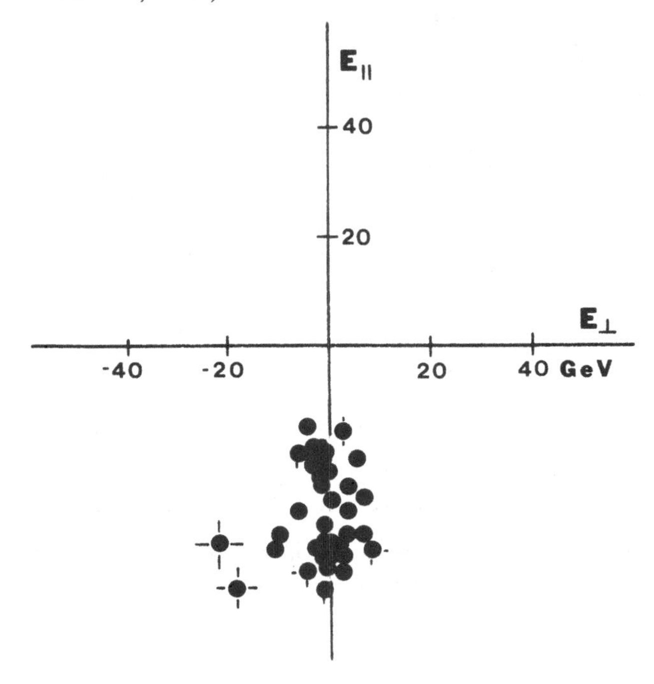

References

[1] A. Marini, I. Peruzzi, M. Piccolo, F. Ronga, D.M. Chew, R.P. Ely, T.P. Pun, V. Vuillemin, R. Fries, B. Gobbi, W. Guryn, D.H. Miller, M.C. Ross, D. Besset, S.J. Freedman, A.M. Litke, J. Napolitano, T.C. Wang, F.A. Harris, I. Karliner, S. Parker, and D.E. Yount, Search for exclusive free-quark production in e^+e^- annihilation, Phys. Rev. Lett. 48: 1649–52, 1982.

[2] R. Seki and C. Wiegand, Kaonic and other exotic atoms, Ann. Rev. Nuc. Sci 25: 241–81, 1975.

[3] F. Farley and E. Picasso, The muon (g-2) experiments, Ann. Rev. Nuc. Sci 29: 243–82, 1979.

[4] G. Kallen, *Elementary Particle Physics,* New York: Addison-Wesley, 1964, Chap. 2.

[5] M. Daum, G. Eaton, R. Frosch, H. Hirschmann, J. McCulloch, R. Minehart, and E. Steiner, Precision measurement of the muon momentum in pion decay at rest, Phys. Rev. D 20: 2692–707, 1979; final results obtained with an improved apparatus are given in R. Abela, M. Daum, G. Eaton, R. Frosch, B. Jost, P. Kettle, and E. Steiner, Precision measurement of the muon momentum in pion decay at rest, Phys. Lett. 146B: 431–6, 1984.

[6] L. Yuan and C. Wu, *Methods of Experimental Physics,* Vol. 5, Part B, New York: Academic, 1963, Sec 2.4.

[7] D. Perkins, *Introduction to High Energy Physics,* New York: Addison-Wesley, 1972, pp. 79–83.

[8] The form we use here is from B. Montague, Elementary spinor algebra for polarized beams in storage rings, Part. Accel. 11: 219–31, 1981.

[9] F. Combley, F. Farley, and E. Picasso, The CERN muon (g-2) experiments, Phys. Rep. 68: 93–119, 1981.

[10] L. Schachinger, G. Bunce, P. Cox, T. Devlin, J. Dworkin, B. Edelman, R. Edwards, R. Handler, K. Heller, R. March, P. Martin, O. Overseth, L. Pondrom, M. Sheaff, and P. Skubic, Precise measurement of the Λ magnetic moment, Phys. Rev. Lett. 41: 1348–51, 1978.

[11] M. Bourquin, R.M. Brown, J.C. Chollet, A. Degre, D. Froidevaux, J.M. Gaillard, C.N.P. Gee, J.P. Gerber, W.M. Gibson, P. Igo-Kemenes, P.W. Jeffreys, M. Jung, B. Merkel, R. Morand, H. Plothow-Besch, J.P. Repellin, J.L. Riester, B.J. Saunders, G. Sauvage, B. Schiby, H.W. Siebert, V.J. Smith, K.P. Streit, R. Strub, J.J. Thresher, and S.N. Tovey, Measurement of Ω^- decay properties in the CERN SPS hyperon beam, Nuc. Phys. B 241: 1–47, 1984.

[12] M. Schardt, J. Frank, C. Hoffman, R. Mischke, D. Moir, and P. Thompson, New measurement of the Dalitz-decay branching ratio of the π^0, Phys. Rev. D 23: 639–48, 1981.

[13] K. Eggert, H. Frenzel, K. Giboni, W. Thome, B. Betev, P. Darriulat, P. Dittmann, M. Holder, K. McDonald, H. Pugh, T. Modis, K. Tittel, V. Eckardt, H. Gebauer, R. Meinke, O. Sander, and P. Seyboth, A measurement of the proton-proton cross section at the CERN ISR, Nuc. Phys. B98: 93–9, 1975.

[14] G. Giacomelli, Total cross sections and elastic scattering at high energies, Phys. Rep. 23: 123–235, 1976.

[15] U. Amaldi, T. Fazzini, G. Fidecaro, C. Ghesquiere, M. Legros, and H. Steiner, Antiproton-proton total cross sections between 0.575 and 5.35 GeV/c, Nuovo Cimento 34: 825–53, 1964.

[16] A. Citron, W. Galbraith, T. Kycia, B. Leontic, R. Phillips, A. Rousset, and P. Sharp, Structure in the pion-proton total cross section between 2 and 7 GeV/c, Phys. Rev. 144: 1101–14, 1966.

[17] L. Fajardo, R. Majka, J. Marx, P. Nemethy, L. Rosselet, J. Sandweiss, A. Schiz, A. Slaughter, C. Ankenbrandt, M. Atac, R. Brown, S. Ecklund, P. Gollon, J. Lach, J. MacLachlan, A. Roberts, and G. Shen, Real part of the forward elastic nuclear amplitude for $\bar{p}p$, pp, π^+p, π^-p, K^+p, and K^-p scattering between 70 and 200 GeV/c, Phys. Rev. D 24: 46–65, 1981.
[18] A. Schiz, L. Fajardo, R. Majka, J. Marx, P. Nemethy, L. Rosselet, J. Sandweiss, A. Slaughter, C. Ankenbrandt, M. Atac, R. Brown, S. Ecklund, P. Gollon, J. Lach, J. MacLachlan, A. Roberts, and G. Shen, High statistics study of π^+p, π^-p, and pp elastic scattering at 200 GeV/c, Phys. Rev. D 24: 26–45, 1981.
[19] R. Fernow and A. Krisch, High energy physics with polarized proton beams, Ann. Rev. Nuc. Part. Sci. 31: 107–44, 1981.
[20] G. Baum, M.R. Bergstrom, P.R. Bolton, J.E. Clendenin, N.R. DeBotton, S.K. Dhawan, Y.-N. Guo, V.-R. Harsh, V.W. Hughes, K. Kondo, M.S. Lubell, Z.-L. Mao, R.H. Miller, S. Miyashita, K. Morimoto, U.F. Moser, I. Nakano, R.F. Oppenheim, D.A. Palmer, L. Panda, W. Raith, N. Sasao, K.P. Schuler, M.L. Seely, P.A. Souder, S.J. St. Lorant, K. Takikawa, and M. Werlen, New measurement of deep inelastic e-p asymmetries, Phys. Rev. Lett. 51: 1135–8, 1983.
[21] V. Hughes and J. Kuti, Internal spin structure of the nucleon, Ann. Rev. Nuc. Part. Sci. 33: 611–44, 1983.
[22] J.J. Aubert, U. Becker, P.J. Biggs, J. Burger, M. Chen, G. Everhart, P. Goldhagen, J. Leong, T. McCorriston, T.G. Rhoades, M. Rohde, S.C.C. Ting, S.L. Wu, and Y.Y. Lee, Experimental observation of a heavy particle J, Phys. Rev. Lett. 33: 1404–6, 1974.
[23] S.C.C. Ting, The discovery of the J particle: A personal recollection, Rev. Mod. Phys. 49: 235–49, 1977.
[24] G. Arnison, A. Astbury, B. Aubert, C. Bacci, G. Bauer, A. Bezaguet, R. Bock, T.J.V. Bowcock, M. Calvetti, T. Carroll, P. Catz, P. Cennini, S. Centro, F. Ceradini, S. Cittolin, D. Cline, C. Cochet, J. Colas, M. Corden, D. Dallman, M. DeBeer, M. Della Negra, M. Demoulin, D. Denegri, A. DiCiaccio, D. DiBitonto, L. Dobrzynski, J.D. Dowell, M. Edwards, K. Eggert, E. Eisenhandler, N. Ellis, P. Erhard, H. Faissner, G. Fontaine, R. Frey, R. Fruhwirth, J. Garvey, S. Geer, C. Ghesquiere, P. Ghez, K.L. Giboni, W.R. Gibson, Y. Giraud-Heraud, A. Givernaud, A. Gonidec, G. Grayer, P. Gutierrez, T. Hansl-Kozanecka, W.J. Haynes, L.O. Hertzberger, C. Hodges, D. Hoffmann, H. Hoffmann, D.J. Holthuizen, R.J. Homer, A. Honma, W. Jank, G. Jorat, P.I.P. Kalmus, V. Karimaki, R. Keeler, I. Kenyon, A. Kernan, R. Kinnunen, H. Kowalski, W. Kozanecki, D. Kryn, F. Lacava, J-P. Laugier, J.P. Lees, H. Lehmann, K. Leuchs, A. Leveque, D. Linglin, E. Locci, M. Loret, J-J. Malosse, T. Markiewicz, G. Maurin, T. McMahon, J-P. Mendiburu, M-N. Minard, M. Moricca, H. Muirhead, F. Muller, A.K. Nandi, L. Naumann, A. Norton, A. Orkin-Lecourtois, L. Paoluzi, G. Petrucci, G. Piano Mortari, M. Pimia, A. Placci, E. Radermacher, J. Ransdell, H. Reithler, J-P. Revol, J. Rich, M. Rijssenbeek, C. Roberts, J. Rohlf, P. Rossi, C. Rubbia, B. Sadoulet, G. Sajot, G. Salvi, G. Salvini, J. Sass, J. Saudraix, A. Savoy-Navarro, D. Schinzel, W. Scott, T.P. Shah, M. Spiro, J. Strauss, K. Sumorok, F. Szoncso, D. Smith, C. Tao, G. Thompson, J. Timmer, E. Tscheslog, J. Tuominiemi, S. Van der Meer, J-P. Vialle, J. Vrana, V. Vuillemin, H.D. Wahl, P. Watkins, J. Wilson, Y.G. Xie, M. Yvert, and E. Zurfluh, Experimental observation of isolated large transverse energy electrons with associated missing energy at $\sqrt{s} = 540$ GeV, Phys. Lett. 122B: 103–16, 1983; Further evidence for charged intermediate vector bosons at the SPS collider, Phys. Lett. 129B: 273–82, 1983.

Exercises

1. Calculate the net precession angle of a 40-GeV/c Λ crossing a 2-m-long, 1-T magnetic field? What is the net precession angle for a 40-GeV/c Σ^-?

2. What is the fractional loss of intensity for a 30-GeV/c proton beam passing through a 25-cm-long liquid hydrogen target?

3. Calculate the value of $d\sigma/dt$ at $t = 0$ for pp elastic scattering between 10 and 100 GeV/c using total cross section measurements. Ignore the real part of the forward amplitude.

4. A polarization experiment uses a polarized pentanol target with polarization 50% and a 60% polarized beam. What is the intrinsic asymmetry of the physical process if 1000 events are measured with the target polarized up and 1040 events are measured with it polarized down?

The following exercises require the reader to go through some of the steps involved in writing a proposal. Find references to the original literature in the Particle Data Tables. Using previously measured values, theory, or reasonable guesses, estimate the cross sections for any relevant processes. Give the required beam energy and intensity, target material and dimensions, and amount of time required for data taking. Specify the detector characteristics, sketch their arrangement, and estimate the acceptance. Give the trigger and possible background processes. Consider possibles inefficiencies that could lead to a loss of events. Estimate the final number of good events and the accuracy of any measurements.

5. Prepare a proposal to search for free quarks at the Tevatron $\bar{p}p$ collider.
6. Prepare a proposal to lower the upper limit on the measurement of the branching ratio for $K^0 \rightarrow e\mu$.
7. Prepare a proposal to study the charged decay modes of the E meson using a K^- beam.

Appendix A: Physical constants

Symbol	Definition	Numerical quantity
c	speed of light in vacuum	2.9979×10^{10} cm/sec
$\hbar = h/2\pi$	Planck constant	6.5822×10^{-22} MeV-s
$\hbar c$		1.9733×10^{-11} MeV-cm
e	electron charge	1.6022×10^{-19} coulombs
$\alpha = e^2/\hbar c$	fine structure constant	1/137.04
N_A	Avogadro number	6.0220×10^{23} mole^{-1}
K_B	Boltzmann constant	1.3807×10^{-16} erg/K
m_e	electron mass	0.51100 MeV/c^2
m_p	proton mass	938.28 MeV/c^2
$r_e = e^2/m_e c^2$	classical electron radius	2.8179×10^{-13} cm
$\lambda_e = h/m_e c$	Compton wavelength	2.4263×10^{-10} cm
$a_0 = \hbar^2/m_e e^2$	Bohr radius	0.52918×10^{-8} cm
$\mu_B = e\hbar/2m_e c$	Bohr magneton	5.7884×10^{-15} MeV/gauss
$\mu_N = e\hbar/2m_p c$	nuclear magneton	3.1525×10^{-18} MeV/gauss
C_E	Euler constant	0.5772
$D_e = 4\pi r_e^2 m_e c^2$		5.0989×10^{-25} MeV-cm^2
ϵ_0	permittivity	8.8543×10^{-12} F/m

Source: Particle Data Group, Rev. Mod. Phys. 56: S1, 1984.

Appendix B: Periodic table of the elements

IA	IIA								
								1 H 1.0079	
3 Li 6.94	4 Be 9.01218								
11 Na 22.98977	12 Mg 24.305	IIIB	•IVB	VB	VIB	VIIB	VIII		
19 K 39.0983	20 Ca 40.08	21 Sc 44.9559	22 Ti 47.90	23 V 50.9415	24 Cr 51.996	25 Mn 54.9380	26 Fe 55.847	27 Co 58.9332	28 Ni 58.71
37 Rb 85.467	38 Sr 87.62	39 Y 88.9059	40 Zr 91.22	41 Nb 92.9064	42 Mo 95.94	43 Tc 98.9062	44 Ru 101.07	45 Rh 102.9055	46 Pd 106.4
55 Cs 132.9054	56 Ba 137.33	57–71 Rare Earths	72 Hf 178.49	73 Ta 180.947	74 W 183.85	75 Re 186.207	76 Os 190.2	77 Ir 192.22	78 Pt 195.09
87 Fr (223)	88 Ra 226.0254	89– Acti- nides	104 (260)	105 (260)	106 (263)				

57 La 138.9055	58 Ce 140.12	59 Pr 140.9077	60 Nd 144.24	61 Pm (145)	62 Sm 150.4	63 Eu 151.96	64 Gd 157.25	65 Tb 158.9254

89 Ac (227)	90 Th 232.0381	91 Pa 231.0359	92 U 238.029	93 Np 237.0482	94 Pu (244)	95 Am (243)	96 Cm (247)	97 Bk (247)

(Particle Data Group, Rev. Mod. Phys. 56: S1, 1984.)

IB	IIB	IIIA	IVA	VA	VIA	VIIA	2 He 4.00260
		5 B 10.81	6 C 12.011	7 N 14.0067	8 O 15.9994	9 F 18.998403	10 Ne 20.17
		13 Al 26.98154	14 Si 28.0855	15 P 30.97376	16 S 32.06	17 Cl 35.453	18 Ar 39.948
29 Cu 63.546	30 Zn 65.38	31 Ga 69.735	32 Ge 72.59	33 As 74.9216	34 Se 78.96	35 Br 79.904	36 Kr 83.80
47 Ag 107.868	48 Cd 112.41	49 In 114.82	50 Sn 118.69	51 Sb 121.75	52 Te 127.60	53 I 126.9045	54 Xe 131.30
79 Au 196.9665	80 Hg 200.59	81 Tl 204.37	82 Pb 207.2	83 Bi 208.9804	84 Po (209)	85 At (210)	86 Rn (222)

66 Dy 162.50	67 Ho 164.9304	68 Er 167.26	69 Tm 168.9342	70 Yb 173.04	71 Lu 174.967	Rare earths (Lanthanide series)
98 Cf (251)	99 Es (254)	100 Fm (257)	101 Md (258)	102 No (259)	103 Lr (260)	Actinide series

Appendix C: Probability and statistics

We briefly summarize here some important results from the theory of probability and statistics. The reader is referred to the references or other texts for proofs and further details [1, 2].

Consider the measurement of some quantity X. In general, measurements of X will give different results, which we denote x. The frequency with which any result for X is obtained is given by a frequency function $f(x)$. The exact form of $f(x)$ depends on the particular process under investigation. Since the quantity X must have some value, the frequency function must have the normalization

$$\int_{-\infty}^{\infty} f(x)\, dx = 1 \tag{C.1}$$

The function $f(x)$ is also referred to as the probability distribution function.

The expectation value for any function $g(x)$ is

$$\langle g(x) \rangle = \int_{-\infty}^{\infty} g(x) f(x)\, dx \tag{C.2}$$

which is just the sum of the various possible values of $g(x)$ weighted by the probability of having that value of x. Two expectation values are particularly important for specifying the characteristics of a distribution. The mean value of x is the expectation value of x itself, or $\langle x \rangle$. This quantity is approximated by the sample mean $\bar{x}$

$$\langle x \rangle \simeq \bar{x} = \frac{1}{n} \sum_{i=1}^{n} x_i \tag{C.3}$$

where n is the number of measurements. This is, of course, a measure of the central tendency of X. The second important expectation value is the variance

$$\sigma^2 = \langle (x - \langle x \rangle)^2 \rangle$$
$$= \langle x^2 \rangle - \langle x \rangle^2 \qquad \text{(C.4)}$$

which is a measure of the spread in the measurements. The square root of the variance is called the standard deviation σ.

Three frequency functions are particularly important for the matters discussed in this book. The binomial frequency function is applicable when there are only two possible outcomes of a given measurement. For example, let A denote that some event has occurred. Suppose the measurement is repeated n times and the result A is obtained x times. The probability of this occurring is

$$f(x) = \frac{n!}{x!(n-x)!} p^x (1-p)^{n-x} \qquad \text{(C.5)}$$

where p is the probability that the event A will occur. The mean and standard deviation are given by

$$\langle x \rangle = np \qquad \text{(C.6)}$$

$$\sigma = \sqrt{np(1-p)} \qquad \text{(C.7)}$$

In the limit that the number of measurements n is large and the mean is small, the binomial distribution approaches the Poisson distribution, where

$$f(x) = \frac{y^x e^{-y}}{x!} \qquad \text{(C.8)}$$

and

$$\langle x \rangle = y \qquad \text{(C.9)}$$

$$\sigma = \sqrt{y} \qquad \text{(C.10)}$$

This distribution is frequently used for the analysis of radioactive decays, since the number of potential decaying nuclei is very large, yet the total number decaying in any short time interval is small.

The normal, or Gaussian, distribution is the limit of the binomial distribution when the number of measurements is large and the probability of the event is not too small. Many types of analog measurements exhibit a Gaussian distribution around the mean value, particularly if the measurement process is subject to random errors. The frequency is given by

$$f(x) = \frac{1}{b\sqrt{2\pi}} \exp\left[-\frac{1}{2}\left(\frac{x-a}{b}\right)^2\right] \qquad \text{(C.11)}$$

and

$$\langle x \rangle = a \tag{C.12}$$

$$\sigma = b \tag{C.13}$$

The full width at half maximum (FWHM) of a Gaussian distribution is related to σ by

$$\text{FWHM} = 2.354\sigma \tag{C.14}$$

References

[1] P. Bevington, *Data Reduction and Error Analysis for the Physical Sciences,* New York: McGraw-Hill, 1969.
[2] A. Melissinos, *Experiments in Modern Physics,* New York: Academic, 1966, Chap. 10.

Appendix D: Cross sections and probability

Consider a flux I_0 of collimated, monoenergetic particles impinging upon a target. The number of atoms per unit volume in the target material is

$$n_a = \frac{N_A \text{ (atoms/mol)} \times \rho \text{ (g/cm}^3\text{)}}{A \text{ (g/mol)}} \tag{D.1}$$

where ρ is the target density, A is its atomic weight, and N_A is Avogadro's number.

In an infinitesimal thickness dx of the target there will be $n_a\, dx$ atoms/cm^2 in the path of the beam. As the beam traverses the target, interactions take place, and the beam intensity is reduced. Let dI refer to the change in flux. This quantity will be proportional to both the incident flux of beam particles and the number of target atoms/cm^2 in the beam's path.

$$dI = -\sigma I n_a\, dx$$

The constant of proportionality σ is referred to as the total cross section and has the units of area. A convenient unit for nuclear work is the barn, where

$$1 \text{ barn} = 10^{-24} \text{ cm}^2$$

Let us assume that the material is homogeneous and that the target is thin enough so that the particle's velocity is not significantly reduced. Then σ is not a function of x, and if we integrate over the target thickness, we find that

$$I(x) = I_0 \exp(-\sigma n_a x) \tag{D.2}$$

Thus, the intensity of particles satisfying the initial conditions drops off exponentially as the beam traverses the target.

If we define Pr to be the probability that a particle interacts in the target, then the probability that the particle does not interact after crossing a thickness L is

$$1 - \mathrm{Pr} = I(L)/I_0 = \exp(-\sigma n_a L) \tag{D.3}$$

The quantity

$$\lambda_I = (\sigma n_a)^{-1} \tag{D.4}$$

is called the interaction length. When $x = \lambda_I$, the beam intensity in Eq. D.2 drops to I_0/e, so λ_I represents the mean free path between interactions. When the target thickness $L \ll \lambda_I$, we can expand the exponential in Eq. D.3 to get

$$\mathrm{Pr} = \sigma n_a L = L/\lambda_I \tag{D.5}$$

The inverse of the interaction length

$$\mu = 1/\lambda_I = \sigma n_a \tag{D.6}$$

is called the attenuation coefficient.

Intuitively, we may consider each atom to present a circular target of area σ to the beam particle. If we assume that the beam particles are randomly distributed over a 1-cm^2 area and that an interaction takes place whenever a beam particle hits one of the circular areas, then $n_a \sigma\, dx$ represents the fraction of the total area in which an interaction will take place, or equivalently the probability of an intersection.

Now consider a scattering experiment using a short target of length $L \ll \lambda_I$ where the scattered particles are only detected in a small solid angle $\Delta\Omega$ around the direction (θ, ϕ). The detected intensity is then given by

$$I(\theta, \phi) = I_0 n_a L \frac{d\sigma}{d\Omega}(\theta, \phi)\, \Delta\Omega \tag{D.7}$$

The constant of proportionality $d\sigma/d\Omega$ is called the differential cross section. The form of the functions σ and $d\sigma/d\Omega$ depends on the dynamics of the scattering process. The probability of an interaction is

$$\mathrm{Pr} = n_a L \frac{d\sigma}{d\Omega}\, \Delta\Omega \tag{D.8}$$

Appendix E: Two-body scattering in the LAB frame

The relations between the kinematic variables are much more complicated in the LAB frame than they are in the CM frame. The exact transformation equations depend critically on the actual masses involved as well as on the relative values of a particle's velocity and the velocity of the CM system in the LAB [1]. The type of transformation is determined by the parameter

$$g_1^* = \frac{s + m_1^2 - m_2^2}{s - m_b^2 + m_t^2} \frac{\lambda^{1/2}(s, m_b^2, m_t^2)}{\lambda^{1/2}(s, m_1^2, m_2^2)} \tag{E.1}$$

and the analogous parameter g_2^* obtained by interchanging the subscripts 1 and 2 in this equation. Recall that b and t refer to the beam and target particles, 1 and 2 refer to the two final state particles, and s is the square of the energy in the CM frame. The function $\lambda(a, b, c)$ was defined in Eq. 1.25. Particles with $g^* < 1$ can be emitted with any polar angle ($0 < \theta < 180°$), whereas particles with $g^* \geq 1$ can only be emitted in the forward hemisphere ($0 < \theta < \theta_{max} \leq 90°$).

In the case of elastic scattering with $m_b = m_1 = \mu$, $m_t = m_2 = m$, and $\mu \leq m$, Eq. E.1 becomes

$$g_1^* = \frac{s + \mu^2 - m^2}{s - \mu^2 + m^2} \leq 1 \qquad g_2^* = 1 \tag{E.2}$$

Since $g_2^* = 1$, the recoil particle is confined to the forward hemisphere in the LAB. The forward particle can be emitted in any direction in the LAB unless $\mu = m$, in which case it is also confined to the forward hemisphere. It is also possible to give explicit relations between p_i and θ_i and between θ_1 and θ_2 [1].

Reference

[1] E. Byckling and K. Kanjantie, *Particle Kinematics*, New York: Wiley, 1973.

Appendix F: Motion of ions in a combined electric and magnetic field

A simple theory of the motion of ions in a region with perpendicular electric and magnetic fields has been derived by Townsend [1]. Consider an ion with mass m and charge q. Let the electric field $\mathscr{E}$ lie along z and the magnetic field B lie along y. The equations of motion are

$$\ddot{x} = \omega\dot{z} \qquad \ddot{y} = 0 \qquad \ddot{z} = f - \omega\dot{x} \tag{F.1}$$

where dots denote time derivatives, $\omega = qB/m$, and $f = q\mathscr{E}/m$. If we assume that the ion is created with a small initial velocity and with a uniform distribution of angles, then the coupled $\ddot{x}$ and $\ddot{z}$ equations have the solutions

$$\begin{aligned} x(t) &= (f/\omega)t - (f/\omega^2)\sin \omega t \\ z(t) &= (f/\omega^2)(1 - \cos \omega t) \end{aligned} \tag{F.2}$$

Let $\{t_i\}$ be the sequence of time intervals between collisions and τ be the mean time interval. The mean displacement of the ion after N collisions is

$$\begin{aligned} \langle x \rangle &= (f/\omega) \sum_{i=1}^{N} t_i - (f/\omega^2) \sum_{i=1}^{N} \sin \omega t_i \\ \langle z \rangle &= (f/\omega^2) \sum_{i=1}^{N} 1 - (f/\omega^2) \sum_{i=1}^{N} \cos \omega t_i \end{aligned} \tag{F.3}$$

The ion traverses a portion of a circular arc between collisions. Townsend showed that the sine and cosine summations over these arcs have the values

$$\begin{aligned} \sum_{i=1}^{N} \sin \omega t_i &= \frac{N\omega\tau}{1 + \omega^2\tau^2} \\ \sum_{i=1}^{N} \cos \omega t_i &= \frac{N}{1 + \omega^2\tau^2} \end{aligned} \tag{F.4}$$

Substituting Eq. F.4 back into F. 3 and taking $N\tau \to t$, we find

$$\langle x(t)\rangle = \frac{q^2\mathscr{E}B\tau^2 t}{m^2\left(1+\frac{q^2B^2\tau^2}{m^2}\right)} \qquad \langle z(t)\rangle = \frac{q\mathscr{E}\tau t}{m\left(1+\frac{q^2B^2\tau^2}{m^2}\right)} \tag{F.5}$$

Note that both $\mathscr{E}$ and B must be nonzero to obtain a net motion along x. An electric field alone causes motion along z. If a magnetic field is also present, the motion along z is decreased. Measurements [2] of the displacement along x as a function of B in a spark chamber have shown that Eq. F.5 gives a reasonable fit to the data for $\mathscr{E} \leqslant 100$ V/cm.

The mean deflection angle and the mean drift velocity follow from Eq. F.5,

$$\tan\theta = \frac{\langle x\rangle}{\langle z\rangle} = \frac{qB\tau}{m} \qquad w = \frac{\langle z\rangle}{t} = \frac{q\mathscr{E}\tau}{m\left(1+\frac{q^2B^2\tau^2}{m^2}\right)} \tag{F.6}$$

Note that a simple estimate of the mean collision time τ can be obtained from a measurement of the drift velocity in a purely electric field.

References

[1] J. Townsend, *Electrons in Gases,* London: Hutchison, 1947.
[2] S. Korenchenko, A. Morozov, and K. Nekrasov, Displacement of spark chamber discharges in a magnetic field, Priboryi Tekhnika Eksperimenta, No. 5, 1966, p. 72.

Appendix G: Properties of structural materials[a]

Material	Composition	Density (g/cm^3)
Aluminum	pure	2.70
Copper	pure	8.93
Iron	pure	7.85
Stainless steel	type 304	8.02
Carbon steel	type 1020	7.86
Brass	70 Cu, 30 Zn	8.5
Lucite	$(C_5H_8O_2)n$	1.18
G-10	glass-filled epoxy	1.82
Glass	Pyrex	2.23

[a] Approximate values at room temperature.
[b] In tension.

Thermal conductivity (mW/cm² · K · cm)	Thermal expansion ($\times 10^{-6}$ K^{-1})	Resistivity ($\mu\Omega$-cm)	Elastic modulus (Mpsi)
2370	25	2.74	10
3980	16.6	1.70	17
803	12	9.8	30
300	17	72	28
1000	12	10	30
2200	19	7	15
2	70	$>10^{20}$	0.4[b]
3	25	$>10^{20}$	3[b]
16	3.2	$>10^{18}$	9.3

Source: Handbook of Chemistry and Physics, 64th ed., Boca Raton: CRC Press, 1983; *Physics Vade Mecum,* New York: American Institute of Physics, 1981; *Modern Plastics Encyclopedia,* New York: McGraw-Hill, 1976; Catalog, Oriel Corp., Stratford, CT.

Author index

Subject index

Printed in the United States
By Bookmasters